OECD Proceedings

AF681582

Innovative Clusters

DRIVERS OF NATIONAL INNOVATION SYSTEMS

OECD

ORGANISATION FOR ECONOMIC CO-OPERATION AND DEVELOPMENT

ORGANISATION FOR ECONOMIC CO-OPERATION AND DEVELOPMENT

Pursuant to Article 1 of the Convention signed in Paris on 14th December 1960, and which came into force on 30th September 1961, the Organisation for Economic Co-operation and Development (OECD) shall promote policies designed:

- to achieve the highest sustainable economic growth and employment and a rising standard of living in Member countries, while maintaining financial stability, and thus to contribute to the development of the world economy;
- to contribute to sound economic expansion in Member as well as non-member countries in the process of economic development; and
- to contribute to the expansion of world trade on a multilateral, non-discriminatory basis in accordance with international obligations.

The original Member countries of the OECD are Austria, Belgium, Canada, Denmark, France, Germany, Greece, Iceland, Ireland, Italy, Luxembourg, the Netherlands, Norway, Portugal, Spain, Sweden, Switzerland, Turkey, the United Kingdom and the United States. The following countries became Members subsequently through accession at the dates indicated hereafter: Japan (28th April 1964), Finland (28th January 1969), Australia (7th June 1971), New Zealand (29th May 1973), Mexico (18th May 1994), the Czech Republic (21st December 1995), Hungary (7th May 1996), Poland (22nd November 1996), Korea (12th December 1996) and the Slovak Republic (14th December 2000). The Commission of the European Communities takes part in the work of the OECD (Article 13 of the OECD Convention).

© OECD 2001
Permission to reproduce a portion of this work for non-commercial purposes or classroom use should be obtained through the Centre français d'exploitation du droit de copie (CFC), 20, rue des Grands-Augustins, 75006 Paris, France, tel. (33-1) 44 07 47 70, fax (33-1) 46 34 67 19, for every country except the United States. In the United States permission should be obtained through the Copyright Clearance Center, Customer Service, (508)750-8400, 222 Rosewood Drive, Danvers, MA 01923 USA, or CCC Online: *www.copyright.com*. All other applications for permission to reproduce or translate all or part of this book should be made to OECD Publications, 2, rue André-Pascal, 75775 Paris Cedex 16, France.

HD
69
.S8
I667
2001

FOREWORD

Innovation through the creation, diffusion and use of knowledge has become a key driver of economic growth and provides part of the response to many new societal challenges. However, the determinants of innovation performance have changed in a globalising knowledge-based economy, partly as a result of recent developments in information and communication technologies. Innovation results from increasingly complex interactions at the local, national and world levels among individuals, firms and other knowledge institutions. Governments exert a strong influence on the innovation process through the financing and steering of public organisations that are directly involved in knowledge generation and diffusion (universities, public labs), and through the provision of inancial and regulatory incentives. They need a sound conceptual framework and an empirical basis to assess whether and how the contribution of public policy to national innovation performance could be improved.

Through a decade of academic research and policy analysis, the National Innovation Systems (NIS) approach has been developed to provide such framework and quantitative information. The OECD Committee for Scientific and Technological Policy (CSTP), and its Working Party on Technology and Innovation Policy (TIP), have contributed to this development through the NIS project, conducted in two phases.

The first phase of the NIS project involved country case studies, the development of internationally comparable indicators and thematic analytical work by six focus groups, including one on clusters. Its results are reported in *Managing National Innovation Systems* (OECD, 1999) and in *Boosting Innovation: The Cluster Approach* (OECD, 1999). This work provided new evidence on the systemic nature of innovation, articulated a new rationale for technology policy and identified broad directions for the improvement of national policies.

The second and last phase of the NIS project was devoted to deepening the analysis on three themes: clusters; innovative firms and networks; and human resource mobility. The work on clusters was, as was the case in the previous phase, led by the Netherlands, whose Ministry of Economic Affairs hosted in Utrecht a workshop entitled "Do Clusters Matter in Innovation Policy?". The present publication builds largely on the presentations and discussions in this workshop.

The work reported in this publication demonstrates the increasing importance of clusters in determining the innovation performance of firms, nations and regions and in structuring international linkages among national innovation systems. Based on the evaluation of concrete national experiences, it provides useful guidance on how the reality of clusters can be better taken into account in designing and implementing innovation policy.

This publication results from a collective work under the efficient leadership of Pim den Hertog of Dialogic, Utrecht (the Netherlands), in close collaboration with Svend Remoe of the OECD Secretariat. It would not have been possible without the competence and efforts of the two other members of the editorial team: Edward M. Bergman of Vienna University of Economics and Business Administration (Austria), and David Charles of the Centre for Urban and Regional Development Studies in Newcastle (United Kingdom).

TABLE OF CONTENTS

Chapter 1. In Pursuit of Innovative Clusters
Edward M. Bergman, David Charles and Pim den Hertog 7

Part I. Learning from Experiences in Advanced Innovation Clusters

Chapter 2. The ICT Cluster: The Engine of Knowledge-driven Growth in Finland
Laura Paija 19

Chapter 3. The Boundaryless Cluster: Information and Communications Technology in Ireland
Roy Green, Imelda Duggan, Majella Giblin, Mike Moroney and Leo Smyth 47

Chapter 4. The ICT Cluster in Denmark
Michael S. Dahl and Bent Dalum 65

Chapter 5. Innovation Dynamics in the Spanish Telecommunication Cluster: Policy Implications
Cristina Chaminade 91

Chapter 6. ICT Cluster Organisations in Flanders: Co-operation in Innovation in the New Networked Economy
Jan Larosse, Patrick Slaets, Jan Wauters, Stephan Bruninx, Peter Simkens and René Wintjes 113

Chapter 7. Innovation in an Adolescent Cluster: The Dutch Multimedia Cluster
Pim den Hertog, Sven Maltha and Erik Brouwer 133

Part II. Learning from Experiences in Mature Clusters

Chapter 8. Innovation Styles in Agro-food Production in Norway
Johan Hauknes 157

Chapter 9. The Construction Cluster in Denmark
Michael S. Dahl and Bent Dalum 179

Chapter 10. Innovation in the Dutch Construction Cluster
Pim den Hertog and Erik Brouwer 203

Chapter 11. An Anatomy of the Swiss Construction Cluster
Patrick Vock 229

Part III. Applied Analyses and Cluster-based Innovation

Chapter 12. Identification of Techno-economic Clusters Using Input-output Data: Application to Flanders and Switzerland
Ludo Peeters, Marc Tiri and Adrian Berwert 251

Chapter 13. Industrial Clusters in the Finnish Economy
Sakari Luukkainen 273

Chapter 14. Knowledge Transfer in Agglomerations: A Regional Approach to Japanese Manufacturing Clusters
Kinji Gonda and Fumihiko Kakizaki 289

Chapter 15. Dynamics of Initial Cluster Formation: The Case of Multimedia and Cultural Content
Michael Peneder 303

Chapter 16. Trade-flow-based Networks in the Finnish Economy: Heterogeneous Practices
Tuomo Pentikäinen 313

Part IV. Policies for Cluster-based Innovation

Chapter 17. Clustering and Economic Complexity: Regional ICT Clusters in the United Kingdom
David Charles and Paul Benneworth 327

Chapter 18. Danish Cluster Policy: Improving Specific Framework Conditions
Mette Holm Dalsgaard 347

Chapter 19. Towards Second-Generation Cluster Policy: The Case of the Netherlands
Victor A. Gilsing 361

Chapter 20. The Cluster Approach in Finnish Technology Policy
Jari Romanainen 377

Chapter 21. Bridging Cluster Theory and Practice: Learning from the Cluster Policy Cycle
Paul Benneworth and David Charles 389

****

Chapter 22. Creating and Sustaining Innovative Clusters: Towards a Synthesis
Pim den Hertog, Edward M. Bergman and David Charles 405

Chapter 1

IN PURSUIT OF INNOVATIVE CLUSTERS

by

Edward M. Bergman, Vienna University of Economics and Business Administration,
David Charles, Centre for Urban and Regional Development Studies, University of Newcastle, and
Pim den Hertog, Dialogic Innovation & Interaction and Department of Innovation Studies, Utrecht University*

Introduction

The enthusiastic reception of *Boosting Innovation: The Cluster Approach* (OECD, 1999) among OECD Member countries and readers was both gratifying and a confirmation that this topic warranted further exploration. While that title could equally well describe the contents of the present volume, *Boosting Innovation* reported accurately the best available thinking on several broad fronts of scholarship and policy by a small community of "first mover" analysts who wished to explore the topic in an internationally comparative context. Its authors had clearly given thought to questions of innovation, but few connected these questions to their analysis of clusters. In retrospect, however, that volume (OECD, 1999) accomplished the valuable and vital task of broadly informing readers about the value-chain cluster concepts and the potential for innovation in clusters that function as reduced-form national innovation systems (reduced-NIS). Perhaps it also created a healthy appetite for the more direct approach to innovative clusters taken in this volume.

The present volume places greater emphasis on the innovation theme pursued by the NIS focus groups through the original contributions of a broad group of analysts from different academic, policy and national backgrounds, all of whom are knowledgeable about clusters and innovation. Genuine improvements in methods, conceptualisations and empirical analyses are included in this volume, although these contributions are now relatively less prominent. The expanded coverage of policies shifts our attention somewhat away from the ways in which clusters are promoted and sustained and more towards their role in supporting innovation. This reflects a conscious editorial effort to adhere more closely to the NIS framework in soliciting and shaping manuscripts for OECD readership. In our view, this readership goes beyond the cluster analyst and advocacy communities to include established policy-making institutions, ministries and organisations that are genuinely concerned with industrial innovation at the national and sub-national levels.

The remaining pages of this introduction examine some of the key features – primarily assumptions and working propositions – that guided the preparation of this volume and that reflect the contributors' collective viewpoints. However, the overall findings and lessons to be drawn from the

* The authors gratefully acknowledge the assistance of Svend Remoe of the OECD, who, in addition to other duties, deserves recognition for his service as a virtual member of our editorial team.

contributions will find their rightful place in the concluding synthesis chapter. Also included is a preview of the major substantive sections.

This volume has a number of key features which include the extension of cluster concepts as reduced-NIS, the expansion of value-chain concepts, the forging of complementary connections with other NIS focus groups, the promotion of multiple cluster innovation policy approaches, and the application of comparative national frameworks.

Reduced-NIS clusters

The national innovation system concepts advanced by Lundvall and others were adopted by this focus group as an endogenous set of interacting agents and resultant practices responsible for the emergence of innovation, a view that departed radically from earlier linear concepts of innovation as the end-stage of a process driven mainly by scientific advances in basic research and one that converged on rapidly developing concepts from the industrial cluster literature.[1] The idea that innovations could result directly from ongoing interactions among scientific, commercial, educational and public institutions gained many adherents, including at the OECD.

This focus group extended that logic by offering the novel hypothesis that an industrial cluster could usefully be thought of as a reduced-NIS,[2] in which the most essential and functional system elements help stimulate the emergence of specific kinds of innovation in various segments of a national economy. This powerful simplification permits a greater focus on the actions and policies that can stimulate processes of innovation.

Although it was originally proposed mainly as an organising metaphor, the reduced-NIS concept has genuine merit that deserves further exploration and testing throughout this volume. Attention will be directed here towards value-chain clusters as key sources of innovation and additive components of a NIS, the recent emergence of regional innovation systems, why size matters in considering clusters as components of NIS, and the role of international trade in large *vs.* small country clusters.

Many of the chapters in this and the preceding volume (OECD, 1999) highlight the fact that a national economy consists of several reduced-form innovation systems, represented by various distinct industrial clusters. This implies that any country's overall innovation system would necessarily include composite features of innovation underway in its few-to-many industrial clusters, depending upon the size and complexity of the national economy. How an NIS approach might blend these different innovation system components is hinted at in a number of the chapters that document policy experiences, but it is clear that this issue deserves more systematic attention.

Initially, the NIS concept was seen by many large OECD countries as requiring a fairly homogenous central policy approach, whose applicability would therefore be greatest in the small national economies represented by – but not restricted to – the Nordic to Belgian/Flemish fringe of Europe. It is therefore no great surprise to see that the majority of the contributors to this volume (and leadership of the related focus groups) represent such countries. Several small countries have sought, or continue to seek, to identify industrial champions that propel key elements of the national economy and are understandably responsive to Porter-like analyses stressing the competitiveness of those champions and their cluster halo. Larger and more complex national economies are sufficiently heterogeneous that a central NIS policy is less compelling than reforms and fine-tuning of core policies to ensure national enabling or framework conditions that are conducive to innovation.

However, the chapters which analyse the situation in the large OECD countries reveal that a *double* reduced-NIS concept may make even better sense, *i.e.* a *regional* innovation system (RIS), consisting of fewer and more locally manageable industrial clusters that share uniquely regional externalities of the type envisioned by Marshall.[3] The idea of RIS has been floating around for nearly a decade now, usually in conjunction with industrial district and local cluster concepts. The second reduction is two-fold: geographic specificity (rather than national generality); and greater distance from national policy frameworks.

Unsurprisingly, the chapters on UK and Japanese clusters in this volume focus almost exclusively on regional clusters, including regional dynamics of clusters of regional industries, similar to the chapter on the United States in the previous volume (OECD, 1999). This implies that regional policy could play a key role in the formulation and implementation of innovation policies in large countries, and perhaps takes its extreme form in the United States, where nearly all cluster-based innovation policy is to be found at the state and regional levels.[4]

International trade among cluster members has completely different implications for large *vs.* small country clusters. A recent study of trade in OECD Member countries (Hummels, Rapoport and Yi, 1998) shows that vertical trade among international members of a value chain is a much higher proportion of total trade in small *vs.* large countries. For example, vertical trade is 25% and 42% for Denmark and Netherlands *vs.* 7% and 14% for the United States and Japan, respectively. The authors consider that these findings reveal a greater likelihood that a cluster's trading partners are within and therefore responsive to a large home country's national and regional policies. Paradoxically, however, it also means that *supra*-national innovation systems (*S*-NIS) may be essential to sound cluster policies, particularly for small countries. Thus, it could well be the case that relevant elements of cluster or innovation policy might logically migrate to the policy frameworks of relevant OECD Member customs unions, such as the EU or NAFTA.

The value-chain cluster concept

As stated in *Boosting Innovation*, "Most of the research on innovation reported in this publication focuses on mutual interdependency and interaction among actors in the value chain… [based upon] …trade linkages, innovation linkages, knowledge flow linkages or on a common knowledge base of common factor conditions." (OECD, 1999, p. 13). We continue to adhere to these basic concepts in this volume, although certain links are emphasised as attention shifts away from clusters *per se* to their emerging role in the core processes of innovation.

The value-chain foundation has proved to be a powerful concept for OECD purposes for several reasons. These include its useful concordance with other interesting but more limited cluster concepts in the literature, an appealing framework for examining cluster-driven trade between OECD Member countries and regions, the potential for integrating related concepts from sister OECD focus groups concerning networking and human capital mobility, and a methodological openness to new analytical approaches for understanding cluster dynamics and policies.

The majority of the chapters in this volume rely upon or refer to value-chain interactions to define cluster members, the majority of which use inter-industry trade data,[5] although inter-industry innovation transfer data are also represented. At the same time, some of the chapters explore more closely key assumptions or changes in clusters that bear upon their potential for innovation. For example, Finnish value chains have been examined here to detect networking intensities and input-sourcing concentration of key cluster industries. These insights reflect upon cluster innovation potential and help to link the implications of the cluster approach to the innovative network approach of a sister OECD focus group.

Other value-chain methods can help to clarify our understanding of – and assumptions about – which elements of the chain are most crucial for the effective diffusion of technology. Since a value chain comprises many actors and elements, all of which are considered *a priori* equally important or essential to the process of innovation, it is more than idle curiosity to inquire if some are of greater value than others. The study of Japanese industry clustering reported in this volume included a survey of SMEs producing intermediate and final goods to detect the most significant sources of innovation in their value chains. The findings from Japan reinforce the evidence accumulating from recent EU and US studies of innovation: the most significant elements of the value chain responsible for innovation are the firms that trade with each other along the value chain as *suppliers or customers* (Bergman and Feser, 2001; Cooke, Boekholt and Toedtling, 2000).

The reasons for this are multiple. Firms with previously embedded technologies in their production processes increase their dependence on value-chain connections for further innovation, particularly in lean-production and process-control technology industries. As examples, such firms procure advanced technological capacities embodied in supplies and equipment, they co-design innovative products and necessary production systems with supply-chain partners, JIT supply chains place logistics managers and industrial engineers in close technical contact, and potential business-to-business suppliers often become ISO and supply-chain certified only after introducing major improvements which have been approved and supervised by customer engineers.[6] Seen in this light, innovation becomes increasingly endogenous to value chains.

While this reasoning and these findings will come as no surprise to analysts who have examined data on innovation transfer and acquisition in many clusters, they do call into question the nature of innovation policies that routinely establish new institutions and practices by public and other non-industrial groups without regard to demonstrable need or opportunity cost.

Questions about the relevance of value-chain institutions that are neither suppliers or customers apply with varying force to different clusters, depending on the length of value chains and the relative maturity of clusters, particularly those with stable "signature" technologies. Very short value chains offer fewer cumulative opportunities for the introduction of product or process innovations of the kind that occur spontaneously along the lengthier value chains of intermediate suppliers, the ICT cluster, for example. Rapid technological progress is endogenously generated within these tightly packed ensembles of innovation and implicit knowledge. Short value chains of other advanced technology clusters, such as those in biotechnology, rely almost exclusively on laboratory-based knowledge and innovations. Mature clusters with stable technologies, such as construction or natural resource industries, are also likely to depend upon relatively exogenous knowledge institutions to organise and distribute state-of-the art innovations as part of their ongoing programmes of education, research and technical assistance. We should therefore expect to find that value chains of varying lengths and technological stability have markedly different degrees of reliance on implicit *vs.* explicit knowledge and on endogenous *vs.* exogenous sources of knowledge.

Complementary connections

If innovation in clusters is to be considered as a reduced-NIS in action, then how do other NIS focus group findings relate to our inquiry into cluster-based innovation?[7] First, the value chain essentially defines regular production interconnections between firms along which innovations and technological learning may occur. This view is closely compatible with, yet distinct from, other views of how innovations arise and become dispersed among a population of firms.

Consider first the question of human mobility, or more precisely, labour mobility. Considerable amounts of human capital and implicit knowledge move between firms as they hire or release key employees. These transfers of human capital and knowledge could occur within the same industrial cluster, but such knowledge is seldom cluster-specific since it can be applied in many different kinds of firms requiring generic skills. Labour is therefore mobile between similar occupational and skill categories, and these can be found in many different sectors or clusters. This indicates that labour mobility can be seen as one way of propagating innovations *across* clusters. Mobility is also a feature of students who become employees, thereby transmitting recently acquired formal knowledge from universities and other knowledge institutions directly to firms. Direct employment of recent graduates may account somewhat for the absence of formal contacts between firms and knowledge institutions noted in earlier studies.

A second question concerns the capacity of individual firms to absorb and embed available innovations or their inherent capacity to generate innovations. The innovative firm focus group isolated six distinct qualities of firms capable of innovative behaviour: *i)* vision and strategy; *ii)* competency base; *iii)* creativity and idea management; *iv)* intelligence; *v)* organisation and process; and *vii)* culture and climate. In building this competence base, the focus group specifically recommended that firms should consider clusters and alliances as a possible platform for developing unique competency mixes, rather than working alone. Working with supply chains or clusters rather than with individual firms was seen as a practical method by which firms can capitalise on "learning through interaction" and promote cultural change. The implications are clear: firms are better able to innovate if they are fortunate enough to be members of a cluster. However, formal clusters should be organised autonomously or by popular demand, rather than by policy fiat.

Finally, innovative networks also link innovative firms who take advantage of the potential for knowledge sharing and further technological learning. Innovative networks can be distinguished from clusters based purely on value chains if one considers evidence of knowledge spillovers: some of the country-specific networks studied by this focus group could not unambiguously be reduced to value-added linkages between R&D-intensive sectors. In short, innovative networks cut *across* various value-chain clusters to include sectors that share similar technologies but produce different goods and services in more than one value chain. The overall effect of programmes to stimulate such networks will not be known in the near future, although one of the impacts anticipated by this focus group is that networked firms will eventually become members of fully-fledged clusters. This implies that innovations created through networks are likely to accumulate in value-chain clusters where their commercial potential can be realised. A final point concerns the better articulation of policy instruments directed towards technology and innovation policy, including co-ordination and complementarity between cluster/network policies and other types of innovation policy.

Comparative case framework

The contributors to this volume were very much aware of the need for representation from OECD countries of different size, industrial structure, economic performance, etc., but the goal was also to be able to draw lessons from a narrower, more readily comparable set of clusters. At the same time, if each country selected a different cluster, it would be impossible to establish useful comparisons of innovation styles and policies.

Therefore, we agreed to examine one or two advanced, innovation-intensive clusters and one or two mature, stable innovation clusters of greatest interest to participants.[8] Negotiations resulted in the ICT, construction and agro-food clusters being chosen, the selection of which necessarily reflected work already underway in participating countries, since funds to support such studies are provided at the country level. These proved to be excellent choices for several reasons.

First, ICT clusters are globally integrated along distant value chains and rely upon a wide variety of distant and local innovation sources. The mature clusters originated in craft or peasant economies, still respond mainly to domestic demand and context, and reflect to varying degrees traditional or national values associated with food and building practices.

Second, ICT clusters are invariably partial in their national composition, *i.e.* key supplies of this globally defined value-chain cluster must be imported or exported before reaching final markets; this contrasts strongly with mature sectors, which are primarily made up of local and national components.

Third, the national and regional policy frameworks must deal with entirely different types and levels of problems confronting distinct clusters, thereby revealing a broader range of potential policy solutions.

Policies and innovation

To the degree that policies were considered in *Boosting Innovation*, most focused on how best to promote and support industry clusters, with only the occasional mention of innovation policies. Innovation was apparently stimulated by an array of NIS-type policies, of which clusters could be one of the possible – but inadequately – defined factors responsible for the successful introduction of innovation measures. And, unlike the view that many policies stimulate and support the emergence of clusters, a view richly documented for ICT clusters in the United Kingdom in this volume, it was previously thought that targeted policies and programmes were essential.

The policy discussions in this volume build upon earlier evidence from many sources, but also lay the groundwork for some new approaches. The richest lode of innovative cluster policies can be found in the many case studies presented here. A key feature of the case studies was to provide sufficient documentation and evidence to enable authors to clearly depict a characteristic style of innovation within types of clusters and particular countries, since consideration was given to the relevant policies for that style in a variety of national and cluster cultures. We are confident that a close reading of these case studies will provide many new and stimulating approaches worthy of careful consideration.

Other approaches take the view of a minister or other official charged with policy responsibilities for clusters, innovation or both. One chapter examines the theoretical framework and analytical policy approach for launching a promising new cluster, taking multimedia as an example. Another chapter examines the evolution between the first- and second-generation policy frameworks of the Dutch Ministry of Economic Affairs for all clusters and related innovation. A third provides analytical insights into national funding of Finnish cluster initiatives, taking into account the significance of clusters for trade flows and regarding social *vs.* private rates of return.

One can also profit much by reading across the various chapters supplied within a given national context, particularly the cases of Finland and the Netherlands. The three Dutch chapters are particularly instructive, since one lays out the previous and present national framework for a portfolio of clusters, while the others present the policy needs of particular clusters within that portfolio.

Profile of contents

In addition to our introductory and concluding chapters, the 20 original contributions included in this book are distributed over an equal number of chapters that reflect experiences in 13 OECD countries. A wide range of national and regional experiences are available for comparison and two or more chapters are available for some countries, thus permitting the interesting intra-country insights noted earlier.

We have elected to group these chapters in four major sections, beginning with *Learning from Experiences in Advanced Innovation Clusters.* These consist entirely of ICT clusters or variants thereof in six OECD countries. In additional to Ireland and Spain, these consist of countries at the Nordic to Belgian/Flemish edge of Europe (Finland, Denmark, Flanders, the Netherlands), all of which possess a distinctive blend of elements from perhaps the most dynamic and internationally oriented value-chain clusters. This cluster has considerable appeal to countries with technological strengths and an interest in new economy growth potentials.

A second major section is analogously entitled *Learning from Experiences in Mature Clusters.* Fewer in number, these case studies examine the significance of mature clusters to four national economies (Norway, Denmark, the Netherlands, Switzerland) and explore how various innovations pulse through or are thwarted by a host of factors. The lessons to be drawn from the agro-food and construction clusters contrast strongly and usefully with those from the ICT cluster.

The five chapters of the third major section, *Applied Analyses and Cluster-based Innovation*, continue to build our useful repertoire of analytical tools and knowledge base, and illustrate their application in distinct national contexts (Flanders, Switzerland, Finland, Japan, Austria). All develop further or apply value-chain concepts in different ways of direct utility to policy analysts in OECD countries concerned with clusters and innovation.

The fourth major section, entitled *Policies for Cluster-based Innovation*, reflects a primary concern of this focus group and the present volume. Although the previous three sections are also written and prepared explicitly to inform questions of policy, this section includes five provocative chapters that examine various policy regimes and options. These range across a number of cluster and innovation policies at both national and regional levels (the United Kingdom, Denmark, the Netherlands, Finland).

The book concludes with the chapter entitled "Creating and Sustaining Innovative Clusters: Towards a Synthesis". This chapter provides a representative sampling of the most useful and striking insights and lessons to be drawn from these studies.

NOTES

1. The distinction between these different schools of literature is discussed in Bergman and Feser (1999).

2. A reduced-NIS can have more than one possible interpretation. Reduced-*scale* NIS implies that the full system is faithfully replicated at the cluster level, which is likelier in small countries dominated by a single cluster around which all NIS activities revolve. The term "reduced-*form*", borrowed from the modelling community, refers to a re-specified system of endogenous relationships that permit one to grasp system essentials, identify their key functions, or to derive practical findings. Seen in this light, cluster concepts retain essential features of the endogeneity and key agents in reduced-form, while leaving certain contextual conditions and detailed interactions of the larger system unexamined.

3. Locational externalities might also arise in a small country NIS. A sampling of recent literature and analyses concerning regional innovation systems is available in Bergman and Feser (2001).

4. The most recent US documents of general relevance were released by the National Science and Technology Council (Rand, 2001) and the Council on Competitiveness (Porter and van Opstal, 2001). The first lays out basic framework conditions for fostering innovation (Clinton White House Office of Science and Technology Policy), while the second business organisation report details changes in national framework conditions and argues strongly for business-driven regional cluster actions.

5. Chapter 12 demonstrates how the application of a common methodology to detect value-chain clusters reveals rather different results for the same clusters in Switzerland and Belgium.

6. A case study of how this supply-chain process unfolds is presented for a mini-cluster surrounding a manufacturer of photocopiers in Rutten (2000, pp. 217-241).

7. Future publications from the focus groups will provide details of broad points summarised in OECD Working Group on S&T (12-13 December 2000) and used as basis for points referenced here.

8. As will be seen in the concluding chapter, there is more innovation underway in mature clusters than was initially assumed when selecting cases.

REFERENCES

Bergman, Edward M. and Edward J. Feser (1999), *Industrial and Regional Clusters: Concepts and Comparative Applications*, University of West Virginia, Morgantown, University of West Virginia Webbook, http://www.rri.wvu.edu/WebBook/Bergman-Feser/contents.htm.

Bergman, Edward M. and Edward J. Feser (2001), "Innovation System Effects on Technological Diffusion in a Regional Value Chain", *European Planning Studies*, Vol. 9.5.

Cooke, P., P. Boekholt and F. Toedtling, F. (eds.) (2000), *The Governance of Innovation in Europe*, Pinter, London.

Hummels, David, Dana Rapoport and Kei-Mu Yi (1998)," Vertical Specialization and the Changing Nature of World Trade", *Economic Policy Review*, pp. 79-99, Federal Reserve Bank of New York, June.

OECD (1999), *Boosting Innovation: The Cluster Approach,* OECD, Paris.

OECD Working Group on Innovation and Technology Policy (2001), "Focus Groups on National Innovation Systems: Reports by Lead Countries", internal working document, OECD Paris, 29 November.

Porter, Michael E. and Debra van Opstal (2001), *US Competitiveness 2001: Strengths, Vulnerabilities, and Long-term Priorities,* Council on Competitiveness, Washington, DC, January.

Rand (2001), *New Foundations for Growth: The U.S. Innovation System Today and Tomorrow,* Science and Technology Policy Institute, Washington, DC, January.

Rutten, Roel (2000), "Innovation in Regional Supplier Networks: The Case of KIC", in F. Boekema, K. Morgan, S. Bakkers and R. Rutten (eds.), *Knowledge, Innovation and Economic Growth: The Theory and Practice of Learning Regions*, Edward Elgar, Cheltenham.

PART I

LEARNING FROM EXPERIENCES IN ADVANCED INNOVATION CLUSTERS

Chapter 2

THE ICT CLUSTER: THE ENGINE OF KNOWLEDGE-DRIVEN GROWTH IN FINLAND

by

Laura Paija[*]
ETLA, The Research Institute of the Finnish Economy, Helsinki

Introduction

Information and communication technology (ICT) is fuelling the ongoing "third industrial revolution". Like steam power and electricity in their time, ICT is a powerful agent of change, influencing virtually all aspects of life. The effects on the demand side (*i.e. application* of the technology) are perhaps more important when considering the overall economic impacts of the technology. However, the supply side of ICT and related services (*i.e.* the *provision* of the technology) has grown to become a sizeable business in its own right.

In what follows, we study the Finnish ICT cluster primarily from the "supply" point of view by exploiting Porter's (1990) cluster application. Given the myriad roles ICT plays in our lives, the cluster composition may be quite different depending on the particular context and point of view. In the Finnish context, telecommunications operation and, in particular, mobile equipment are at centre-stage.

The chapter begins by identifying the ICT cluster and continues with a review of the economic relevance of ICT in domestic and foreign markets. Then, after an overview of cluster history, we dig deeper into the cluster through an analysis of the factors of competitive advantage, *i.e.* firm strategy, structure and rivalry; factor conditions; demand conditions; supporting and related industries; government; and coincidental factors. Next, we attempt to shed some light on the characteristics of innovation patterns in the Finnish ICT context (acknowledging, however, the constraints related to the generalisation of this very case-specific phenomenon). Finally, we conclude by assessing the role of government in the development of a world-class industrial complex.

---

* Holtron Ltd, a Finnish seed and start-up investor, provided a background study for the discussion on the capital market developments. The author is indebted to Petri Rouvinen (ETLA) for his invaluable support and assistance.

Cluster identification

The ICT cluster is depicted in Figure 1. In Finland, the cluster centres on telecommunications equipment manufacturing and service provision as the key industries. The universities and research institutes have been successful in producing competent human resources and world-class R&D to support cluster development. The electronics industry, in turn, has become highly specialised in the needs of the key activities. Digitalisation of content, provided by related industries, is seen as the most prominent factor in boosting the future demand for telecommunications infrastructure. The venture capital market, as an example of associated services, has emerged as a new and important source of funding that has greatly enhanced the preconditions for cluster growth.

Figure 1. **ICT cluster chart**

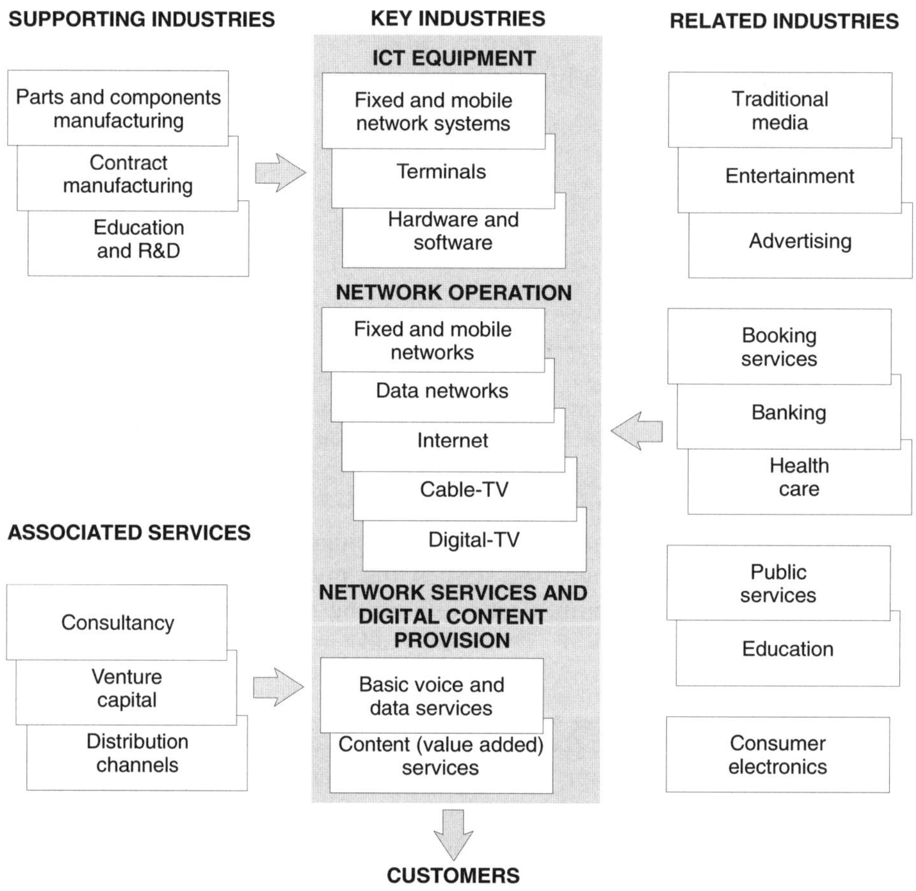

In recent years, however, it has become increasingly difficult to place firms on the cluster chart. Three trends, namely *convergence* of networks, terminals and services, *digitalisation* and *deregulation* have dramatically complicated the clear-cut cluster chart that we had a few years ago.[1] The cluster actors are gradually penetrating each other's domains, blurring the competitive environment. In the near future, for example, both terminal suppliers and operators will compete in providing the same technical solutions (*e.g.* user authentication) or service access (*e.g.* mobile portal).

Furthermore, cluster actors merge vertically (*e.g.* content providers, packagers, distributors and service providers; or, business consultants, IT integrators and new media) to encompass a wider range of the value chain. A number of firms in different communications fields have taken part in the global wave of cross-sectoral mergers.

Cluster mapping

Economic relevance in the domestic market

The key figures of the ICT cluster for 1998 are presented in Table 1.[2] The gross value of the cluster was EUR 17.5 billion. Manufacturing of equipment and their electronic components dominated the cluster, accounting for two-thirds of the value, while telecommunications services represented one-fifth of the turnover. The significance of software supply and other IT services is underestimated in the table since ICT equipment includes an important amount of software, and the construction of telecommunications networks involves IT services that are included in sales of equipment manufacturers.

Table 1. **Key economic indicators of the ICT cluster, 1998**

	ICT manufacturing		ICT services				Cluster (total)	
			Telecom services		Software, IT services			
	EUR (million)	Share of production	EUR (million)	Share of production	EUR (million)	Share of production	EUR (million)	Share of production
Production	11 631	100%	3 408	100%	2 500	100%	17 538	100%
Value added	3 728	32%	2 045	60%	1 724	69%	7 497	43%
Labour cost	951	8%	682	20%	706	28%	2 339	13%
Exports	9 543	82%	110	3%	932	37%	10 585	60%
Imports	1 694	15%	150	4%	578	23%	2 422	14%

Source: Statistics Finland, Ministry of Transport and Communications.

Figure 2 reveals the breakthrough of communications products in domestic production. Since Nokia's recovery (from 1992 onwards), value added in ICT manufacturing has grown at the average annual rate of 35%. In 1998, the share of the cluster in GDP was 6.6%.[3]

Figure 2. **The share of ICT value-added on GDP**

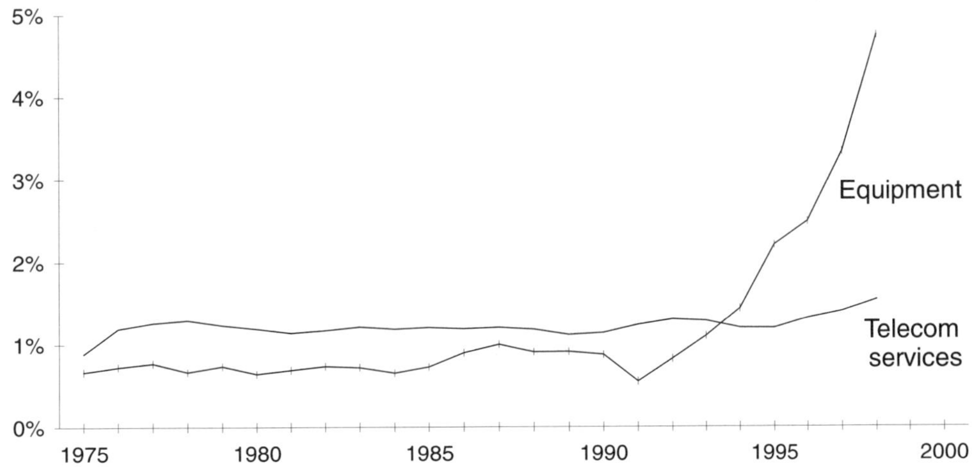

Note: Excludes software and IT services as well as computers due to lack of data. There is a slight discontinuity in the data between 1994-95 due a change in statistical classification.
Source: Statistics Finland, Ministry of Transport and Communications.

With its 75 000 employees, the ICT cluster accounted for 3% of total national employment in 1998. Nokia alone employed 21 000 persons in Finland and thus accounted directly for almost 30% of employment in the cluster. According to estimations, Nokia indirectly employed an additional 14 000 persons through its first-tier subcontractor firms (Ali-Yrkkö *et al.*, 2000). As production networks go further to sequential tiers, the employment effect of the major firm is significant, but cannot be readily quantified. However, without the chronic shortage of skilled labour, the employment potential of the cluster would allow much higher recruitment.

Figure 3. **Annual employment growth in the ICT cluster and in the economy as a whole**

Note: The ICT cluster lacks complete data for computer manufacturing (which employed approximately 3 000 persons on average between 1995-98).
Source: Statistics Finland.

Despite the relative importance of the cluster in the economy, the share of ICT firms is under 2% (4 000) of the total. In addition, the firms are relatively small from the global perspective. In terms of sales, Nokia is a leader in its own class in Finland, although in the global market it still lags many of its peers. Sonera, despite its rapid growth and aggressive penetration in new markets, is still a minor company in the international playground (see Table 2).

Foreign trade and international market position

In 1998, the export share of total ICT cluster production was over 60% (Table 1). In telecommunications services, exports represented still an insignificant share, while in equipment manufacturing, the share was about 85%. In 1999, ICT product exports represented 20% of total exports, while in 1990, the share was only 5%.

Figure 4 shows the trade balance in cluster products and illustrates the dominance of telecommunications equipment in Finnish ICT cluster trade. On top of telecommunications equipment exports, there is a margin of other ICT exports that has kept constant in value over time. In imports, the electronics industry is dependent on standard components (semi-conductors).

Figure 4. **Foreign trade in ICT and telecommunications products**

In EUR millions

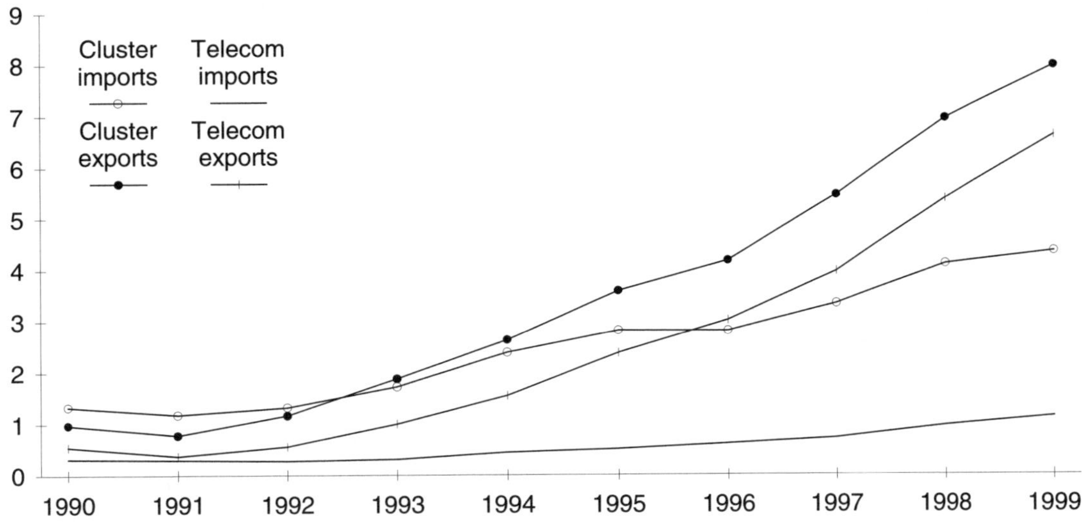

Source: National Board of Customs.

The pace and intensity of the growth of the Finnish electronics industry has been extraordinary throughout the 1990s. It has led to an industrial restructuring in the former forest and metal-based economy, in which knowledge has replaced capital, raw materials and energy as the dominant factor of production (Figure 5). During the past decade, Finland has become the world leader in high-technology trade surplus (ratio of high-technology exports/imports) among indigenous high-technology producers.

Figure 5. **Export shares by industry groups, 1960-2000**

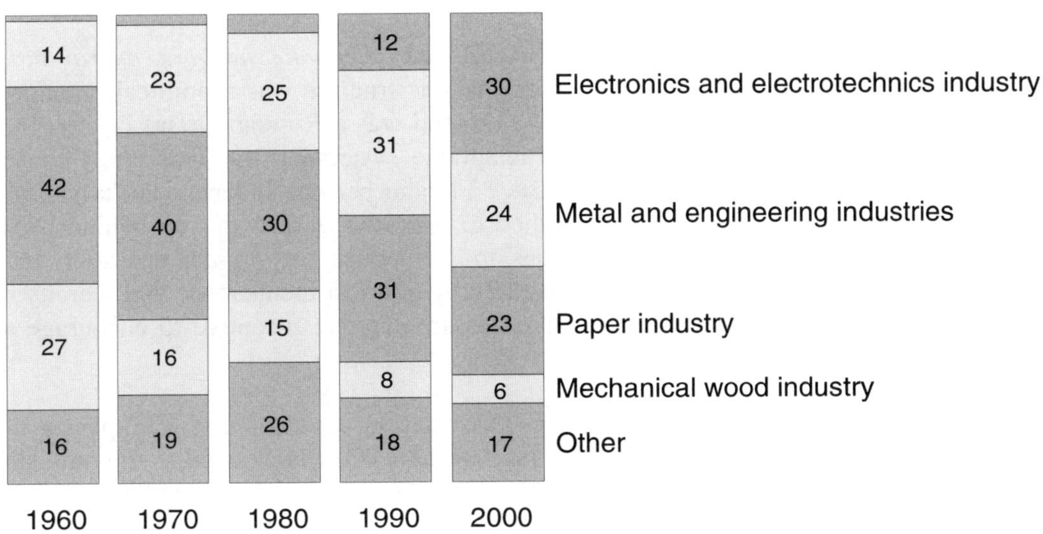

Source: National Board of Customs, ETLA.

In an OECD comparison for 1997, in ICT exports specialisation, Finland ranked joint second with the United States, while Japan ranked first (Figure 6, left-hand panel). Limiting the comparison to telecommunications exports reveals that Finland ranked first in specialisation in 1998 (Figure 6, right-hand panel). During the 1990s, Japan lost its lead to the two Nordic countries, which have been racing for the lead position.

Figure 6. **Export specialisation in 1997**

RSCA index

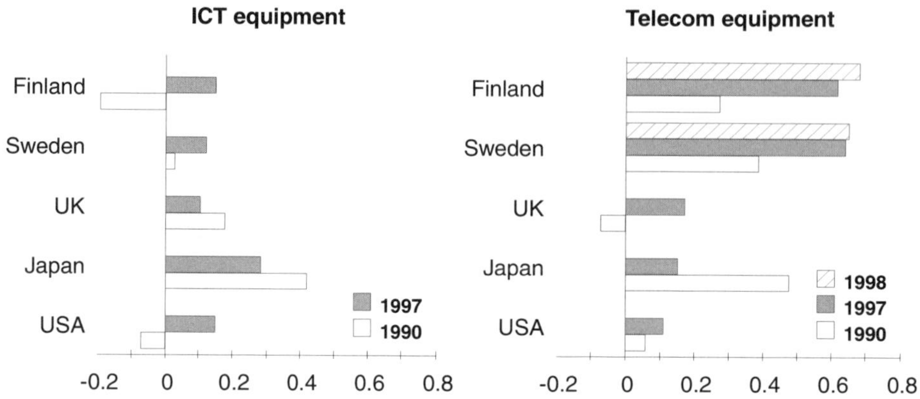

Note: 1998 data was not yet available for all countries. If the RSCA index equals zero, the country is as specialised in ICT cluster exports as the OECD countries on average. Thus, positive values indicate specialisation in cluster exports. See Hatzichronoglou (1996) for the definition of the index.
Source: OECD.

Components of the cluster system

The evolution of the ICT cluster

Unlike most of the European countries, in Finland *telephone network operation* was not monopolised by the state. Initially this fragmented market structure was a political outcome. At the time the first cables were being laid, in the 1800s, Finland was a Russian Grand Duchy. In order to complicate the potential seizure of the national telephone system by the Tsar, the Finnish Senate granted many licences in telephony operation.[4] Once Finland became independent, a national public telecommunications operator (PTO) was established to operate the network left behind by the Tsar. There were several plans throughout the decades to nationalise the private operation in view of harmonising the infrastructure, but they were frustrated by political incoherence and scarcity of public funds.[5] Indeed, the threat of nationalisation worked as an effective incentive to encourage technical upgrading.

In 1921, the private companies founded the Association of Telephone Companies aiming at administrative co-operation and joining forces in face of the PTO, which acted as the regulatory body authorised to redeem poorly performing operators. The Association dominated local operation in growth centres, while the PTO had monopoly over long-distance and international calls. Over the years, the Association grew to become a powerful opponent to the PTO, giving rise to a duopolistic market structure.

In the 1970s, the Nordic Telecom Conference, consisting of the national Post and Telegraph Administrations, engaged the industry in a research project on an automatic cross-border Nordic mobile telephone (NMT) network, which was intended to set the foundation for consumer-oriented mobile communications. Based on their experience, the Conference played an active role in initiating the Groupe Spécial Mobile (GSM) in 1982, and in the design of the pan-European digital mobile network.

The introduction of the NMT in 1981-82 made the Nordic countries (Denmark, Finland, Norway and Sweden) the world's largest mobile market in the early 1980s. The market, expanding at an unanticipated rate, started to attract the private sector. Private licence applications were, however, rejected by the regulator-PTO that pleaded for the economies of scale of natural monopoly. As a countermove, the Association founded a joint venture, Radiolinja, to operate a private GSM network. The private GSM licence application gave the decisive stimulus to swift deregulation and full liberalisation of the telecommunications market, finalised in 1994. The Telecommunications Services Act of 1987 separated the administrative and operational functions of the PTO, transferring the regulatory authority to an independent body under the Ministry of Transport and Communications.

As the winner of the regulatory battle, in 1991, Radiolinja was the first operator in the world to launch commercial GSM services. The liberalisation meant fundamental organisational and regulatory changes for the PTO, which was transformed into a public corporation. It started to actively improve its service and price efficiency and launched its GSM service soon after Radiolinja, among the very first in Europe. In 1994, the post and telecommunications functions were separated. The government began to reduce its ownership and indicated that further privatisation would take place in due course. In 1998, the name of the company was changed to Sonera to pinpoint the change in strategic focus and the change in direction towards mobile and media services.

The Finnish *telecommunications equipment market* differed from many foreign markets by allowing competition. Up until the 1980s, the market was dominated by leading foreign manufacturers such as Siemens, Ericsson and ITT. Attracted by the multi-operator market, they had set up production facilities in Finland. The leading resource-intensive foreign companies put pressure on the emerging domestic industry. As an illustration, in 1970, the turnover of the Siemens Group was EUR 2 billion – almost equalling the total Finnish state budget of EUR 2.5 billion (Mäkinen, 1995).

The seeds of the Finnish radiophone industry were planted in three companies: Salora, Suomen Kaapelitehdas and Valtion Sähköpaja in the 1920s. New radio technology was typically developed in the sideline of main activities by fervent engineers who often had to endure suspicion and opposition from their more conservative colleagues. During a complex organisational evolution process, finalised in 1987, the three companies merged under Nokia's roof.

Salora (established in 1928) was a manufacturer of TV and radio sets, with a strong brand-name beyond national borders and accumulated experience in serial production and marketing, which proved valuable in the later mobile phone business development. The development of radiophones, initiated in 1964, was based on pioneering experiments conducted in addition to core activities.

Suomen Kaapelitehdas (literally Finnish Cable Works, founded in 1917), in turn, was a producer of telecommunications cables. Trade with the Soviet Union, which originated during the deliveries of war indemnities, was decisive to the development of the company's technical skills.

Valtion Sähköpaja (literally State Electric Works, established in 1925) of the Ministry of Defence was founded to strengthen national development and production of strategic radio technology. After the wars, the activities were industrialised and merged with the R&D unit of the PTO. The company was renamed Televa and, in 1976, it became a state-owned limited company serving mostly public establishments for which it was the prime, although not exclusive, provider.

In 1963, the Army gave a decisive stimulus to the domestic industry by putting out an invitation for tenders for a radiophone. This was the first in a series of orders by which the government provoked companies to exceed their capacity to meet demanding technology requirements. Rather than a business opportunity, firms regarded the order as a chance to give physical form to the know-how accumulated "backstage". Ultimately, the Army did not have the funds to redeem the phone, but the prototypes served the bidding firms in developing new portable phones, which soon found their way to export markets.

In 1966, Suomen Kaapelitehdas was merged with Suomen Gummitehdas (literally Finnish Rubber Works) and Nokia, a 100-year-old wood grinding mill that gave its name to the new corporation. The merger secured R&D investments in telecommunications, which was now regarded as one of the strategic business areas of the company.

In the 1970s, it became apparent that the market was too small and resources too scarce for parallel development of digital exchanges in both Televa and Nokia. Consequently, the companies combined their R&D and marketing efforts on digital transfer technology in a joint venture, Telefenno, and finally, in 1982, introduced the first domestic digital exchange – only shortly after Ericsson, Alcatel, ITT and Siemens. It was the first fully digital exchange installed in Europe and thus served to convince the market of domestic competence *vis-à-vis* foreign manufacturers. For years, the exchange was the most successful export article of Nokia.

In 1979, Nokia and Salora, in turn, joined their complementary resources. Mobira, owned 50-50 by the two companies, was set up to market and develop radio technology, and especially the NMT terminal. The design of the NMT standard brought the Nordic telecommunications administrators and companies into close co-operation. While active in terminal development, the Finnish industry was not yet able to contribute to network specifications. Fierce pressure from the Finnish PTO's side to engage the domestic industry in cellular exchange development materialised finally, in 1981, in the base station supplied by Mobira. In the retrospect, it turned out to be crucial in maintaining the company's position in the emerging market.

The introduction of the NMT in 1981-82 marked the start of a fast-expanding new industry. The specifications were open to promote competition in equipment provision. No less than ten manufacturers entered the Nordic market.

Following its vision of global mobile communications, Mobira took substantial risks in technology development and market expansion.[6] By 1985, it held a leading position in a number of foreign markets, and the average annual growth rate of sales was 50%. Mobira allied with established foreign actors, teaching the company, among other things, the importance of branding – which later served to distinguish a Nokia from other mobile phones in the challenging consumer market.[7]

Finally, in 1986-87, Finnish telecommunications know-how was organised under one management when Nokia obtained full ownership of Mobira and Telefenno. In the search for rapid growth and global presence, Nokia ran into a crisis that almost destroyed the company. Severe external shocks, *i.e.* the collapse of Russian bilateral trade and the abrupt economic recession, aggravated the downturn. However, the crisis also provided the stimulus for a drastic dismantling of the firm's business sectors – which varied from tissue paper and rubber boots to cable machines and consumer electronics – leading to an exclusive concentration on telecommunications activities

(Figure 7). The structural changes were coupled with an important redesign of the company's management.

At the same time, the world witnessed a wave of telecommunications liberalisation. The boost in global demand for digital mobile equipment, coupled with Nokia's global position built up since the 1980s, saved the company from a dive that would probably have destroyed it.

Figure 7. **The structure of Nokia's sales, 1967-2000**

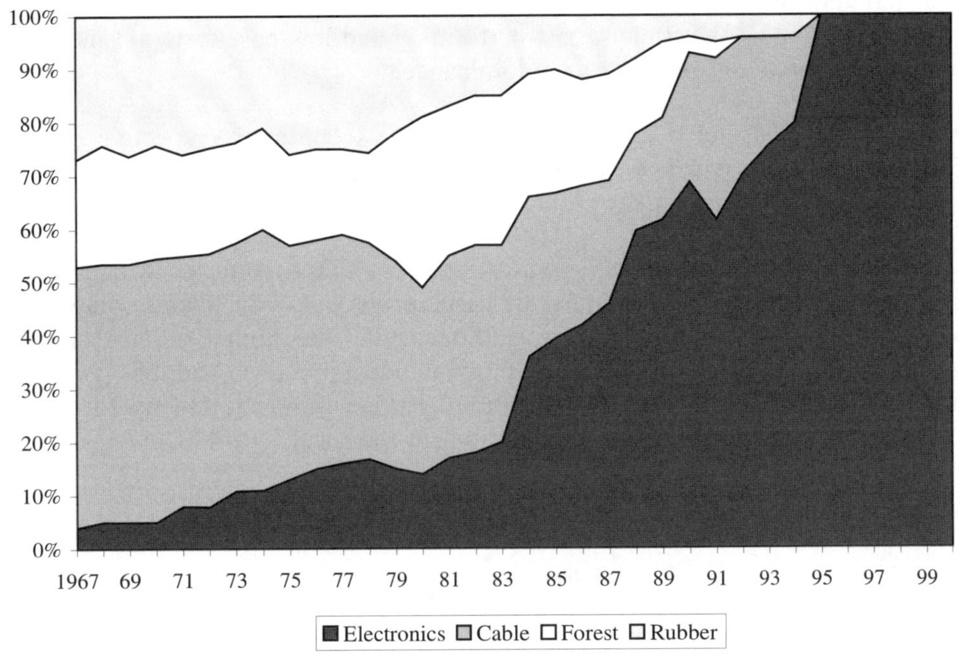

Source: Lemola and Lovio (1996), updated by ETLA.

Owing to the recession which hit consumer demand hard, it became crucial to put an end to the luxury image of the portable phone. With softer aesthetic design and a more user-friendly customer interface, Nokia invented the key to the consumer market. Since the first consumer-targeted model in 1994, Nokia has highlighted the life-style feature of communications in its brand building – a strategy that explains an important share of its breakthrough. In 2000, Nokia was the fifth most valuable global brand (Interbrand).

The facets of competitive advantage

Firm strategy, structure and rivalry

Nokia dominates the ICT cluster both by size and effect. The share of the company's domestic sales in the cluster turnover was close to 50% and its share in cluster exports was 66% in 1998. However, Nokia is not alone. There are several Finnish firms in the *equipment industry* that have important shares in global niche markets, especially in wireless and Internet technology.

For example, Comptel has attained the position of world leader in mediation device solutions (subscriber data management solutions for operators). Tecnomen was the first company to develop a unified messaging system and now leads the market for enhanced network service systems. In spring 2000, Iobox, one of the myriad of new technology-based start-ups, was named as one of the 100 most important companies in the world by *Red Herring* (a technology business magazine).[8] The company offers localised mobile content services over different platforms. SSH Communications Security, F-Secure and Softstone have established their positions in narrow but fast-growing niches in the highly fragmented data security industry.[9] In fact, network security solutions are becoming the backbone of the Finnish software industry. Benefon – founded by an ex-manager of Nokia – has attacked a global niche with analogue and GSM/GPS navigator phones. Linux is in a league of its own in the history of Finnish ICT. With its reputation as an efficient, fast-performing and low-cost system, Linux has been suggested as a possible alternative to Microsoft's predominance.[10]

The small home market and global competition define the scope and perspective of company strategies. The Finnish market provides a valuable development and test ground, but the actual business environment and reference groups are global from the outset.

There are some 120 *telecommunications operators* in the market, some of which operate on leased network capacity. Despite the large number of participants and fully liberal competition, the market is dominated, in equal shares, by two actors, Sonera and the Finnet Group (the renamed Association of Telephone Companies, an alliance of 46 private local operators and their jointly owned subsidiaries), which generate 95% of telecommunications service turnover. Owing to the lack of technically and economically viable solutions, competition in the local loop has not taken place as anticipated.

Figure 8. **Operator market shares in 1999**

In EUR millions

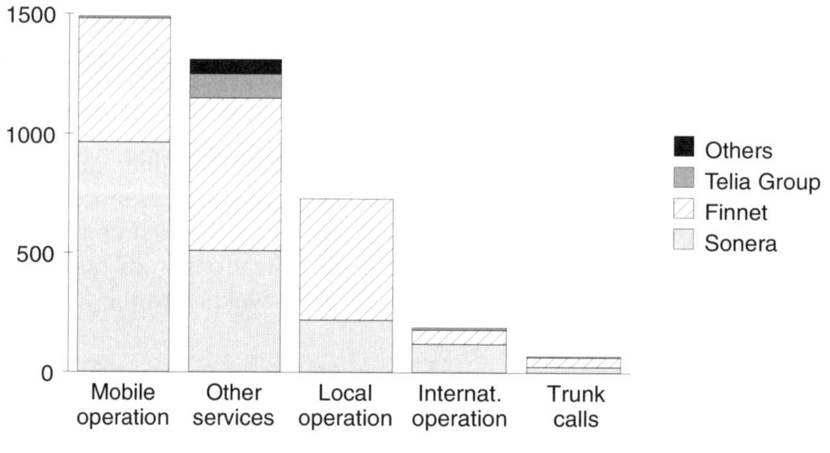

Note: "Other services" include data transmission incl. ISP, cable-TV, business solutions, equipment sales, international calls of the "others" (RSL Com Finland, Globetel, Global One, Telenordia) and Telia's advertising and equipment sales units.
Source: Ministry of Transport and Communications.

In mobile services, Sonera's main competitor, Radiolinja, is owned by the largest private operator Elisa Communications. Swedish Telia has not succeeded in eroding any significant share of the GSM market since its entrance in 1997.[11] The third national GSM network was launched in 2001 by a group of private operators.

Table 2. **Some Finnish ICT cluster firms in 1999**

Firm	Line of business	Sales (mEUR)	Personnel
F I N N I S H F I R M S			
NOKIA OYJ*	phones and network systems	19 620	51 177
SONERA OYJ*	telecom operator	1 849	9 270
TIETOENATOR OYJ*	IT solutions	1 220	11 058
ELISA COMMUNICATIONS OYJ*	telecom operator	1 070	5 489
ELCOTEQ NETWORK OYJ*	electronic manufacturing services	747	4 733
NOVO GROUP OYJ	IT solutions	310	2 100
PERLOS OYJ*	mobile phone enclosures	282	1 378
ASPOCOMP OYJ*	printed circuit boards	199	1 886
PKC GROUP OYJ*	communications cables	113	730
SCANFIL OY	mechanics and electronics manuf.	106	756
JOT AUTOMATION GROUP OYJ*	industry automation	99	565
EIMO OYJ*	mobile phone enclosures	77	681
DATATIE OY	data network services	71	255
TECNOMEN OY	enhanced network service systems	51	430
SAMLINK OY	electronic banking systems	45	229
BENEFON OYJ	mobile phones	39	296
F O R E I G N A C Q U I S I T I O N S (acquierer/new name in parentheses**)			
NOKIA DATA OY (**ICL INVIA OY*)(**UK)	information technology	353	1 902
MARTIS OY (**TELLABS OY***)(USA)	network access and transfer systems	*351*	*902*
NK CABLES OY(DRAKA HOLDING NV)(NL)	communications cables	*275*	*1 221*
KYREL EMS OY(**FLEXTRONICS FINLAND**)(USA)	electronic manufacturing services	*145*	*532*
SALORA-LUXOR OY(**SEMI-TECH TURKU)**(USA)	hotel TVs	*120*	*700*
NOKIA MAILLEFER OY (**NEXTROM OY**)	communications cables	*117*	*390*
LK-PRODUCTS OY (FILTRONIC PLC) (UK)	RF filters, access products, antennas	*83*	*849*
SOLITRA OY (**ADC TELECOM. OY**) (USA)	integrated RF solutions	*64*	*802*
ENVISET OY(**ESSEX COMMUN. EMS OY**)(USA)	electronic manufacturing services	*59*	*278*
F O R E I G N F I R M S W I T H R & D			
HEWLET-PACKARD FINLAND OY (USA)	information technology	*375*	na
ICL INVIA OY (UK)	information technology	353	1 902
SIEMENS OY (GER)	phones and network systems	298	1 329
ERICSSON OY (SWE)	phones and network systems	192	1 056

Note: Figures in *italics* are 1998 data. * Consolidated figures. ** Figures refer to the companies in **bold**.

Finland was the first country in the world to grant UMTS licences for all four GSM operators free of licence charges in early 1999. In the same year, licences in digital-TV operation were also granted. It is expected that the analogue TV network could be waived by the end of 2006.

Liberalisation and new business opportunities in the global arena have motivated some operators to increase their international operations. Foreign subsidiaries, international joint ventures and R&D co-operation are strong indications of the swift revolution in the once monopolised and regulated market.[12] It is, however, expected that there will finally be room for four to five global telecommunications companies, and a pitched battle is currently being waged for market positions in the global market. Governments are moving away from their traditional domain to free the hands of the national players. In Finland, the largest operators, Sonera and Elisa Communications, are looking for global partners with whom to share the uncertain future.

Factor conditions

Liberalisation of the capital market at the end of 1980s remodelled the institutional environment of corporate funding. In particular, the emergence of venture capitalists during the 1990s opened up unparalleled opportunities for innovative technology-based start-ups to enter the market at a very early phase of product development. Successful investment stories in the early 2000s have boosted the amount of available venture capital, and venture capital has actually become the most common source of capital for start-ups. The share of the ICT cluster in total capital investments has been around 30% over the last few years.

The availability of venture capital has reshaped the role of public funding, traditionally the prime resource for risky enterprises. New kinds of investment syndicates have come into existence, where the public sector carries the technology risk and a venture capitalist the commercial risk of an enterprise (based on Puhakka *et al.*, 2000).

The strong growth in the capital market has been one of the most noteworthy contributors to recent ICT cluster growth and increased versatility (together with the sizeable rise in public R&D funding; see section on "Government" below). In fact, a phenomenon like Nokia would not have been possible for a small country like Finland in the absence of foreign capital investments. Foreign ownership (around 90% in Nokia) shares risk in a country that is increasingly dependent on the unsettled ICT sector.

The bottleneck, however, is skilled labour. There is a structural mismatch in available skills not only at the macro level, but also *within* the cluster, notably in the software industry owing to the fast pace of technological development. Owing to the lengthy lead-time in education, increased intake has not yet alleviated the problem (see section on "Government"). In fact, there is a danger of watering down the education system by excessive intakes coupled with inefficient allocation of educational resources. In addition, the lack of employees draws both students and personnel from higher education institutions into industry, severely eroding the generation of future resources.

Demand conditions

Finland has been the leader in mobile penetration since 1996. In 1998, mobile subscribers outnumbered wired subscribers; 60% of households have both terminals, while 20% rely solely on mobile communications. Furthermore, mobile phone replacements have exceeded the number of new subscriptions (Ministry of Transport and Communications). In terms of the Internet host penetration rate, Finland ranked first, with 121 hosts per 1 000 inhabitants in 1999 (EITO, 2000).

Since liberalisation, the general telecommunications price level has declined by about 25% in real terms (Ministry of Transport and Communications). Although Finland has lost its relative price efficiency – at its highest in the mid-1990s – it was still leading in lowest prices for data (including Internet) and digital mobile services in the OECD in 1998.[13] Low prices – together with the introduction of cheaper portable phones to replace car phones – were major factors behind the breakthrough of consumer mobile communications.

The enthusiasm of the Finns in adopting the mobile phone has been explained by the "technology-oriented" Finnish character. In addition, the generally high level of basic education has been seen as a factor in high mobile diffusion. All in all, the home market has provided technology developers with a fruitful ground on which to experiment future products and services. After the global breakthrough of Nokia and Sonera, the Finns seem to have adopted the mobile phone as a national symbol.

In the early 1990s, Nokia utilised the advanced home market to develop consumer-oriented terminals. New models were designed to reflect individual life-styles and consumer aspirations investigated in polls – an innovation that was soon to be imitated by the firm's competitors. Today, the Finnish market has lost its importance in absolute terms (2% of Nokia's total revenue) in global operations, but the market still serves as an important laboratory since domestic demand anticipates future trends in foreign markets.

Supporting and related industries

In the recent years, the domestic *supporting sector* has become specialised in the needs of original ICT equipment manufacturers (OEM). The growth in production volume *and* in the share of outsourcing, together with increasingly sophisticated needs of the key companies, have generated an ever-growing number of new suppliers. By the same token, established firms have fine-tuned their product lines towards the specific needs of telecommunications growth companies.

The Finnish supplier sector has focused on highly customised inputs, while in standard components – which require large-scale and effective distribution channels – Finnish OEMs rely on imports. In Finland, special expertise can be found in contract manufacturing of parts and components (*e.g.* ASIC, rf-filters, printed circuit board production and surface mounting technology, hybrid circuits, silicon wafers), electronic manufacturing services (EMS), automation and precision mouldings.

Many supplier relationships have been stretched to OEMs' foreign markets. Suppliers' global presence has become increasingly important for efficient outsourcing. This has created enormous growth opportunities, but also risks for relatively small Finnish firms since foreign investment decisions cannot rely on one customer relationship.

The ICT key industries have a number of *related industries*, producing complementary or value-adding services to the infrastructure. The industries with the most promising prospects are those producing digital content. However, non-fixed definitions and backward statistical classifications severely complicate the valuation of the economic relevance of digital services, ranging from transaction and information services to education and entertainment.

The digital content industry in Finland is in its early phase of evolution. There are numerous signs of an emerging industry, but innovative enterprises often lack the skills crucial to professional business development and large-scale market penetration. For example, in 1999, there was a group of some 50 firms in the games and entertainment software production, generating EUR 10-12 million. Despite advanced technological skills, this group still operates on the fringe of the software sector, lacking sectoral concentration and volume (Autere *et al.*, 1999). However, technology leadership and new business models enabled by the Internet (digital mass distribution) provide new opportunities for the Finnish digital entertainment industry.

The favourable development and global competence of the Finnish content industry rank high on the national agenda. In 1999, the government initiated the Content Finland Programme, an inter-ministerial agenda for 2000-03 to improve the preconditions for Finland's development as a lead country in the provision – in addition to telecommunications technology – of content industrial products. The programme involves eight ministries under which new content products for a wide range of application areas will be developed in co-operation with businesses and other financiers (Ministry of Trade and Industry, 1999, http://www.minedu.fi/opm/hankkeet/sisu/index.html).

Government

The original Finnish cluster study was co-ordinated by ETLA in the early 1990s (Hernesniemi *et al.*, 1996). The cluster approach dominated the policy guidelines outlined in 1993 in the Ministry of Trade and Industry's National Industrial Strategy. The approach guided the construction of a novel platform for interaction between the parties involved (ministries, public and private research units, companies and relevant users).

However, rather than providing a radical approach towards industrial policy, the study served to convince stakeholders of the relevance of the policy direction adopted in Finland in the past: after an era of interventionist industrial policy in the 1960s-70s, there had been a shift towards more liberal public involvement in the 1980s. Knowledge, technology and competition were put at the top of the policy agenda in order to improve competitiveness, a necessary precondition for accessing international markets.

The government's adoption of the cluster framework was not an attempt to centralise its implementation or control. Instead, the approach has been applied as a conceptual tool in policy design by a number of national institutions and firms. Perhaps most importantly, "cluster thinking" has been effective in creating a dialogue between the public and private parties to the clusters by providing them with a common language (Rouvinen and Ylä-Anttila, 1999). (For an in-depth discussion of the Finnish cluster-policy framework, the reader is referred to the chapter by Romanainen in this volume.)

Government performance is now considered through policies concerning *technology, education* and *competition,* the core of the new Finnish industrial policy.

Technology policy. Dating back to the 1970s, the statutory separation between science and industry was terminated by the change in industrial policy orientation of the 1980s. The crucial role of science in technological development was recognised and, consequently, science policy was intertwined with technology policy in the newly established Council of Science and Technology, which co-ordinated the co-operation between the Ministries of Education and of Trade and Industry. This re-orientation of technology policy witnessed the foundation of the National Technology Agency (Tekes) in 1983, which became the main executor of technology policy.

The new industrial policy materialised in, for example, continuous growth in the share of R&D in GDP. Between 1985-99, the share doubled, reaching EUR 3.75 billion at the end of the period – over 3% of GDP. With this share, Finland ranks second in the world in terms of R&D input.

High R&D intensity reflects the government's 1996 decision to systematically increase R&D funding in line with the National Industrial Strategy (Figure 9). The target for 2001-04 is to increase the funding in line with the GDP growth rate. The share of public R&D funding was targeted to reach 40%; however, due to intense growth in the private share, it has so far attained only 30%. The additional appropriation for research went contrary to the trend towards significant downsizing of general public expenditure which accompanied the severe recession.

The allocation of additional funds was made, with specific attention being paid to cross-sectoral diffusion of knowledge. Thus, a share of these funds was directed to sectoral ministries' cluster programmes (see below). It should be noted, however, that the ICT cluster was not among the selected sectors as it was already covered by existing technology programmes.[14]

Figure 9. **R&D expenditure in some OECD countries**

As a percentage of GDP

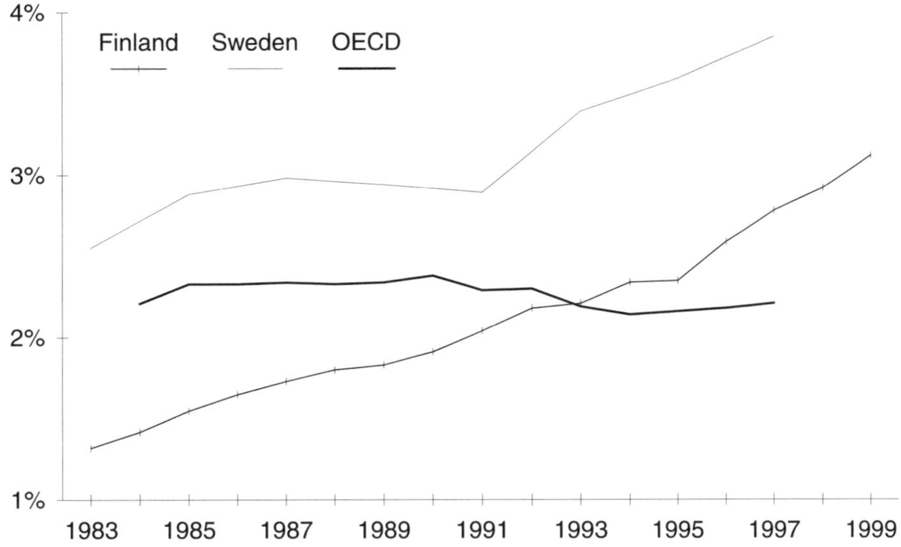

Source: OECD, *Main Science and Technology Indicators 1999.*

The convergence of ICT technologies and globalisation has meant that public funding has had to be redirected away from a technology orientation (beginning of the innovation chain) towards a market orientation (end of the chain). This new approach has given rise, for example, to a series of digital media technology programmes, which, contrary to established technology-oriented practice, have allocated R&D funding also to *service* development. This has supported the creation of export-oriented digital content service production.[15]

Education policy. The rapid growth of the electronics industry has exhausted the resources of available skilled labour. The Finnish Government has reacted by increasing openings in higher education institutions. Between 1993-98, the total intake in universities nearly doubled, and in polytechnics it nearly tripled. However, this has not eradicated the chronic lack of educated labour in the cluster. In early 1998, the government adopted a programme aimed at increasing further education in the information industry fields between 1998-2002.

Since 1999, there are 12 postgraduate schools in information technology. At present, the universities, together with the industry, have taken the initiative to establish an ICT-oriented university exclusively for foreign students in order to attract labour resources from overseas to Finland.

Competition policy. The regulatory approach in telecommunications policy is based on pro-competitive policies, light-handed regulation and technology-neutral competition. The market is subject to general competition and consumer protection legislation. The telecommunications authorities pursue a minimum interference policy, intervening mainly in cases of insufficient competition. The approach is less interventionist than in many other OECD countries. Some mandatory EU requirements have been enforced in Finland, rather reluctantly as they are considered to go against the liberal functioning of markets (Ministry of Transport and Communications).

Despite the policy objective of enhancing high-speed transmission capacity in Finland, the government decided in 2000 – unlike in Sweden – not to engage in direct infrastructure provision in order to ensure technology neutrality and free functioning of the market. In line with its liberal policy principles, Finland granted third-generation mobile network licences free of charge in comparative tendering, being among the few countries to date to support free distribution of new technology.[16]

By mandate from Parliament, the government intends to withdraw from the telecommunications business. However, there has been some political debate as to the share the state should ideally retain in this nationally important sector.

Coincidental factors

The 1990s saw several external incidents which had significant repercussions on the Finnish ICT cluster without which, it is fair to say, the average 30% annual growth rate of the electronics industry would not have materialised.

Following the agreements within the EU and the WTO, the traditionally monopolistic telecommunications equipment and service markets were gradually liberalised. The opening of the east European market gave an additional boost to demand for mobile equipment. The effects of liberalisation were momentous. Between 1990-98, the value of OECD exports of telecommunications equipment grew by a multiple of some 2.5, reaching USD 110 billion at the end of the period (OECD, 1999, and OECD trade statistics). Correspondingly, 96% of the OECD market, as measured by telecommunications revenue, was open to competition by the beginning of 1999.[17]

In contrast, the collapse of the Soviet Union, together with the severe recession in Finland, severely hit ICT cluster demand in the early 1990s. Without the counterbalancing effects of market liberalisation, Finland's path to economic recovery would have been somewhat different.

Cluster innovation patterns

R&D activities

In an OECD comparison for 1997, the share of private ICT-related R&D in total manufacturing R&D was highest in Finland.[18] In fact, between 1991-97, Finland moved from being a below-average investor to a leader. The share of the electronics industry accounts for two-thirds of total private R&D expenditure, of which Nokia alone accounts for an estimated 40%.

Firms in the ICT service sectors were the most involved in innovation activities in Finland in 1996. The electrotechnics industry, although it includes a host of non-ICT firms, was among the most innovative industries (after the paper and the chemical industries).[19] Innovation in ICT is reflected in Finnish patent counts. Almost 30% of the patents granted to Finland in the United States were in the field, which was the highest ICT patent share in the OECD in 1998 (OECD, 1999b).

During the 1990s, a number of multinational ICT manufacturers (*e.g.* ICL, IBM, Siemens, Hewlett Packard and Ericsson) extended their R&D activities in Finland, and sought intensified co-operation with local firms.[20] Finnish-based R&D units have become strategic knowledge centres, supplying the multinationals with information on new communications technologies (see also Pajarinen and Ylä-Anttila, 2001). The share of foreign affiliates in ICT-related R&D, however, is not significant compared to the domestic expenditure.[21]

Finland has remained the main R&D base for Nokia despite its network of 52 research centres in 14 countries. According to one estimate, around 60% of the company's R&D input are spent in Finland (Ali-Yrkkö *et al.*, 2000).

Industrial co-operation

Since the mid-1990s, there has been an increasing tendency towards intensified outsourcing and collaboration in the ICT cluster. According to our study on network relations in the cluster in 2000, vertical relationships play a crucial role in innovation and firm upgrading through multidimensional interactions. There is a strategic shift towards generating increased value added from vertical relationships, pertaining especially to complex and strategic components or sub-systems. From the network perspective, these kinds of relationships seem to be more common in the proximity of the original equipment manufacturer (OEM), than in the lower tiers of the production value chain, where supplier relationships still have the more traditional features of standard outsourcing.

Nokia has acted as an effective catalyst and forerunner in creating "new-generation" relationships. Starting in the 1980s with standard outsourcing to buffer demand fluctuations, in the 1990s the company started refining its supplier strategy towards value-adding partnerships, to induce innovation and to dedicate in-house resources to core competence building. At present, Nokia is extending its supplier co-operation to R&D activities, which is a sign of its moving to the next generation in supplier relationships.

The number of Nokia's first-tier high-technology partnerships has been estimated to total some 300 companies in Finland. Partners are selected through careful screening and audition procedures to ensure a match in objectives and firm cultures, and to gain a thorough knowledge of the managerial and operative capabilities as well as the development potential of candidates.

Owing to the critical role of knowledge transfers and real-time information flows in co-operation, inter-firm communication has been characteristically brought down to operative levels to enable direct interaction between the corresponding functions of firms. Nokia, particularly, is famous for its decentralised decision-making system, allowing rapid advancement in, for example, product development.

As Nokia holds exclusive knowledge on certain technologies developed in house, co-operative projects occasionally start with a period of intensive knowledge transfer to the supplier. Suppliers with a strategic position in the network are typically engaged in early-phase product and production process design to ensure, for example, enhanced solutions and efficient product life-cycle management.

The efforts to improve co-ordination and interoperability within the network have induced operational improvements on both sides of the firm interface. Internet-based network solutions are gradually replacing operative point-to-point communication, and harmonising the plethora of firm interfaces within the network.

The networked production paradigm, enhanced by co-operative long-term relations, can be seen behind much of the superior performance of Nokia and the Finnish ICT sector in general. Companies have been able to improve their focus, increase flexibility, share risk and exploit the expertise of other specialised firms. Our data indicate that the great majority of firms can apply the technical skills acquired in customer projects in other contexts, too.

Although Nokia is an exemplary case, strategic features in firm interaction have become increasingly common in the cluster as a whole. There are, however, still plenty of opportunities to be reaped in the cluster. Our study implies that suppliers would be willing to go deeper into the client relationship and take more responsibility in the design phase of product development. This is in line with the intentions expressed by manufacturers, who look for improved efficiency in outsourcing through fewer supplier interfaces. There is also an outspoken desire to increase the share of risks and rewards of suppliers in product development. However, the relatively small firm size limits the extent to which greater responsibility can be allocated to Finnish suppliers.

Although technology development is often client-dependent, as indicated by our study, co-operation has not overpowered independent R&D activities. As Nokia-dependence tends to become excessive in some firms, individual product development is often regarded as an important means of increasing the independence of the client and improving competitiveness through technological distinctiveness.

Industry-science interaction

The share of all Finnish firms having co-operation arrangements with universities or public research institutions was 40%, the second highest in the OECD area, in 1994-96 (OECD, 1999b). After an era of legislative separation which lasted until the turn of the 1980s, co-operation between the industry and academia in R&D has begun to strengthen (see chapter by Romanainen in this volume).

In fact, the interaction between the industry and ICT-oriented universities (in Helsinki, Lappeenranta, Oulu and Tampere) dates back to the early phases of the industry. An important party in the science-industry dialogue is the Technical Research Centre (VTT, founded in 1941), the largest public research unit in Finland. Intrinsic enthusiasm towards new technology, and in many cases personal relationships built up during studentship, were strong drivers in co-operation. The somewhat exceptional legislation, which protects the intellectual property rights of academia on their inventions, has contributed to a number of start-ups based on academic research. In addition, ICT academics have connections to the industry through a number of board memberships.

The shortage of skilled employees has further activated the discourse between universities and the industry. As the skills currently required by companies cannot always be found in textbooks, "real-time" knowledge transfer from the industry has been regarded as an important means of targeting education towards the demands of companies. In early 1998, the government adopted a programme for increasing education in the information industry fields between 1998-2002. The industry has committed itself to its implementation by providing internships (avoiding recruitment of under-graduates), and by aiming at increased participation in training and education.

Conclusions: government policies and cluster dynamics

The Finnish ICT cluster has been evolving for 100 years. The cluster as we see it today looks like the product of a master plan in industrial policy; an internationally attractive and vigorous industrial system boosting innovation and external trade. Considering the role of the government in the cluster evolution process, the primary motives and initiative in implementing telecommunications sector-related policies have varied over time. However, the great majority of actions have served as the "appropriate" measures of government identified by Porter (as in Rouvinen and Ylä-Anttila, 1999), *i.e.* measures that create a context which encourages upgrading and establishes a stable economic and political environment.

Market structure-driven dynamics

Among the most influential factors behind the evolution of the ICT cluster has been the competitive environment, which has given rise to multidimensional and far-reaching dynamic impacts.

At the outset, in the late 19th century, the dispersed telephony market structure was motivated by the Finnish Senate's objective to ensure sovereignty in telecommunications under the Tsar's reign. The division of infrastructure between public and private ownership promoted technical improvements to the infrastructure, and created the preconditions for equitable competition in the later (1994) liberalised market.

The multi-operator market, coupled with another local peculiarity, *i.e.* free equipment supply, provided an upgrading mechanism for the emerging industry. Domestic firms had to catch up with foreign competitors on leading technology. Furthermore, the variety of competing technologies available developed operators' skills in interface technology and enabled them to become sophisticated customers in technology development.

The government also played the role of a demanding customer. Domestic firms with incubating know-how on radio technology were motivated to introduce physical products which ultimately served as prototypes for exportables. Finally, technological co-operation between the Nordic telecommunications administrations and companies enabled the industry to become the vanguard of international mobile markets. The Nordic market was an invaluable training field for future challenges.

In telecommunications operation, the decisive stimulus for liberalisation came from private operators. The manoeuvre to penetrate the state's overpowered domain, initiated in the late 1970s, had wider repercussions than was initially anticipated. At the end of the day, after lengthy political arm-wrestling, the whole telecommunications field was opened to competition, among the first in the world.[22]

Competition stimulated rapid penetration of mobile services. For the Finnish equipment industry, the delivery of the network for Radiolinja, the first GSM operator in the world, was a reference that lifted it to the international stage. In the subsequent international wave of liberalisation, Nokia was well equipped to take off.

Once started, the swift liberalisation of telecommunications signalled the modern industrial policy orientation of the Ministry of Transport and Communications, reflected in the problem-solving attitude and active communication with the parties involved. Since the GSM licence dispute, politics have not played a role in the administration of the industry; rather, decisions have been directed by competitive and technological objectives (Häikiö, 1998).

Apparently, there was a pinch of good luck – rather than well-thought-out policy objectives – behind the perfect timing of liberalisation, the implications of which were momentous. The Finnish telecommunications sector got a head-start in the exploding GSM market, and its stimulus was substantial in supporting the revival and restructuring of the economy.

Cluster-oriented policy

After an era of interventionist policy-orientation, the 1980s marked a change in the government's approach to economic and industrial policy. Public ownership and the regulatory framework were seen in a new, more liberal light. The intertwining of technology, science and economy was regarded as the prime driver of societal change, and the idea was communicated by the adoption of the concept of a national innovation system in the policy outline (see Romanainen in this volume; Jääskeläinen, 2001).

The National Industrial Strategy of 1993 was shaped on the basis of the cluster approach and served to clarify the new role of the government as the creator of favourable framework conditions. The cluster approach has had a clear impact on subsequent government actions emphasising inter-organisational co-operation as well as the accumulation and transfer of know-how.

The major tool in cluster-policy implementation was the government's additional appropriation for research, totalling EUR 540 million. Implemented in 1997-99, it was allocated through, *e.g.* selected sectoral ministries to establish forums for wide-ranging co-operation between science, industry, users and financiers. The major channels of distribution were, however, the National Technology Agency (Tekes) and the research sector (Figure 10). Despite the fact that the cluster-specific programmes did not target the ICT industry, due to the good coverage of contemporary ICT-related programmes, the injection of additional R&D resources was allocated to the cluster through other channels.

The evaluation report of the effects of the additional appropriation concluded that the cluster-oriented research funding had been "highly successful".[23] The policy tool, as noted in the report, has had perceivable positive impacts on private research investments and on growth in both productivity and employment, arising from intensified R&D activity. Further, the report underlines the increase in networking – between industry and science, and between large firms and SMEs – which is "internationally recognised as one of Finland's strong points and has shown marked improvement over the past two decades" (p. 47). Yet, it is emphasised, the full effects of the funding will be manifested over the years to come.

Figure 10. **Breakdown of allotment of additional appropriation of research, 1997-99**

In EUR 540 millions, by source of funds

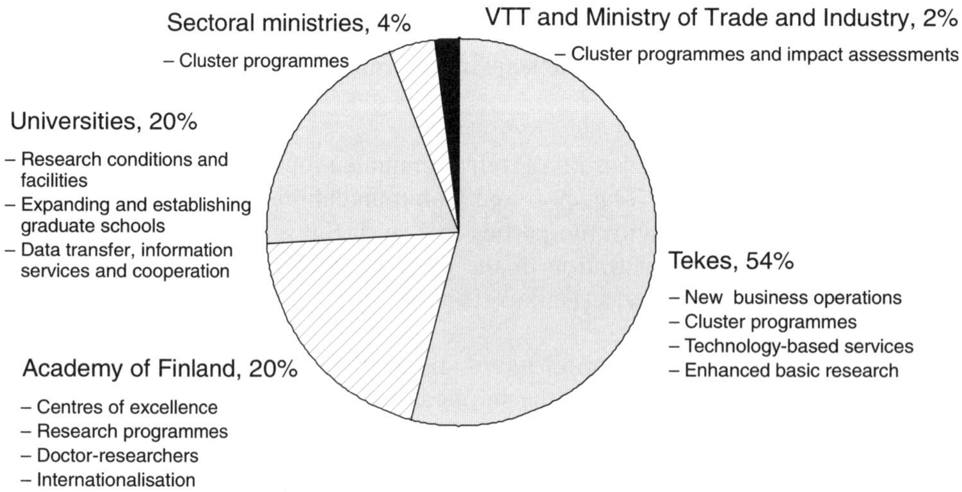

Note: VTT refers to the Technical Research Centre, while Tekes denotes the National Technology Agency.
Source: SITRA – Finnish National Fund for Research and Development, 2000.

Despite calls for an extension of cluster programmes to the ICT sector (Pentikäinen, 2000), the report considers that for the public sector to engage in a field as industry-driven as the ICT sector would be too active an intervention. According to the report, public involvement can be better justified in efforts related to, *e.g.* development of standards and promotion of new firms.

The policy options for the future as presented in the evaluation report, can be regarded as an initiative for the second generation of cluster-oriented policies. The main proposals highlight the following application areas for policy:

- *Strengthen the conditions for basic research.* A world-class research base is necessary to attract and maintain global companies' activities in Finland. In addition, the boundary between basic and applied research is becoming blurred, calling for attention to be paid to the focus of industrial research.

- *Improve the cluster approach.* The greatest potential is foreseen in areas where completely new linkages are created, especially in fields where improvement of public services can be implemented with innovative industrial products and processes.

- *Integrate the new and the old economies.* New business and management concepts for traditional industries.

- *Focus more on innovation.* An intensified shift away from a technology-push type of innovation policy to a customer-driven policy is needed. Increased cultural and commercial know-how is an important ingredient for successful R&D and a prerequisite for future success in global competition.

- *Develop the future competencies of the workforce.* The mismatch between educational resources and requirements calls for a structural- and content-related change in the system.

The above proposals reflect the rapid shift from a capital- and resource-based economy to one founded on knowledge and competence. As competition in global markets has taken on the form of learning competition, as noted in the report, there is no time to take a break.

The proposals – although expressed in general terms with reference to a range of sectors in the Finnish economy – draw essentially from the future prospects and requirements of the ICT cluster. In a sense, they also provide an indication of the mechanisms by which industrial policy is made in Finland today.

The emergence of the Finnish ICT cluster is the outcome of dynamic interaction within the innovation system, in which the government has assumed the role of facilitator and co-ordinator. The setting for industrial policy design is characterised by intensive and informal communication between government, industry, academia and the labour market. The system is based on a clear definition of the role of each actor, as well as on a shared view of the policy objectives and tools – excluding subsidies or other direct supports.

This interaction has been institutionalised in the Council of Science and Technology. Chaired by the Prime Minister and represented by key participants, it has high status in the industrial policy system – a situation which remains uncommon on the international stage. Industrial associations are another institution with a important role in the Finnish policy arena, acting as influential intermediaries between the industry and the public sector.

Through online communication, the government keeps itself abreast of changes in the economic and technological environments. As suggested by Castells (1996), in a globalising society, modern industrial policy needs to move away from a planning and control role and to adapt to the shift in decision making away from public institutions to networks of actors (Jääskeläinen, 2001). This new role will be best dealt with through the creation of favourable conditions conducive to straightforward interaction between the various stakeholders.

Annex

NACE CODES UTILISED IN THE CALCULATION OF ECONOMIC INDICATORS FOR THE ICT CLUSTER

ICT Manufacturing

30020 Manufacture of computers, etc.

31300 Manufacture of insulated wire and cable

32100 Manufacture of electronic components

32200 Manufacture of radio transmitters, etc.

32300 Manufacture of radio receivers, etc.

ICT Services

Telecom services

64201 Telephone communication

64202 Other telecommunications

64203 Data transmission services

Software and IT services

72100 Hardware consultancy

72200 Software consultancy and supply

72300 Data processing

72500 Maintenance of office machinery, etc.

NOTES

1. Luukkainen and Mäenpää (1994) carried out the telecommunications cluster study as part of the national cluster identification project, co-ordinated by ETLA (Hernesniemi *et al.*, 1995).

2. Clusters do not follow sectoral boundaries. Sectoral data inevitably includes firms that are not active in the cluster, and alternatively, excludes many important actors. For example, national statistics do not yet enable quantification of digital content production which is, however, largely included in the data on telecommunications operation and software production. Further, it is necessary to combine the data for electronic components (inputs) and ICT equipment (outputs), since many of the input suppliers are classified under the sector code of their main clients. Despite classification problems, the national data applied here cover the crucial business sectors of the cluster. See Annex for the NACE codes used.

3. In contrast, the share of the largest cluster, *i.e.* the forest cluster, was 9.6% of GDP in the same year (Lammi, 2000). Owing to higher growth rates in the telecommunications sector, the ICT cluster is about to become the largest industrial cluster in Finland.

4. In the 1930s, there were no less than 815 private local telephone companies. Between 1950-65, the number of operators declined dramatically as structural regulations – with a view to general network development – forced minor companies to merge either with larger firms or with the PTO.

5. State redemption of the long-distance operation in 1934 was an exception to the rule. There were also occasional acquisitions of operators by the state, motivated by national defence and technical concerns.

6. Mobira manufactured equipment for five standards adopted in different countries. Only Motorola supported an equal amount of standards.

7. Mobira obtained access to the US market under an OEM agreement with Tandy Corporation which offered an extensive distribution channel. The alliance with Alcatel and AEG for marketing and system development opened the doors of the French and German PTOs and gave credibility to the emerging mobile manufacturer. Co-operation was gradually terminated once the company became capable of independent supply of a GSM system in 1991.

8. Nokia was the only other Finnish firm ranked in the list.

9. For example, in 1998, SSH Communications Security was awarded the EC Esprit programme's European IT Prize for representing "Europe's strengthening position in information technology and telecommunications". Its cryptography and authentication technology (SSH Secure Shell) for Internet has become a *de facto* standard for log-ins.

10. The kernel (the central part of the operating system) of a UNIX-like operating system was developed by Linus Tordvals at the University of Helsinki in Finland. As a publicly open and free system, extendible by any contributor, it soon gained supporters from all over the world. Linux comes in versions for all the major microprocessor platforms, and it is commercialised by a number of companies.

11. Telia owns networks only in major cities; for nation-wide services, it has a leasing contract with Radiolinja. There are other mobile service providers in the market, as well.

12. Sonera has 12 joint ventures in mobile operation, and it is actively looking for more. The newly established mobile technology units (Sonera SmartTrust and SoneraZed) are targeted to international

markets. Sonera has a number of alliances with leading global ICT companies with which it develops new solutions for mobile communications. The largest private operator, Elisa Communications, in turn, has advanced through acquisitions in the German city carrier market.

13. In digital mobile service prices, Finland ranked first in residential services but seventh in business services (Ministry of Transport and Communications, 1999; OECD, 1999a).

14. A small ICT-cluster-based programme was established promote the exploitation of information networks in SMEs.

15. Digitaalisen Median Sisältötuotteet (Digital Media Content) Programme initiated in 1996 (TEKES, 1999).

16. The choice of a third-generation technology standard is not restricted by the licence.

17. ITU (1999). The remaining six countries are committed to liberalisation in coming years.

18. OECD ANBERD data was used in the calculation of the RSCA index (see Appendix 2 for the formula) for the ICT R&D specialisation.

19. In telecommunications services, IT services, and the electrotechnics industry, 86%, 68% and 58% of firms, respectively, engaged in innovation activities between 1994-96, while, on the average, 34% of all firms were innovative. Community Innovation Survey 2 (Statistics Finland, 1998).

20. For example, in 2000, Hewlett Packard set up the *Mobile E-Services Bazaar*, an innovation centre in Finland to co-ordinate mobile service development in Europe. The objective is to gather together communications technology developers from different fields to co-operate in innovative e-services development. IBM has expressed its intentions to establish a similar kind of innovation centre in Finland.

21. The share of foreign affiliates in total manufacturing R&D was around 10% in 1997, compared to the most "internationalised" country, Ireland where the share was 68% (in 1993) (OECD, 1999b).

22. With the exception of the NMT network that is being gradually closed down.

23. SITRA (2000). The expert group for the evaluation was established by the Ministry of Trade and Industry and the Ministry of Education

REFERENCES

Ali-Yrkkö, J., L. Paija, C. Reilly and P. Ylä-Anttila (2000), *Nokia – A Big Company in a Small Country*, ETLA – The Research Institute of the Finnish Economy, Series B162, Taloustieto, Helsinki.

Autere, J., I. Lamberg and A. Tarjanne (1999), *Ohjelmistotuotteilla kansainväliseen menestykseen*, TEKES – The National Technology Agent, Technology Report 74/99.

Castells, M. (1996), *The Information Age: Economy, Society and Culture,* Vols. I-III, Blackwell Publishers, Oxford.

Hatzichronoglou, T. (1996), "Globalisation and Competitiveness: Relevant Indicators", *STI Working Papers 1996/5*, OECD, Paris.

Hernesniemi, H., M. Lammi and P. Ylä-Anttila (1996), *Advantage Finland – The Future of Finnish Industries*, ETLA – The Research Institute of the Finnish Economy, Series B113, Taloustieto, Helsinki.

Hienonen, R. (2000), *Elektroniikan- ja sähköalan kehitysnäkymät 2000...2005*, VTT Automaatio, Helsinki.

Häikiö, M. (1998), *Alkuräjähdys*, with English summary: *The Big Bang of GSM – Mobile Phone Revolution*, The Story of Radiolinja, Finland, 1988-98, Edita, Helsinki.

Jääskeläinen, J. (2001), *Klusteri tieteen ja politiikan välissä – Teollisuuspolitiikasta yhteiskuntapolitiikkaan* [Cluster – Between Science and Policy. From Industrial Policy to Social Policy], with English summary, ETLA – The Research Institute of the Finnish Economy, Series A33, Taloustieto, Helsinki.

Lammi, M. (2000), *Metsäklusteri Suomen taloudessa*, ETLA – The Research Institute of the Finnish Economy, Series B161, Taloustieto, Helsinki.

Ministry of Trade and Industry (1993), *National Industrial Strategy for Finland*, Publication 3/1993, Industry Department, Helsinki.

Ministry of Trade and Industry (1999a), *Elektroniikan sopimusvalmistus*, Report of the Electronic Manufacturing Services (EMS) Industry, http://ktm.tt-tietopalvelut.fi/ktm/fin/select1.html.

Ministry of Trade and Industry (1999b), *Sisältötuotanto – sisältö- ja kulttuuriteollisuus*, Report of the Content Industry, http://ktm.tietopalvelut.com/ktm/fin/ssisalto1.html.

Ministry of Transport and Communications (1999), *Suomen telemaksujen hintataso*, Publication 14/99, Helsinki.

Ministry of Transport and Communications (2000a), *Case Mobile Finland*, Publication 16/2000, Helsinki.

Ministry of Transport and Communications (2000b), *Televiestintätilasto 2000*, Helsinki.

Mäenpää, K. and S. Luukkainen (1994), *Teletekniikasta monimuotoiseen viestintään – Teleklusterin kilpailukyky*, ETLA – The Research Institute of the Finnish Economy, Series B96, Taloustieto, Helsinki.

Mäkinen, M. (1995), *Nokia Saga*, Gummerus, Jyväskylä.

OECD (1999a), *OECD Communications Outlook 1999*, OECD, Paris.

OECD (1999b), *Science, Technology and Industry Scoreboard 1999: Benchmarking Knowledge-based Economies*, OECD, Paris.

OECD (2000), *Mobile Phones: Pricing Structures and Trends,* OECD, Paris.

Paija, L. and P. Ylä-Anttila (1996), *The Impact of Structure and Competition on Employment in the Telecommunications Cluster – Case Finland*, ETLA – The Research Institute of the Finnish Economy, Discussion Papers No. 549, paper presented at the OECD Workshop on Economics of the Information Society, Istanbul, 14-15 December 1995.

Pajarinen, M, P. Rouvinen and P. Ylä-Anttila (1998), *Small Country Strategies in Global Competition. Benchmarking the Finnish Case*, ETLA -- The Research Institute of the Finnish Economy, Series B144, SITRA 203, distributed by Taloustieto, Helsinki.

Pajarinen, M. and P. Ylä-Anttila (2001), *Maat kilpailevat investoinneista – teknologia vetää sijoituksia Suomeen* [Countries Compete for Investment – Finnish Technology Attracts Foreign Firms], Taloustieto, Helsinki.

Pentikäinen, T. (2000), *Economic Evaluation of the Finnish Cluster Programmes*, Working Papers No. 50/00, VTT, Group for Technology Studies, Espoo.

Porter, M. (1990), *The Competitive Advantage of Nations*, The Macmillan Press Ltd., London.

Puhakka, M., P. Rönkkö and D. Steinbock (2000), *Growth and Internationalisation of Technology-based, New Companies within the Finnish ICT Cluster*, in the forthcoming ICT cluster study report of ETLA – The Research Institute of the Finnish Economy.

Pulkkinen, M. (1996), "Miten jättiläisiä horjutetaan?", in T. Lemola and R. Lovio (eds.), *Miksi Nokia, Finland?*, WSOY, Juva.

Pulkkinen, M. (1997), *The Breakthrough of Nokia Mobile Phones*, Doctoral Dissertation, Helsinki School of Economics and Business Administration, A-122.

Rouvinen, P. and P. Ylä-Anttila (1999), "Finnish Cluster Studies and New Industrial Policy Making", in OECD, *Boosting Innovation: The Cluster Approach*, OECD, Paris.

SITRA – Finnish National Fund for Research and Development (2000), *Assessment of the Additional Appropriation for Research*, SITRA Report Series 2.

Statistics Finland (1998), "Innovaatiotutkimus 1996", *Tiede ja teknologia* 1998:3.

Statistics Finland (2000), *Joukkoviestimet - Finnish Mass Media 2000*, Helsinki.

Tavaila, P. (1989), "Suomen teletoimen organisaation kehittyminen", in Ministry of Transport and Communications, *Televiestintätilasto 1989*.

TEKES – The National Technology Agent (1999), *Suomalainen uusmedia – Eväät kasvuun ja kansainvälistymiseen*, Technology Programme Report 3/1999.

Toivola, K. (1992), *Poimintoja teletoimen historiasta*, Vol. 4, TELE Matkaviestinverkot.

Turpeinen, O. (1996), *Yhdistämme*, Vols. 1-2, Edita, Helsinki.

Chapter 3

THE BOUNDARYLESS CLUSTER: INFORMATION AND COMMUNICATIONS TECHNOLOGY IN IRELAND

by

Roy Green, James Cunningham, Imelda Duggan, Majella Giblin, Mike Moroney and Leo Smyth
School of Business and Economics, National University of Ireland, Galway

Introduction

As an emerging knowledge-based economy, Ireland has one of the highest concentrations of information and communications technology (ICT) activity and employment in the OECD. This activity comprises primarily electronics hardware manufacturing, such as personal computers (PCs); software products and services, especially business application products and "localisation"; and call centres. It has largely been driven by foreign direct investment (FDI), although recent evidence suggests that indigenous firms are now growing at a much faster rate than the multinational sector. It is also a major factor in high-skill, high-wage job creation since the early 1990s and comprises a key element of Ireland's national innovation system, which is also a regional one in the European context. Significantly, policy accompanying progress towards a single market in the European Union (EU), far from stifling innovation and diversity in the regions, has mobilised it as a powerful new source of competitive advantage.

This chapter identifies the main characteristics of the Irish ICT sector, with particular reference to the apparently spontaneous development of a software cluster in and around the city of Galway, and evaluates the respective contributions of public policy and local initiatives. The chapter suggests that this regional cluster is "boundaryless", in the sense that its global character defies traditional stereotypes of domestic rivalry and collaboration. However, it also finds that this is not inconsistent with Michael Porter's observation that "the enduring competitive advantages in a global economy lie increasingly in local things – knowledge, relationships, motivation – that distant rivals cannot match" (1998, p. 78). Finally, the chapter reflects concern that, while Ireland is a leading exporter of ICT products and services, it is a net importer of new technologies with an historically weak but improving innovation performance in key areas, most notably research and development (R&D), as well as in the take-up and diffusion of ICT across the economy as a whole.

Clusters in context

Over the past decade, Ireland has experienced the fastest growth rate of output and employment of any country in the OECD, with annual average growth rates of 8% and the number of jobs across the economy increasing by nearly 50% between 1990-2000 (Green, 2000) (Figure 1). While the services sector was the main source of this increase, manufacturing employment also grew rapidly, for the most part in high-value-adding, knowledge-based activities (Figure 2). The scale and composition of this growth have given rise to the term "Celtic Tiger", and initial doubts about its sustainability have been met with evidence of far-reaching cumulative structural change and adaptation.[1]

Figure 1. **Employment change, 1990- 2000**

Percentages

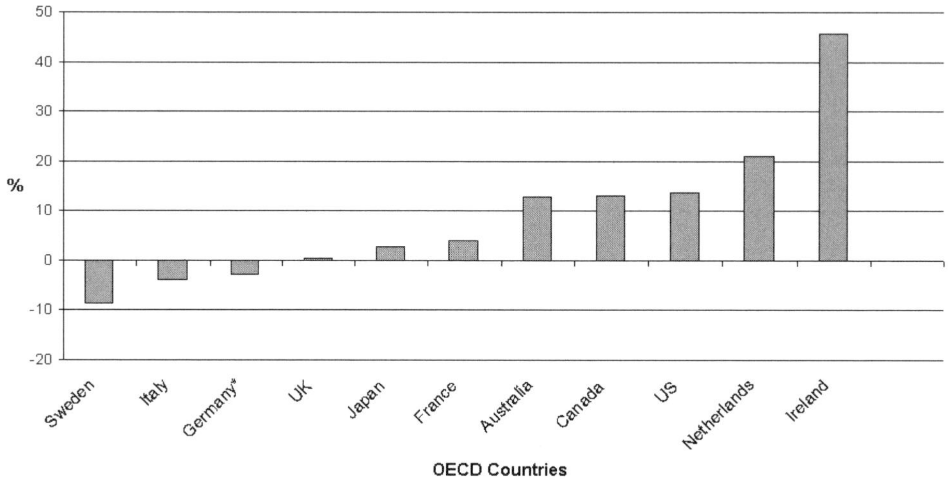

Note: Germany: employment growth, 1991-2000.
Source: OECD.

Figure 2. **Manufacturing employment change, 1990-97**

Percentages

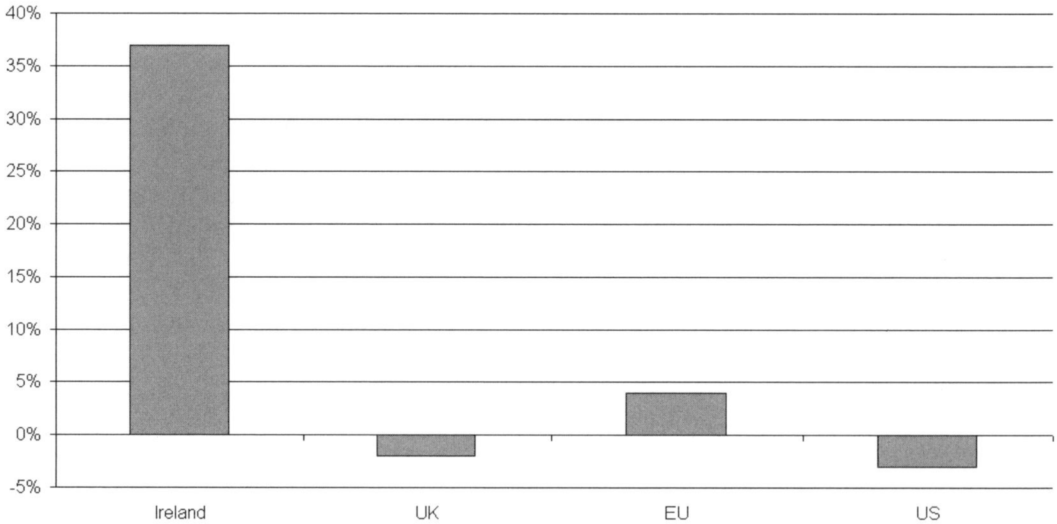

Source: OECD and Eurostat.

Irish growth is essentially export-led, with over 90% of GDP sold abroad and an average increase in export volumes of more than 12% a year during the 1990s. Indeed, export growth shows signs of accelerating even further this year, despite the worldwide slowdown. Ireland has consistently recorded a trade surplus over this period in excess of 10% of national production, due to a very large extent to the impetus provided by a world-scale, export-oriented and increasingly sophisticated ICT sector. This surplus has been a key ingredient in Ireland's improved capacity to overcome the balance of payments problems that characterised its disappointing performance in the 1980s.

For the purposes of analysis, the ICT sector in Ireland may broadly be disaggregated into: *i)* a large electronics hardware industry which produces a standardised output of PCs, PC components and office machinery (NACE 3002, 313, 321, 322, 323); *ii)* a smaller but more rapidly growing computer software industry which combines elements of standardisation and differentiation in a constantly evolving product mix (NACE 72); and *iii)* a teleservices or call centre industry associated with PCs (NACE 30) and software (NACE 72), providing a spectrum of services from routine inquiries to advanced problem-solving with a reach across domestic and European markets. These elements of ICT are increasingly related through producer-user linkages as well as shared innovation processes and skills pools. There are also a number of content sectors relating to software and multimedia developments in the areas of entertainment, education and tourism (NACE 221, 222, 74.4, 92.1, 92.2, 92.4, 92.51) and related telecommunications services (NACE 64.2). However, these areas are more difficult to quantify due to data limitations, and are discussed in this chapter in conjunction with the software industry.[2]

Ireland is now the *fifth largest exporter of computers* in the world, despite its small size (OECD, 2000). These account for more than one third of Irish exports, with a third of the PCs sold in Europe manufactured in Ireland. Recent data also indicate that Ireland has the highest proportion of high-technology industries represented in its manufacturing exports of all OECD countries (OECD, 1999). In addition, with 34% of the global market, Ireland is the *biggest exporter of software products* in the world, having overtaken the United States in 1998. Over 40% of packaged software and 60% of business application software sold in Europe is produced in Ireland and, while much of this localises or simply replicates software originating in the United States, an increasing proportion is research-intensive, high-value-added production. International demand is the main factor in continuing export growth not only for multinational companies in Ireland but also for the indigenous software industry which exports almost 60% of its output (Travers, 1999; Clancy *et al.*, 2001).

The key drivers of Ireland's trade performance in ICT have been the immense scale, high quality and local "embeddedness" of inward investment, with FDI accounting for two-thirds of manufacturing output and over 80% of manufacturing exports. This has been a deliberate strategy of the Industrial Development Agency (IDA), at least since the influential Telesis report of 1982, supplemented by the role of Enterprise Ireland in the promotion of indigenous supply chains (Forfas, 1996). In 1998, Ireland attracted FDI inflows of USD 6.8 billion, which makes it one of only four countries in the OECD – along with Finland, Sweden and the Netherlands – where FDI amounts to more than 8% of GDP. Moreover, with only 1% of the EU population, Ireland gained *23% of all FDI projects in Europe* in 1997, covering manufacturing, software, teleservices and shared services projects. This is a remarkable achievement, but one which, as we shall see below, is not wholly positive.

In computer software, Ireland has the largest market share of FDI in Europe with 55% of the total, more than twice the market share of the next most successful country (France at 21%). Similarly, Ireland's market share of FDI in teleservices (over 28% in 1994-97) and in shared services such as back-office activities (37% over 1996-97) is also the highest in Europe. In electronics, Ireland's market share is second only to the United Kingdom at around 22%, and the gap is closing. In the manufacturing sector, too, Ireland's market share of FDI increased to 13% of the total by 1997, behind the United Kingdom and France. Finally, it is noteworthy that 19 of the top 25 computer firms in the world have manufacturing operations in Ireland, with Microsoft recently committing itself to major European "hub" facilities in Dublin.

While the tax regime, telco facilities and EU membership have all played their part in attracting high levels of FDI (Gunnigle and McGuire, 2000), Ireland's human resource base has also been an important factor for the ICT sector. This goes beyond use of the English language. It is recognised that "the ability of a country to attract, successfully absorb and benefit from foreign direct investment, and the transfer of technology which it may bring, depends to a large extent on its own technological capabilities, of which the skills and technical knowledge of its workforce are critical components" (ILO, 1999). In the 1998 IMD *World Competitiveness Report*, Ireland was ranked first in the world for the "fit" between its educational system – with its high output of third-level graduates in computer science and engineering – and the needs of a competitive economy. According to recent analysis, "[P]ast national strategies for investing in education and training have paid off in terms of faster productivity growth and higher levels of productivity at the aggregate level, and higher earnings and employability at the individual level" (OECD, 1998b).

Yet the economy's dependence on FDI has also encouraged the use of *imported* rather than locally generated technologies. This is reflected in the very large deficit in Ireland's "technology balance of payments", which measures flows in knowledge and "disembodied" technologies between countries (Figure 3) and which in turn is due to Ireland's traditionally low R&D intensity, as indicated by low levels of business expenditure on R&D as a proportion of GDP (at 1.5%) and a low government share of R&D spending (5.5% compared with an OECD average of 8%). It is only in the most recent phase of ICT expansion that this problem is being addressed, with significant support from EU programmes targeted at Research, Technological Development and Innovation (RTDI) (Grimes, 1999a). R&D business spending as a proportion of GDP has more than doubled over the 1990s, with ICT industries now contributing a third of the total (OECD, 2000). The major part of this contribution is in electronics, although software companies, especially in the indigenous sector, are gaining ground, with companies undertaking R&D allocating 20% of the value of sales to this function (National Software Directorate, 1996). Some recent developments suggest that a "critical mass" has been reached in indigenous private sector R&D, although the latest economy-wide data have yet to be released. The public sector is also increasing its investment in research, as we shall see below, through the *National Development Plan 2000-2006*.

Figure 3. **Technology balance of payments, 1997**

As a percentage of GDP

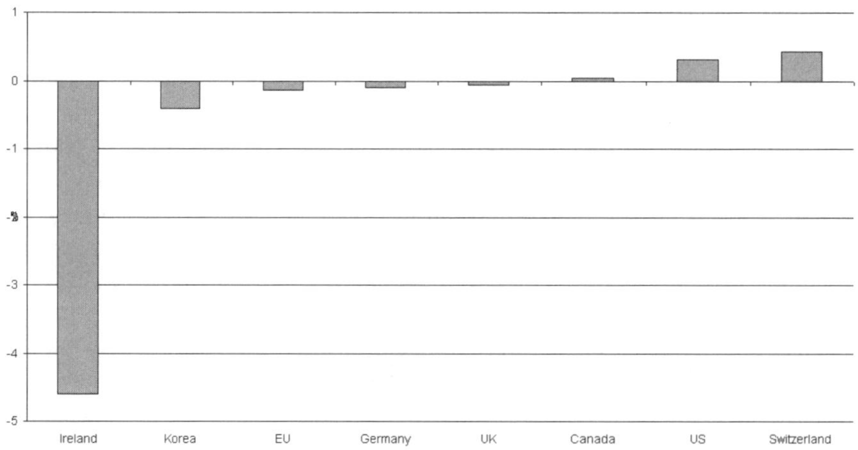

Source: OECD, 1999.

ICT cluster characteristics

Ireland's participation in global ICT activity was characterised at the initial stages of development by relatively low-value-adding manufacture and assembly of electronics hardware, as well as by call centre investments, but these have been overtaken by more complex integrated manufacturing and software operations, including those of a "boundaryless" indigenous software industry cluster as assembly operations relocate to lower-cost countries. During the 1990s, indigenous firms achieved growth rates of 11% a year for employment, 25% a year for the value of sales and almost 40% a year for the value of exports. While the electronics sector continues to be dominated by large multinational companies – with significant technology and skills transfer – employment in software products and services is more evenly divided between overseas and Irish companies, which consist mainly of small and medium-sized enterprises (SMEs).

Ireland's ICT cluster is primarily a self-contained group of *producers* rather than a significant market for ICT products. The Irish ICT market is very small at 0.18% of the OECD, and ICT intensity – measured by ICT expenditure as a proportion of GDP – is low at 6%, compared with an OECD average of 7%. However, this is changing rapidly with the increasing importance of e-business, which builds on Ireland's production capabilities. At the opposite end of the spectrum, Australia's ICT intensity is 10% but that country ranks poorly as a producer, with ICT accounting for only 1% of GDP, thus severely restricting its global options (Green and Genoff, 1998). Ireland is also ranked sixteenth out of 19 OECD countries for diffusion of ICT equipment, such as number of phone lines per 100 inhabitants and use of PCs. It is only recently that the historical underinvestment in advanced telecommunications, including fibre optics, has begun to be addressed. Previously, the take-up of ICT in the "old economy" was seen as a spill-over effect of globally targeted production rather than as a strategic objective in its own right.

It is as a producer, however, that Ireland has constructed the foundations for long-term growth, both domestically and in international markets. In the period 1991-97, turnover in electronics hardware increased by 25% a year and numbers employed in the sector by 14% a year to more than 35 000, taking it beyond chemicals and pharmaceuticals to become Ireland's largest manufacturing employer. Although there are still three times as many people working for the Irish subsidiaries of multinationals as for indigenous companies, the latter are expanding considerably faster, at over 17% a year. The electronics sector is almost entirely geared to exports, with three-quarters of all firms exporting and more than 80% of turnover accounted for by exports. These comprise mainly PCs, computer components and office equipment, and their value more than doubled in the period 1990-98 to IEP 13.4 billion – just lagging the United Kingdom and the Netherlands, but ahead of Germany and France (Figure 4).

While the low level of business expenditure on R&D generally in Ireland has been a matter for concern, at least until more recent favourable data began to emerge, the electronics sector is increasing its own contribution, already the largest of any industry, by almost 30% a year. This reflects the trend for MNCs to devolve the R&D function to "nodes" of creativity and expertise in established clusters encompassing their subsidiaries, their suppliers and even in some cases their competitors, as shorter product life cycles, "virtual" proximity to markets and the seemingly insatiable demand for high-level skills redefine first-mover advantage in the industry.

A similar trend is discernible in the Irish software sector which is experiencing even more spectacular growth, with exports of IEP 3.3 billion in 1998. In the period 1993-97, the number of firms in the sector increased by over 60%, employment doubled to 18 000 and revenue more than doubled to IEP 4.5 billion. In software, by contrast with electronics hardware manufacturing, indigenous firms not only account for 83% of all firms in the industry but also *more than half of total employment.* However, the firms are generally small, with just 27 in 1997 employing more than 50 people, and they generate only 12% of total revenue. The MNC subsidiaries are again much larger and, given the limited size of the domestic market, it is not unexpected that they would export their output, in some cases in a close and evolving association with computer hardware manufacturers. On the other hand, 80% of indigenous firms are also exporters, creating new benchmarks of global penetration and competitiveness.

Figure 4. **ICT and software goods: leading exporters, 1998**

In USD billions

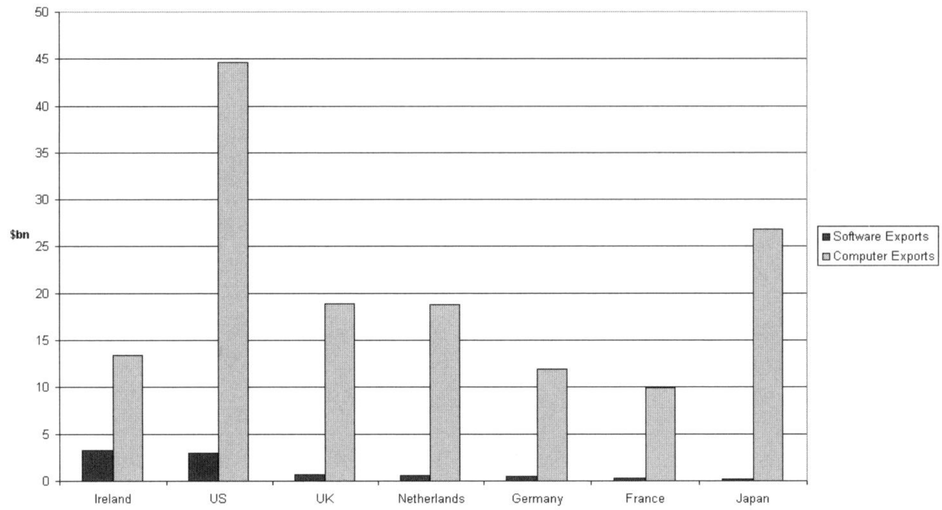

Source: OECD, *Information Technology Outlook 2000.*

While call centres employ 6 000 people in Ireland, only a small proportion of these operate in the ICT sector. They draw upon local skills and infrastructure but have no significant impact on innovation, value adding and cluster development within the sector. Over 60 global companies have established pan-European teleservices operations in Ireland in recent years, including large ICT firms such as Gateway and Dell, and these focus mainly on sales, technical and customer support for computer system hardware and software. Although English remains the dominant language for these operations, they also employ a range of other European languages. What impetus exists in relation to the cluster is not so much through direct employment or company expansion as through the generation of local supply chains with expertise in call centre technologies and skills. Certainly, call centres have had neither the creative input or the impact of software growth in Ireland over the last ten years.

Most of the software multinationals in Ireland are packaged software or product companies selling to mass markets, although the growing emphasis on localisation and product development, especially in the financial sector, requires a higher level of software engineering skills and is more reliant on outsourcing and indigenous supply chains, including translation, fulfilment, packaging, manual printing, transport and technical support. Irish public policy has recognised that "the full benefit of the presence of foreign production firms depends on the extent to which they can be integrated into their environment. Such relationships are not only beneficial for local suppliers that benefit from technology transfer [...] Foreign firms will be anchored to the regional economy, merging local and global interests, and making sudden divestiture less likely than before" (OECD, 1998a).[3]

As a result, indigenous software producers tend to be more specialised in terms of both types of products and types of customers. According to a recent survey, half or more of the sales of about a third of these firms go to the Irish subsidiaries of multinationals (O'Gorman *et al.*, 1997). However, a further third have little or no linkage with the MNC sector and have targeted niche markets both locally and globally. Indigenous firms also provide software services such as programme development, consultancy and technical training, which tend in many cases to lead to the development of new software products. Irish software start-ups, particularly Iona, SmartForce, Baltimore Technologies, Trintech, Parthus and Riverdeep, have contributed to the "critical mass" achieved by the industry, but they have now been joined by an array of dynamic new operators such as Datalex, Flexicom, Macalla, Piercom, Managed Solutions Corporation and Peregrine Systems.

Figure 5. **Share of ICT in total business expenditure on R&D, 1997**

Percentages

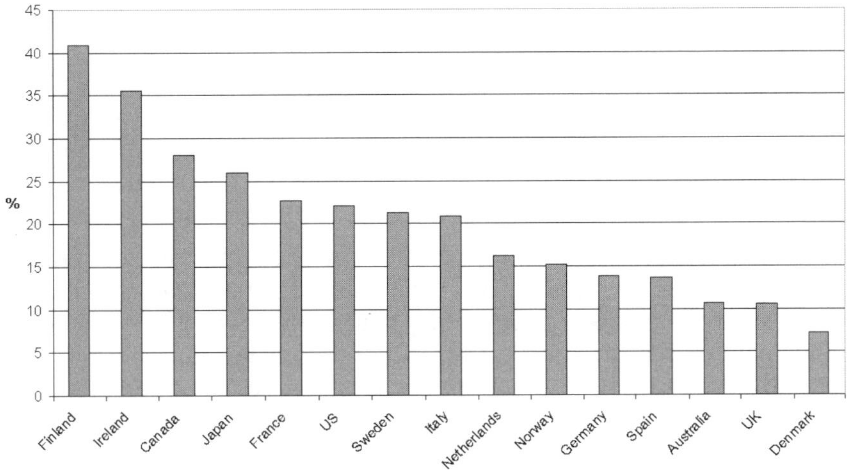

Source: OECD, *Information Technology Outlook 2000.*

A major source of competitive advantage for a growing number of these firms is ingenuity and speed of adaptation. They have demonstrated a capacity to meet these requirements and in most cases to sustain their momentum, not only by benefiting from technology transfer but also by originating software applications, including for e-business. This capacity is based on institutional support, training and education, workplace flexibility, increased venture capital availability and the remarkable growth of R&D expenditure in the software sector during the 1990s of more than 40% a year, with over three-quarters of this growth accounted for by indigenous firms. The ICT share in total business expenditure on R&D is, at 35%, much higher than any other country apart from Finland (Figure 5). However, this may also reflect to some degree the lower levels of R&D in other sectors. Nor is very much known other than on an anecdotal basis about the nature, quality and direct and indirect outcomes of R&D in the ICT sector and the social organisation needed to foster "collective and cumulative learning" in this context (O'Sullivan, 2000).

Galway regional cluster

While manufacturing employment, including electronics, tends to be widely dispersed in Ireland, the software industry is concentrated largely in the Dublin area, with smaller regional clusters in Cork, Limerick/Shannon and Galway itself (O'Gorman *et al.*, 1997). The Galway regional software cluster comprises three large North American companies, the biggest being Nortel, and over 50 small and medium-sized companies, many of them indigenous start-ups.[4] These companies span a range of activities from localisation to multimedia, and competitive advantage is driven by factor conditions specific to Ireland, such as English-language proficiency and the high level of computer and engineering skills. While domestic demand is important for many indigenous companies, predominantly through linkages to major software, computer and telecommunications firms, the multinational sector itself and an increasing proportion of indigenous companies are oriented almost entirely to global markets. These companies specialise in niche products and services and are less likely to face local competition (Clancy *et al.*, 2001).

There is now a policy commitment to "foster the development of clusters of new knowledge-intensive enterprise in regional centres" (Enterprise Ireland, 2000, p. 2) in recognition of the increasingly accepted finding that "clusters of industries […] are the driving force behind economic development [...] Their mutually reinforcing character energises innovation, fosters upgrading, spawns new companies and new industries, and stimulates demand for local industries" (Porter, 1991). This finding has been incorporated into both EU funding guidelines and the operational methodology of Enterprise Ireland. Indeed, over the 1990s, they first became interdependent and then mutually reinforcing. The cluster dynamic is supplied in the case of Irish regions by a unique mix of inter-firm collaboration, interaction and rivalry, by the development and constant replenishment of common pools of skilled labour, by the localised support of research and educational institutions, by the commitment of local business organisations and unions and by the strategies of national and regional development agencies.

This dynamic also broadly follows the pattern identified by previous OECD research which found that "industrial clusters with links to local and regional innovation networks have been associated with accelerated diffusion of technology and know-how. The pace at which technologies are diffused within national innovation systems depends on the country's industrial structure and technological specialisation, institutional set-up, corporate governance regimes, degree of economic openness and the flexibility of firms' organisational and managerial structures" (OECD, 1997). This "holistic" interpretation of regional growth may be illustrated by the development over the past decade of the Galway software cluster, which had its origins not so much in conscious policy making, although ultimately policy and planning played a role, but in the closure of a large computer manufacturing facility owned and operated by the Digital Electronics Corporation (DEC).

Digital established its manufacturing operation in 1971 and, by 1977, employed 1 000 people in the production of the PDP 8E computer, mainly for the European market. The numbers employed rose to 1 800 in the 1980s with further expansion of the operation, the development of Digital's first European Software Distribution Centre (ESDC) and the establishment of European Systems and Support Engineering. The role of the ESDC was to redistribute and "localise" software products sourced in the United States to Europe, including both software programs and technical documentation. At this point, Digital had become not simply the major employer in the local economy, with an estimated "net worth" to local business and the community of IEP 100 million a year, but also a provider of advanced training and development for its workforce, a large employer of third-level graduates and a source of broader research linkages and collaboration. For example, Digital contributed in the early stages to a high-profile research project on Computer Integrated Manufacturing with Renault, Compaq, Poletecnico di Torino and NUI Galway.[5]

During the 1980s and early 1990s, Digital was the focus, perhaps an overly dominant one, of a shallow but functional regional innovation network in Galway. Its closure in 1993 decimated this network, including at least 40 direct suppliers to the company such as Pulse Engineering (Tuam) and Cable Products (Castlebar), but it may be seen in retrospect that it also facilitated the development of a new, more open and diversified set of linkages (WESTBIC, 1999). The closure resulted in 760 redundancies in the Digital operation itself, after a period of already severe downsizing, whose scale and impact became a focus of national as well as local concern. The reasons for closure were complex, but included: *i)* the worldwide economic downturn; *ii)* intense competition in computer manufacturing, with Nixdorf, Wang and IBM building their manufacturing presence in Europe;

iii) failure to manage the shift in technology and consumer preferences from large mainframes to personal desktop computers; *iv)* a preference, in contrast with Microsoft, for "closed" rather than "open" operating systems; and *v)* lack of support within the company for local R&D initiatives such as a powerful new microprocessor chip which was ultimately sold to a competitor, Intel (Needham, 1999, Chapter 4).

However, the response of Digital, the local business chamber and national and regional development agencies to the closure of the manufacturing operation set the pace for European regional policy. In particular, it created new opportunities in the Galway region for the pool of skills and professionalism within Digital's workforce. In addition to provision of generous redundancy packages, Digital itself established in-house programmes for job search, career change, new business start-ups and relocation. This approach was supplemented by an Inter-Agency Task Force established by the Minister for Enterprise and Employment, Mr Ruairi Quinn TD, comprising the IDA, FAS (*Foras Aiseanna Saothair* – the national employment agency) and Enterprise Ireland as well as local government, business, trade unions, *Udaras na Gaeltachta* (the development body for the Irish language area) and WESTBIC (business innovation centre) with the support of the National University of Ireland, Galway, and institutes of technology. The most significant outcome of these discussions was the establishment of the Galway Technology Centre, the provision of additional training support and advisory services and funding for business start-ups, including via the conversion of tax on redundancy pay into a seed capital grant (Keating, 1994).

No single measure could be identified as responsible for the subsequent growth of Galway's software cluster, but business support, training and "incubator" facilities, together with informal networks among key ex-Digital staff, all played their part, and continue to do so, in an ongoing transformation of the local economy.[6] Nor did all ex-Digital staff enter the software sector. It required some high-profile successes, however, exemplified by Toucan Technology, to create the momentum necessary to encourage other start-ups and to attract major new investors such as Siebel Systems while at the same time persuading existing ones such as Nortel and Compaq (which absorbed the software and computer engineering areas of Digital) to expand their operations. These investors have in turn generated local supply chains, and have had to co-operate as well as compete for skills, infrastructure and market opportunities, both with each other and with the simultaneous emergence in Galway of Europe's leading medical instruments cluster, which is anchored by the presence of Medtronics AVE and Boston Scientific.

On the other hand, there are also impediments to the future growth and sustainability of the Galway cluster, some of which relate to the ICT sector as such and others to its regional status. While, in principle, distance from markets should not be an obstacle to ICT companies, especially in the software sector, the reality remains that scale and proximity are factors in competitive advantage. These problems can be addressed to some degree by the cluster approach which combines the positive attributes of scale with small-firm flexibility and which creates "virtual" proximity through "international recognition associated with establishing a world-leading position in the market" (Enterprise Ireland, 2000, p. 5). However, in the short term, the problems are compounded by poor access to venture capital, and the propensity of "dotcom" firms to position for a buy-out of intellectual property rather than expansion through an Initial Public Offering (IPO).

Other impediments common to ICT worldwide include shortages of high-level skills. Employment in the Irish software sector, for example, is expected to double by 2003. Recent estimates indicate a supply of 6 100 employees a year but an average demand for 8 300, leaving a shortfall of 2 200 (Expert Group on Future Skill Needs, 1999). Even if the demand for core technological capabilities can be met, the ICT cluster generally has a narrow skill set and is weak in functional areas of sales and marketing, management of human resources, corporate finance and strategy. This is an area where business and engineering schools, including those at NUI Galway, will be required to play a more active role as part of regional innovation systems. Finally, there are still severe infrastructure deficits in the Border-Midlands-West (BMW) region, mainly affecting transport and communications, although high bandwidth is in the process of being rolled out and connected with the new Global Crossing transatlantic cable.

Innovation processes and policies

The Galway ICT cluster comprises attributes that are unique to the regional innovation system and others with a more universal application. The unique attributes include the withdrawal of Digital from the region, with the loss of many previously secure jobs, and the precise configuration of the response of policy makers, development agencies and the local community. While these are instructive, there are also broader processes of regional business growth and innovation at work which have major implications for national policy frameworks across the OECD. The first point to make in this context is that the Porter "diamond" model of competitive advantage, particularly the requirement that it comprises only indigenous firms, may not be applicable to ICT cluster formation in a small regional economy (O'Donnell, 1998; Green, 1999; O'Gorman and Kautonen, 2000; Clancy *et al.*, 2001). The globalised nature of Irish ICT, the influence of the multinational sector and the niche operations of indigenous firms suggest the need for a new, more outward-looking theoretical approach to the advent of the "boundaryless" cluster.[7]

Second, the presence of at least one large ICT operation provides a useful catalyst and focus for cluster development, even when, as in the case of Galway, the operation withdraws from the region at a later stage. This presence affords the opportunity to build local capacity in new technologies and skills both within the operation itself and more widely in the emerging cluster through outsourcing, vertical supply chains and, ultimately, horizontal inter-firm linkages. However, the extent to which the potential benefits are realised in practice will depend not just on company strategy but also on the policy environment created by the national and regional innovation system. The greatest benefits flow from global, high-value-adding, research-intensive investments that are nevertheless embedded in the location; while the least are associated with low-value-adding operations, such as assembly manufacture and call centres, which also tend to be more "footloose".

Third, the development of the regional skills base is cumulative and parallels the scale and sophistication of the industry cluster, whose growth patterns are themselves path-dependent. The large ICT operation may act as a catalyst, establishing a demand for skills that are met through graduate employment, attraction of skilled personnel and through the creation of complex networks which supply training, consultancy and proprietary products and services. As well as employing skills, however, the operation may also play a role in the diffusion of research and technological competences which in turn become a springboard for entrepreneurship and participation in global niche markets. This was the experience of the Galway software cluster, which, after the closure of Digital's manufacturing operation, was required very rapidly to demonstrate an independent capacity to participate in these markets. The regional skills base was a necessary but not sufficient condition for doing so.[8]

Fourth, the Galway experience would suggest that appropriate regional business support structures are the final major local ingredient in successful cluster development. Without such structures, skilled personnel would be unemployed or underemployed, or alternatively would emigrate. As previously indicated, the traditional concept of "regional aid" has been superseded by more cost-effective approaches to the promotion of long-term growth and jobs, including business start-up assistance, industry network brokering and incubator facilities and collaborative approaches to new technology, R&D, training and project management. In Galway, a key role was played by Enterprise Ireland, in association with WESTBIC and the Technology Centre, in developing new enterprises with ideas, products and technologies that contribute to critical mass in the ICT cluster. Indeed, the demand for its sliding-scale fee for service facilities is such that the Technology Centre can impose a rigorous system of entry and review, ensuring high-quality outcomes that feed into future demand for the facilities and generate the resources to improve them.

Fifth, the policy framework (and to some extent the funding) for regional business support and innovation are settled at national and, increasingly, EU level (Grimes, 1999b). There is wide recognition in Ireland of the need for inward investment but, given the success of the IDA in attracting MNCs in recent years, especially in global growth sectors, the emphasis is now shifting to the quality and location of investment. This approach has been complemented over the last decade by Enterprise Ireland in "developing an indigenous, entrepreneur-driven technology sector", with a newly announced commitment to "clusters of new knowledge-intensive enterprises in regional centres" (Enterprise Ireland, 2000, pp. 2, 13). The most recently developed instrument of intervention will be a series of technology hubs known as "Webworks", whose task will be to "generate a critical mass of high-potential start-ups in the regions – companies that are high R&D and export performers" (p. 17). The first Webworks facility is to be established in Galway.

Sixth, the sustainability of the ICT cluster will derive from constant innovation, which in turn must be based on leading-edge research and research training. This has been a weakness in the past. However, it is now being addressed not only by firms themselves but by government through increased support for third-level education and expansion of research funding as part of the *National Development Plan 2000-2006*. In particular, the RTDI strategy is based on the recognition that "there is a strong link between investment in the research and innovation base of the economy and sustained economic growth... [T]he accumulation of "knowledge capital"... will facilitate the evolution of the "knowledge-based" economy" (Government of Ireland, 1999, para. 6.35).[9] Substantial resources have been earmarked for internationally peer-reviewed research on ICT and biotechnology to be administered by a new body, Science Foundation Ireland, and further rounds of third-level research funding are also in preparation as part of a total allocation of IEP 1.9 billion.

Conclusion

The evolution of Ireland's ICT sector has been driven not only by market forces but by the conscious design and implementation of public policy in the context of EU framework and cohesion programmes and, over the last decade, social partnership (O'Donnell, 2000). This has comprised measures to attract knowledge-intensive inward investment through IDA Ireland, support for indigenous companies and networks through Enterprise Ireland, promotion of education and training at all levels, especially universities and technical colleges, development of a world-class telecommunications infrastructure, increased funding support for research in third-level institutions and strengthened linkages between companies and the education sector. A major feature has been the emergence of internationally competitive clusters that are positioned to take advantage of local skills and expertise, on the one hand, and global scale, on the other.

However, the experience of the Galway software cluster demonstrates that the presence of a dynamic skills base is a necessary but not sufficient condition for success in international ICT markets. There is also a role for active industry and regional policies that contribute to business growth, technological innovation and cluster development – with an increasing emphasis on capacity-building in research and commercial applications. This is the approach of the "flexible developmental state", which is "defined by its ability to nurture Post-Fordist networks of production and innovation, attract international investment and link these local and global technology and business networks together in ways which promote development" (O'Riain, 1999). It is this strategic approach to the role of government and markets, in conjunction with a unique evolution of the boundaryless cluster, that lies at the heart of Ireland's knowledge-based economy and the broader canvas of European regional innovation policy.

Annex

ICT COMPANIES IN THE GALWAY AREA

Name of company	Activity/Product	Size of company
ADC Software Systems	Telecommunications billing system	M
Advance Design Centre Ltd.	Computer software	S
Aimware Ltd.	Software	S
Amicus Technology	Software – GIS/Mapping	S
Armon Group Ltd.	Software development	S
AND Data	Software – GIS/Mapping	M
Aro	Internet business development	S
Asita	Network back-up & retrieval	S
Baurer Software & Consulting Gmbh	Software	S
Bigback	Software	S
Blue Tree Systems	Computer software	S
BSM Ireland Ltd.	Software products and services	S
Celtic Technologies Ltd.	Multimedia software	S
Compaq Computer Corporation	Software	L
Cormorant Telematic Systems	Software	S
CT Solutions	Software	S
Cuan AV	Computer software, Web pages	S
Data Dimensions	Computer consultancy services	M
D.B.A. Ireland Ltd.	Database consultancy and development	S
DER Ireland	Shared services	M
Digital Filing	Document management technology	S
Direct Marketing Technologies Ltd.	Software	S
Distributed Software Consultancy Ltd.	Software	S
Europa Technology Ltd.	Internet technologies, Web design	S
European Software Centre, Digital Equipment Int BV	Software	L
Ex Tech Ltd.	Software	S
FKM Software Services	Software services	S
Fotonation	Software	S
Gcom Teoranta	IT services: Internet, multimedia	S
Graham Technology	Software	S
GP Care Systems Development Ltd.	Software	S
Infopoint Systems Ireland	Multimedia software	S
Infoscience Software	Software	S
Integrity Data Associates Ltd.	Software	S
Intra Fish	Software	S
Marine Computation Services	Software	S

Name of company	Activity/Product	Size of company
Martec Computer Services Ltd.	IT services: Internet, consulting, training	S
Menlo Park Technologies	Software	S
Multis Ltd.	Computer re-manufacturing	S
Netaccess	Web-page design	S
Nortel Networks	Telecommunication equipment	L
PlanNet21 Communications	Web-page design	S
PMC Sierra	Software	S
Q-Set Ltd.	Software/training	S
Quick-Tech	Manufacturing	S
S&P Media Ireland*	Software	S
Saville Ireland	Software/telecommunications	M
Siebel Systems	Shared services	M
Silicon & Software Systems	Software	M
Start Amadeus	Software	M
Storm Technology Ltd.	Software	S
SyberNet Ltd.	Software	S
Toucan Technology Ltd.	ASIC design	S
Unisoft Systems Ltd.	Software	S
Visual Computergrapic	Software	S
World Wide Web Marketing	WWW marketing/design	S
Yak	Web-page design	S

Key: S = small (1-50 employees); M = medium (51-200 employees); L = large (201+ employees).
Note: Total number of ICT companies = 57.
Source: Enterprise Ireland, Galway Chamber of Commerce & Industry, NUI Galway, 2001.

NOTES

1. This evidence is not yet conclusive. Ireland is now ranked fifth by the World Economic Forum, after the United States, Singapore, Luxembourg and the Netherlands, in terms of "growth competitiveness", which measures factors contributing to growth. However, Ireland remains 22nd in the world for "current competitiveness", encompassing factors in the *continuation* of growth, mainly due to an infrastructure deficits in a number of areas (*Irish Times*, 7 September 2000).

2. NACE codes have been used wherever possible to generate data, but in some cases alternative sources, such as the National Software Directorate, have been preferred.

3. This cluster dynamic may be further reinforced by leading-edge research, such as that envisaged for the MIT's MediaLab Europe facility to be sited in Dublin. By contrast, the lack of "embeddedness" in local supply chains may a factor in disinvestment by MNCs, even if not the catalyst for it, as we shall see below in the case of Digital in Galway in the early 1990s.

4. These figures are derived from unpublished databases held by Enterprise Ireland and the Galway Chamber of Commerce and Industry. For an overview of ICT companies in the Galway area, see Annex.

5. Building on this experience and a complement of former Digital staff, Compaq's super computing engineering team subsequently designed the Alphaserver SC in Galway, which integrates Alpha processors in a "massively parallel system" to deliver super computing power (used, for example, to sequence the human genome).

6. A survey conducted in 1999 found that 90% of the redundant staff had returned to full-time employment, with 10% starting their own businesses in Galway and 70% of these earning the same or better pay than prior to redundancy (Needham, 1999, Chapter 7). There are parallels here with closures in more traditional areas of manufacturing, but also substantive differences (O'Neill and Green, 2000). In retrospect, it may be seen that the prospects of the Galway workforce would prove more favourable due to the nature of their skills, the emerging opportunities in the ICT sector and the coherence of national and regional intervention, which matched available skills to the opportunities.

7. According to a recent study, "the suggestion that foreign-owned MNEs should be included as potential contributors to the competitive advantage of host-country industry clusters seems likely to have merit in the Irish case...Even in the absence of fully developed Porterian clusters, there generally are appreciable benefits... arising from... groupings of connected or related companies or industries, and from interactions between them" (O'Malley and van Egeraat, 2000).

8. In this sense, Robert Reich's thesis in the *Work of Nations* (1991) is incomplete and even misleading. The futility of a world-class skills base without relevant and expanding local employment opportunities has been demonstrated in many places, including Ireland itself in the 1980s, as well as in industrial and rural regions elsewhere (Green, 1998; ESC, 1999).

9. The objectives of the RTDI are to: *i)* develop intellectual infrastructure to "root" overseas companies here through more extensive use of research based in Ireland; *ii)* persuade and encourage companies to develop their own research activities; *iii)* develop a world-class research environment in our higher education institutions and state research institutions; and *iv)* ensure a vibrant and dynamic pool of high-quality, technically literate graduates from the graduate to postdoctoral levels to service the needs of these companies and to start their own companies (Government of Ireland, 1999, para. 6.35).

REFERENCES

Clancy, P., E. O'Malley, L. O'Connell and C. van Egeraat (2001), "Industry Clusters in Ireland: An Application of Porter's Model of National Competitive Advantage to Three Sectors", *European Planning Studies*, Vol. 9, No. 1.

Enterprise Ireland (2000), *Opportunities for Ireland's High-Technology Internationally Traded Services (ITS) Sector to 2007*, Enterprise Ireland, Dublin.

ESC (Employment Studies Centre) (1999), *Clever Region: Using Knowledge and Skills to Compete in the Global Economy*, report for Department of Employment, Workplace Relations and Small Business, University of Newcastle.

Expert Group on Future Skill Needs (1999), *Ireland's Information Technology Skill Requirements*, Forfas, Dublin.

Forfas (1996), *Shaping our Future: A Strategy for Enterprise in Ireland in the 21st Century*, Forfas, Dublin.

Government of Ireland (1999), *Ireland: National Development Plan 2000-2006*, Stationery Office, Dublin.

Green, R. (1998), "Unemployment and Structural Change: The Regional Experience", in B. Cass and R. Couch (eds.), *Divided Work, Divided Society*, Research Institute of the Humanities and Social Sciences, University of Sydney.

Green, R. (1999), *What Makes Global Regions: The Role and Significance of Regional Innovation Systems*, National Innovation Summit, Federal Department of Industry, Science and Resources, Canberra.

Green, R. (2000), "Unemployment: Perspectives and Policies", *International Journal of Manpower*, Vol. 21, No. 5, September.

Green, R. and R. Genoff (1998), "What is Industry Policy?", in *Manufacturing Prosperity: Ideas for Industry, Technology and Employment*, Federation Press, Sydney.

Grimes, S. (1999a), "The Information Society in Ireland – A Late Developer Catches the Wave", National University of Ireland, Galway, paper presented to the European Regional Science Association Conference, Dublin, August.

Grimes, S. (1999b), "Information Society Technology Programmes, Structural Funds and European Cohesion: The Case of Ireland", National University of Ireland, Galway, paper presented to Royal Geographical Society Conference, Brighton, January.

Gunnigle, P. and D. McGuire (2000), "Why Ireland? A Qualitative Review of the Factors Influencing the Location of US Multinationals in Ireland with Particular Reference to the Impact of Labour Issues", University of Limerick, paper presented to Conference of the Irish Academy of Management, 7-8 September.

Keating, S. (1994), *The Galway Task Force Report*, Galway Task Force, Galway.

ILO (1999), *World Employment Report 1998-99*, International Labour Organisation, Geneva.

National Software Directorate (1996), *Irish Software Industry Survey 1995: Results*, National Software Directorate, Dublin.

National Software Directorate (1998), *Irish Software Industry Survey 1997: Results*, National Software Directorate, Dublin.

Needham, L. (1999), "The Impact of the Closure of Digital Equipment Corporation International Ltd, Galway on the Workers and the Local Economy", research dissertation, Faculty of Commerce, National University of Ireland, Galway.

OECD (1997), *National Innovation Systems*, OECD, Paris.

OECD (1998a), *Inward Investment and Regional Economic Growth*, OECD, Paris.

OECD (1998b), *Technology, Productivity and Job Creation*, OECD, Paris.

OECD (1999), "The Knowledge-based Economy", Meeting of the Committee for Scientific and Technological Policy at Ministerial Level, OECD, Paris, June.

OECD (2000), *Information Technology Outlook 2000*, OECD, Paris.

O'Donnell, R. (1998), "Post-Porter: Exploring Policy for the Irish Context", in *Sustaining Competitive Advantage: Proceedings of a NESC Seminar*, National Economic and Social Council, Dublin.

O'Donnell, R. (ed.) (2000), "The New Ireland in the New Europe", in *Europe: The Irish Experience*, Institute of European Affairs, Dublin.

O'Gorman, C. and M. Kautonen (2000), "Developing Clusters of SMEs in New Knowledge-intensive Industries: The Cases of Dublin, Ireland, and Tampere, Finland", seminar paper presented at National University of Ireland, Galway, October.

O'Gorman, C., E. O'Malley and J. Mooney (1997) *Clusters in Ireland: The Irish Indigenous Software Industry*, National Economic and Social Council, Dublin.

O'Malley, E. and C. van Egeraat (2000), "Industry Clusters and Irish Indigenous Manufacturing: Limits of the Porter View", *Economic & Social Review*, Vol. 31, No. 1, January.

O'Neill, P. and R. Green (2000), "Global Economy, Local Jobs", in P. McManus P. O'Neill, R. Loughran and O.R. Lescure (eds.), *Journeys: The Making of the Hunter Region*, Allen & Unwin, Sydney.

O'Riain, S. (1999), "The Flexible Developmental State: Globalization, Information Technology and the 'Celtic Tiger'", unpublished manuscript.

O'Sullivan. M. (2000), "The Sustainability of Industrial Development in Ireland", *Regional Studies,* Vol. 34, No. 3.

Porter, M. (1991), *The Competitive Advantage of Nations*, Macmillan, London.

Porter, M. (1998). "Clusters and the New Economics of Competition", *Harvard Business Review*, November-December.

Reich, R. (1991), *The Work of Nations*, Vintage Books, New York.

Travers, J. (1999), "The Role of Information Technology in Attracting Foreign Investment, Creating Industrial Zones and Developing Human Resources: Case Study Dublin", paper presented to World Competitive Cities Congress, World Bank, Washington, DC, May.

WESTBIC (1999), *WESTBIC Activity Report*, WESTBIC Business & Innovation Centre, Galway.

Chapter 4

THE ICT CLUSTER IN DENMARK

by

Michael S. Dahl and Bent Dalum
DRUID/IKE-Group, Department of Business Studies, Aalborg University

Introduction

What are usually called "high-tech" or R&D-intensive industries are those industries that develop and produce information and communication technologies (ICT), pharmaceuticals/biotechnology and aerospace. Originally, the major focus in analysis of high-tech industries was on physical output in terms of hardware. However – and especially in the case of ICT – it has become increasingly evident that the hardware and software aspects are fundamentally intermingled; in many cases, they are two sides of the same coin. This has led many analysts and researchers to use in the place of the electronics industry, the concept of an ICT sector which, taken in a narrow sense, comprises the electronics industry as well as software development and IT services. In a broader perspective, it may, however, include such industries as broadcasting, electronic and printed media, publishing, movie production and advertising, all activities which are to an increasing degree converging towards what have traditionally been conceived as the "core" ICT industries. This chapter takes as its point of departure such a broad specification of the ICT mega-cluster (ICT-MC) as defined by the Danish Agency for Trade and Industry (EFS).[1]

The purpose of cluster studies is typically to study the general economic development (growth of value added, employment and export market shares) of the cluster. However, this approach sets out from the outset to incorporate the structural characteristics of the cluster as a whole, with a focus on important interactions between the use of new technologies and their development and manufacture. Given that the application of new technologies is very often located in other industries than those that developed them, the *combination* of the user and producer industries is emphasised. In the context of ICT, it may even be harder than for other mega-clusters to draw the boundaries, taking into account the very fast development of the underlying basic technologies. This makes it rather difficult to undertake international comparisons, especially for the various service industries involved in the ICT-MC. In the Danish context, a project has been launched by the EFS to further examine how best to undertake international comparisons at the cluster level, although the results are not yet ready for use. However, a recently published benchmarking study of a series of Danish mega-clusters (EFS, 2000) contains some international comparisons which can be drawn upon.

The next section *identifies* the cluster, while the following one *maps* it. The *cluster dynamics* as well as *innovation style and pattern* are then dealt with, followed by an analysis of the geographic pattern. The final section presents some *policy recommendations* relevant for the Danish ICT cluster.

Identifying and defining the ICT mega-cluster

The analytical ancestors of the current EFS cluster studies can be traced back to the Lundvall *et al.* (1984) cluster-type analysis of the Danish "agro-industrial complex". The microeconomic foundations were stated in Lundvall (1985), with explicit reference to the emerging discipline of the "economics of innovation". This approach became one of the central foundations for the Danish contribution to Porter's major competitive advantage of nations project 1987-90, published in Danish as Moeller and Pade (1988). There was a clear Scandinavian influence behind two of the components in the "diamond model" (Porter 1990, Chapter 4), *i.e.* "demand" and "related and supporting industries".

In parallel, but closely related to the proliferation of cluster studies undertaken during the 1990s and surveyed in Porter (1998, Chapter 7), a stream of literature has emerged under the heading of "systems of innovation". Taking as its point of departure the concept of national systems of innovation (Freeman, 1987; Lundvall, 1992; Nelson, 1993; Freeman, 1995), research has been carried out on sectoral systems (Breschi and Malerba, 1997; Malerba, 2001), technology systems (Carlsson and Jacobsson, 1997) and regional systems (Braczyk *et al.*, 1997).[2] The common thread of this work is the emphasis on interactions among actors, leading to a conceptualisation of innovation as a process which is often deeply embedded in a given social context.

The innovation systems literature can be seen as a more specialised line of research – in terms of focusing on the economics of innovation – which remains fully compatible with the fairly large body of cluster studies published in the recent decade.[3] Further, during the 1990s, a visible part of the discipline of economic geography, such as Storper (1992), Saxenian (1994) and Maskell and Malmberg (1999), has clearly converged with these lines of research. From the point of view of the economic geographers, clusters are often geographically embedded in cities and/or fairly small regions (see also the broad survey of the entire field by Clark, Feldman and Gertler, 2000).

The tradition in Denmark during the 1990s has been to undertake broad analyses of business sectors using a cluster-based approach.[4] The basic idea is to include important interactions which may not be captured by traditional industrial classification schemes. The explicit goal is to promote discussions of policy implications at the cluster level. Recently, the EFS launched a second round of cluster studies. A so-called "benchmarking" report has been published which provides an overview of the basic statistical trends of the clusters.[5] This second round contains a broad ICT cluster (the ICT-MC), which was not present in the first round. From the outset, the ICT-MC has been seen as a collection of interconnected statistical sectors. The "mega" level indicates that it may also be relevant to focus upon "small-scale clusters".[6]

The EFS has recently broken down the ICT-MC into a matrix of four horizontal segments and three vertical fields (see Annex Table A2): IT/Electronics, Telecommunications, Broadcasting and Information/Entertainment. The principle behind the vertical dimension of each of the four segments is to focus on various levels of the value chain. The first, "equipment/infrastructure", basically consists of manufacture, wholesale and repair of hardware. "Transport/mediation" and "content/application", concentrate on the next stages of the value chain. In Table A2, a few segments are empty, indicating that it would require a more detailed level of aggregation than the four-digit NACE codes in order to distinguish adequately between, especially, transport/mediation *vs.* content/application.

Mapping the ICT mega-cluster

Relative size of the economy and segmentation

The ICT-MC accounts for between 10% and 15% of the Danish private sector in terms of number of firms, employees, turnover, salaries and exports, as specified in Table 1, which also shows the four horizontal segments of the ICT-MC.[7]

Table 1. **ICT-MC segments and their share of the Danish private sector, 1997**

	Percentage of the economy	IT/ Electronics	Telecomm-unications	Broadcasting	Information/ Entertainment
Number of firms	8%	50%	1%	12%	37%
Number of full-time employees	13%	47%	12%	12%	29%
Turnover	12%	51%	11%	17%	21%
Exports	11%	55%	9%	22%	14%
Aggregate salaries	16%	49%	13%	9%	28%

Note: The last four columns sum to 100%. The definition of the segments is shown in Annex Table A2.
Source: Statistics Denmark (The Danish Mega Cluster Statistics).

Nearly half of ICT-MC employment belongs to the IT/Electronics segment and 29% to Information/Entertainment. Broadcasting accounts for 22% of exports but only 12% of employment. The Telecommunications segment is concentrated on very few firms (1%), but accounts for 12% of employment, mainly due to heavy concentration in telecommunication service provision. In Table 2, 1997 data on employment and exports are assigned to the two dimensions of the EFS segmentation, the four horizontal segments as well as the vertical fields.

Table 2. **The EFS segments of the ICT mega-cluster, 1997**

Segment	Types	Number of full-time employees	Share of ICT employment	Exports (DKK 000)	Share of total exports
IT/electronics	Transport/mediation	18 706	13%	3 291 001	7%
	Equipment/infrastructure	49 656	34%	22 819 135	48%
Telecommunications	Transport/mediation	15 242	10%	1 596 634	3%
	Equipment/infrastructure	2 409	2%	2 766 739	6%
Broadcasting	Content/application Transport/mediation	1 115	1%	166 156	0%
	Equipment/infrastructure	15 970	11%	10 163 731	21%
Information/ entertainment	Content/application	26 227	18%	5 652 539	12%
	Transport/mediation	2 695	2%	227 883	0%
	Equipment/infrastructure	13 656	9%	917 313	2%
Total ICT-MC		**145 676**	**100%**	**47 601 131**	**100%**

Note: The definition of the segments is shown in Annex Table A2.
Source: Statistics Denmark (The Danish Mega Cluster Statistics).

The IT/Electronics segment is dominated by "equipment/infrastructure" manufacturing activities, which are relatively export-intensive compared to other segments. Exports from this segment are closely related to wholesale of IT electronics. Only a small part of the Telecommunications segment consists of manufacturing; the "transport/mediation" dimension of telecommunications clearly dominates the segment. Telecommunication service provision is basically not export-oriented. In the Broadcasting segment, the "equipment/infrastructure" activities are clearly dominant, with considerable exports (mainly of consumer electronics equipment). Two-thirds of the employment in the Information/Entertainment segment is within the "content/application" field, with newspaper and journal publishing and advertising the main activities. The general printing industry (with the exception of newspapers) dominates employment in the "equipment/infrastructure" field. The main export industries within the Information/Entertainment segment are newspapers and general printing.

Economic development

Figure 1 presents various domestic economic indicators for the ICT-MC based on the EFS benchmarking project (EFS, 2000).

Figure1. **Economic indicators for the ICT-MC, 1992-98**

Index: 1992 = 100

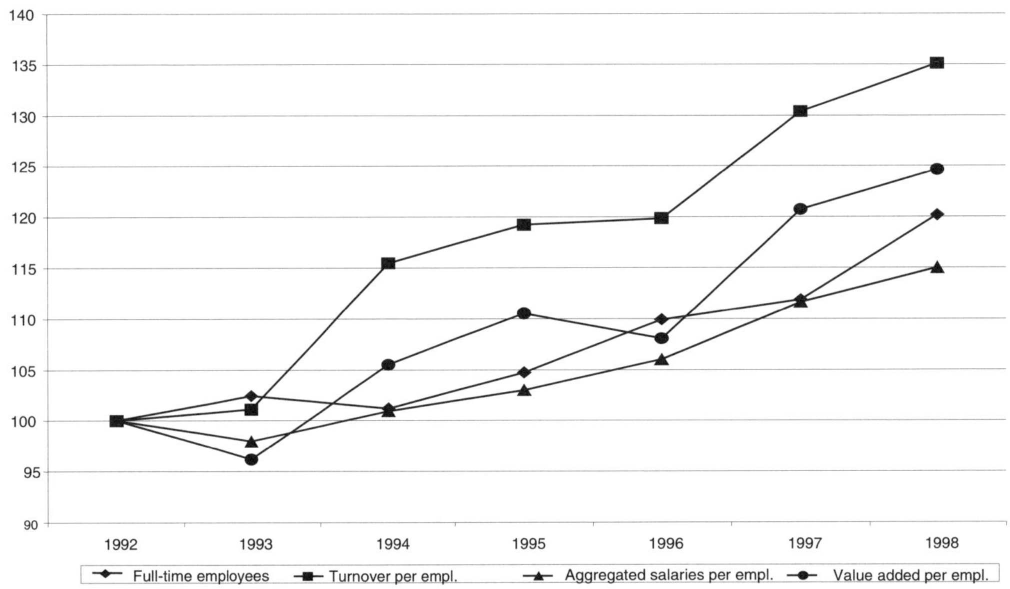

Source: 1992-96: EFS (2000). Updated to 1998 using data supplied by EFS. All data are in current prices.

The 1992-98 pattern appears to be one of substantial growth, especially in terms of turnover per employee (35%), but also in employment (20%) salaries and value added per employee (25%), the latter two in current prices. In general, value added per employee has increased faster than employment, indicating increased productivity (measured in current prices). The increase in the number of employees can primarily be assigned to ICT services, as ICT manufacturing has shown below-average growth. No internationally comparable data set for the ICT-MC is available along the lines shown in Figure 2. For a group of 15 OECD countries, the OECD (2000b, Table 1) has published value added for the ICT industries as a percentage of GDP for 1980, 1990 and 1997.[8] In 1997, only one of the 15 countries, Iceland, showed a share lower than the Danish one for the ICT industries.

Danish ICT manufactures lost market shares on the OECD markets in 1988-96, as indicated in Figure 2, which also draws on EFS (2000). The data sources used are, however, the result of approximations due to the lack of international data directly comparable with the Danish mega-cluster definitions.[9] OECD imports of ICT equipment have grown by more than 120%, while Danish exports of ICT goods have increased by 68%.

Figure 2. **Danish ICT export performance of manufactured goods on the OECD market**

Index: 1988 = 100

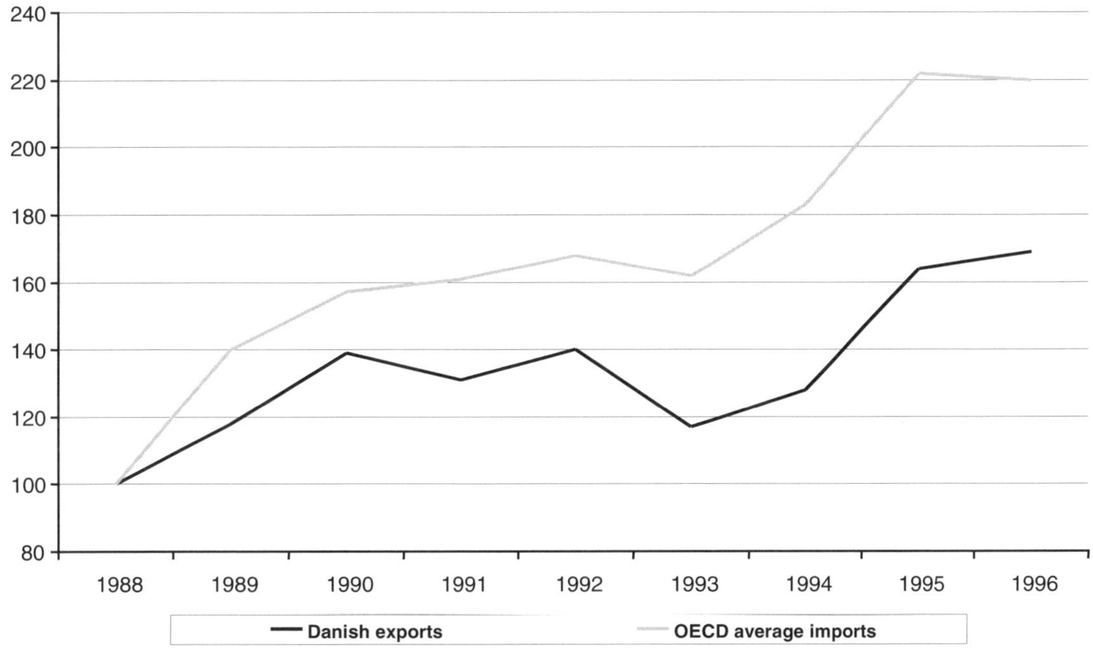

Note: OECD average imports of ICT goods are a weighted average of 22 OECD Member countries.
Source: EFS (2000).

A slightly different data set is used to illustrate the structure of Danish ICT exports compared with the other OECD countries (Table 3). Export specialisation provides a convenient measure for analysing the relative export structure of a country *vis-à-vis* an average pattern – *i.e.* total OECD. Specialisation for two ICT manufacturing industries is shown for a sample of small OECD countries, for the United States and for Japan.

Among the countries shown, Ireland was the only small county to be export-specialised in high-technology industries in 1990.[10] During the 1990s, this pattern changed, with Sweden and the Netherlands joining this group, closely followed by Finland.[11] In the Swedish and Finnish cases, two ICT companies, Ericsson and Nokia, were the major force behind the increased high-tech specialisation (especially due to very rapid increase in mobile communications), although pharmaceuticals was also of importance for Sweden. The Dutch case has been a combination of pharmaceuticals and the computer industry. For the Dutch ICT industry, this can be related directly to Philips in terms of non-Dutch production (and thus exports) of consumer electronics and lack of success in telecommunications hardware. Ireland is an outstanding case, being by far the most high-tech export-intensive OECD exporting country in 1996 (44.3% of manufacturing exports).[12]

Table 3. **Export specialisation, 1990-96**
Selected countries compared to the OECD average

	High-tech		Pharma-ceuticals		Computers		Radio, TV & communic.		*Low-tech*		Food processing		Wood and furniture	
	1990	1996	'90	'96	'90	'96	'90	'96	'90	'96	'90	'96	'90	'96
Finland	0.5	0.9	0.4	0.2	0.3	0.5	0.9	1.4	2.3	1.9	0.3	0.4	3.8	3.3
Sweden	0.8	1.1	1.6	1.9	0.6	0.3	1.0	1.5	1.3	1.1	0.3	0.3	2.7	2.7
Netherlands	0.8	1.1	0.8	1.1	1.2	2.1	0.7	0.8	1.5	1.4	2.6	2.6	0.6	0.5
Ireland	2.2	2.5	3.1	3.1	5.0	4.9	1.2	1.8	1.5	1.1	3.1	2.3	0.3	0.2
United States	1.8	1.5	0.9	0.7	1.8	1.6	1.4	1.3	0.8	0.8	1.0	1.0	0.8	0.7
Japan	1.5	1.4	0.2	0.3	1.8	1.6	2.6	1.9	0.2	0.1	0.1	0.1	0.1	0.1
Denmark	**0.8**	**0.7**	**2.6**	**2.7**	**0.5**	**0.5**	**0.8**	**0.6**	**2.1**	**2.1**	**3.7**	**3.5**	**3.0**	**3.2**

Note: The export data only contain manufactured goods (in current USD). "Specialisation" is the relative share of national exports from a given industry compared to the OECD average. If the indicator is above 1.0, the country has an above-average export share, *i.e.* is "specialised".
Source: OECD (1999b), Appendix Table 12.1.1.

At the same time, all of the small countries shown have the necessary characteristics for becoming specialised, although mainly decreasingly so, in exports from the low-tech industries. However, it is a well-known fact that the so-called low-tech industries of these countries often operate with very advanced process technologies. The technology intensities used in the standard OECD classifications are international averages which may hide substantial national differences.

In this context, the lack of Danish specialisation in ICT hardware is striking. This can mainly be attributed to the lack of "domestic" multinationals of the Ericsson-Nokia-Philips type. Among the high-tech industries, Denmark is only specialised in pharmaceuticals. However, telecommunications may become visible in a more fine-grained export analysis.[13] Given that no satisfactory data are available for an international comparison of the ICT service industries, the preliminary conclusion from this statistical study is that the Danish ICT industry is rather weak.

However, the notion underlying the cluster approach is to integrate some of the service industries and related manufacturing industries in the analysis. A closer inspection of the export data behind Table 2 above indicates that there is a rather large export of "content" from the Information/Entertainment segment concentrated on printed material. Finally, a flourishing movie industry, which can now boast an international reputation, has emerged during the 1990s, although it is difficult to analyse in an international context with the available statistical data.

The ICT consumption pattern

A substantial amount of data on the *use* of ICT equipment indicates that the markets of the small, rich countries – as well as the US market – represent the most sophisticated demand. This is typically illustrated by per capita penetration ratios in such fields as Internet connections, mobile phones, PCs in homes, etc. (see, for example, OECD, 1999a, 2000b, Appendix Table 4-6).

Among the small countries, the above especially holds true for the Nordic area. The mobile communications field represents a now classical case of close interaction between advanced demand characteristics, institutional set up, regulation and international competitiveness of the supplier industries – *i.e.* the very essence of the factors emphasised in cluster analysis.

The establishment of the Nordic Mobile Telephony System (NMT) in 1981, through co-operation between the Nordic telecommunication service providers and regulatory authorities, created the first cross-border market for public mobile telephony based on a common standard. This was the decisive factor behind the Nordic firms (Ericsson, Nokia and the US-Danish Storno)[14] becoming world leaders in mobile communications equipment from the early 1980s. This strong competitive position was strengthened when the pan-European GSM standard, now the dominant world standard, was implemented in 1992.[15] In the late 1980s, the GSM standardisation process was to a substantial degree influenced by Nordic companies and authorities, with Nokia and Ericsson the leading architects. Motorola, the largest producer of mobile terminals at the time, acquired Storno in Denmark in order not to be excluded from this decisive process.

The story has to a large extent been repeated in the specification from 1997/98 of the third-generation mobile communications system, UMTS, which also happened to be dominated by Ericsson and Nokia. The development of the mobile communications industry is probably one of the most clear-cut examples of "textbook" cluster dynamics. From this perspective, it may, however, be a cause of some concern that the first third-generation mobile communications system to come into operation will be in Japan in May 2001. In the past, the Nordic countries were the "lead-users", and this will no longer will be the case. Allocation of UMTS licences in Europe was very turbulent during 2000, with telecommunication service providers facing rocketing costs to become licence holders. Denmark has not yet entered this phase; the auction will take place during 2001. The country runs the risk of losing the advantage of being an early advanced user.

Given the advanced nature of demand in the Nordic region concerning use of ICT products and services, it is to be expected that a similar mechanism could emerge in the new software development and IT service industries. On the international scene, the Nordic area has been characterised as "leading Europe in the technology revolution that will dominate the early years of the 21st century".[16]

In a consultancy report by PLS Ramboell and Boersens Nyhedsmagasin (2000a), ten Western countries were benchmarked according to an extensive set of criteria combining statistical data and peer judgements by 170 experts from the ten countries. The data provides information on ICT application as well as R&D efforts, expenditure on IT education, macroeconomic stability and performance, characteristics of the capital and labour markets, etc. A group of "front runners" – the United States, Sweden and Finland – are leading the race. Denmark belongs to a group of "followers", along with the United Kingdom, Norway and Netherlands, while the "pack" consists of France, Germany and Spain. With all the caution to be used with such kinds of methodology, the conclusion fits quite well with the pattern described above. Denmark appears to be fairly well positioned to exploit the business opportunities in ICT from the point of view of demand. However, this potential has not been realised to the extent shown by Sweden and Finland.

To a certain extent, this might be due to a "lack-of-Ericsson-Nokia" effect. However, this kind of explanation is too simplistic. The Danish Government recently published a major Network Report (Ministry of Research, 2000), which contains policy reviews as well as a series of internationally comparable data on, for example, ICT consumption patterns. In terms of high-speed Internet access – from ISDN (2 lines from 124 Kbit/sec.) and upwards (the various xDSLs) – for households and small firms, Denmark ranks in the middle of the group of advanced countries.[17]

Knowledge and innovation in the ICT mega-cluster

Education and R&D

Internationally, the quality of the Danish higher education system (and not least the system of education for skilled labour) is generally considered to be high. Table 3 presents some basic characteristics from the *Education at a Glance 2000* report (OECD, 2000d).

Table 4. **Percentage of population aged 25-64 by level of educational attainment, 1998**

	Primary and secondary education		Post-secondary tertiary education		Graduates in science and engineering (% of employment)
	Below upper	Upper	Education of at least two years, focusing on practical skills	Education of at least three years theoretical duration	
Finland[1]	32	39	17	13	0.08[2]
Sweden	24	48	15	13	0.07[3]
Denmark	22	53	20	5	0.04[2]
EU	46	57	10	12	0.12
OECD	38	44	8	14	0.09

1. 1997.
2. 1995.
3. 1996.

Source: OECD (2000c), Appendix Table 7. Data from OECD (2000d), *Education at Glance 2000.*

Denmark has a high proportion of the population with upper secondary education as well as with post-secondary practical education. However, the shares of graduates with at least three years of study and of graduates in science and engineering are rather low. These data are disputed. *Education at a Glance 1998* used a different classification: non-university and university level education. Whatever the classification used, Denmark performed better than the OECD average (see the summary table in OECD, 1999b, Appendix Table 2.6.1.).

The efforts by the Nordic Statistical Institutes (1998) to produce comparable data on ICT activities represent a step forward in international comparisons, as shown in Table 5. Unfortunately, no data are available for Finland. The high levels of university educated employees in the Scandinavian ICT industries are striking. The close similarities in the education systems of these countries are no doubt a major factor behind this pattern. In this perspective, there does not seem to be a substantial difference in the level of tertiary ICT-related education between the ICT industries in Denmark and Sweden.

The two technical universities, the Technical University in Copenhagen (DTU) and Aalborg University (AAU), produce *engineers* in electronics, bachelors (diploma engineers), masters and PhDs. DTU has enjoyed an international reputation in electrical engineering for two centuries. AAU, founded in 1974, rapidly became established in electronics, granting half of Danish MScs in electronics. *Computer scientists* are produced at the Copenhagen, Aarhus and Aalborg universities. There is also a system of decentralised engineering schools producing BScs in electrical engineering, created to provide higher education opportunities for skilled workers who follow a tailor-made admission programme. At present, there are six of these schools, two of which are in the process of being integrated with the new Southern Danish University.

Table 5. **Percentages of employees in ICT with university degrees, 1996**

	Denmark	Norway	Sweden
ICT manufacturing	17	22	18
ICT services	20	18	17
Wholesale of ICT products	16	11	10
Telecommunications	13	16	10
Consultancy services	31	31	31
Total manufacturing	8	6	7
Total services activities	11	9	10
Total private sector	9	-	8

Note: University degree level of education is equal to ISCED levels 6 and 7. Total manufacturing industry (NACE 15-37), total services activities (NACE 50-74, 92) and total private economy (NACE 15-37, 45, 50-74, 92, 93).
Source: Own elaboration, based on Nordic Statistical Institutes (1998).

Table 6 shows the relative distribution of research expenditure in an OECD context as well as a division of manufacturing into four levels of R&D intensity.

The Danish (and Dutch) emphasis on food-processing-related R&D is outstanding, with a share three times higher than the OECD average of 2%. The relative specialisation in pharmaceuticals R&D (some 3%) represents 30% of Danish R&D expenditure in manufacturing. Conversely, Danish R&D efforts in ICT manufacturing are generally very small when viewed from the international perspective. A more direct indicator is the OECD (2000b, Table 14) measure of ICT research in total business-enterprise expenditure on R&D. In 1977, Denmark lagged way behind the 14 most developed OECD countries, with a share of 7.2%. The country was also at the low end in terms of the share of business sector R&D for communications services (telecommunication services).

Table 6 clearly illustrates the high degree of concentration on a few industries in the small, high-income OECD countries. In Finland, the increase in specialisation from 1.3 to 3.1 in telecommunications equipment and semiconductors reflects the fact that this industry covered 50% of Finnish manufacturing R&D in 1998, mainly due to the very rapid growth of Nokia. In the Swedish case, the dominance of Ericsson is less outstanding, although the firm contributes to more than 20% of Swedish manufacturing R&D.[18]

Table 6. **R&D specialisation in manufacturing for selected small OECD countries**

	OECD-14[1]		Denmark[2]		Sweden		Netherlands		Finland	
	1990	1995	1990	1998	1990	1995	1990	1996	1990	1998
Food, beverages, tobacco	2.0	2.0	3.5	3.1	1.0	0.7	2.9	3.7	3.4	1.2
Textiles and clothing	0.6	0.6	1.0	0.2	0.6	0.4	0.8	0.8	1.6	1.2
Wood and furniture	0.2	0.4	1.8	0.8	0.5	1.0	0.0	1.4	5.1	2.3
Paper and printing	11.5	1.3	0.5	0.3	3.6	2.4	0.4	0.6	8.7	3.2
Chemicals	21.2	21.6	1.4	1.6	0.9	0.9	1.8	1.5	1.0	0.6
Pharmaceuticals	7.6	9.6	3.1	3.1	1.8	1.7	1.1	1.2	0.7	0.4
Non-met. mineral prod.	1.2	1.1	1.8	0.7	0.5	0.5	0.3	0.5	1.8	0.9
Basic metals	2.1	1.8	0.7	0.3	1.0	0.8	0.7	1.2	1.7	0.8
Metals and machinery	71.1	70.7	0.7	0.7	1.0	1.0	0.7	0.8	0.8	1.1
Non-electrical mach.[3]	6.0	7.1	2.6	3.2	2.3	1.7	0.5	0.5	2.3	1.7
Computers	***9.8***	***7.1***	***0.3***	***0.2***	***0.3***	***0.2***	***0.4***	***0.7***	***0.3***	***0.1***
Telecom & semiconductors	***14.5***	***16.3***	***0.7***	***0.6***	***1.9***	***1.4***	***1.1***	***0.9***	***1.3***	***3.1***
Shipbuilding	0.1	0.2	21.5	13.0	2.9	0.5	1.0	1.1	10.2	1.4
Motor vehicles	12.6	14.4	0.0	0.0	1.3	1.3	0.4	0.5	0.1	0.0
Aerospace	14.4	10.7	0.0	0.0	0.4	0.5	0.1	0.2	0.0	0.0
Scientific instruments	4.9	7.3	2.4	1.2	0.2	1.1	0.2	0.2	1.1	0.6
Other manufacturing	0.6	0.6	14.4	12.1	0.4	0.2	0.0	0.2	0.8	1.0
Total manufacturing	100.0	100.0	1.0	1.0	1.0	1.0	1.0	1.0	1.0	1.0
High-tech industries	46.5	43.8	0.8	0.9	1.1	1.1	0.7	0.8	0.6	1.3
Medium-high technology	40.7	44.1	0.9	0.9	1.0	1.0	1.4	1.1	1.0	0.6
Medium-low technology	9.0	8.0	1.9	1.8	0.6	0.6	0.8	1.0	1.7	1.1
Low technology	3.7	4.2	2.3	1.7	1.6	1.2	1.8	2.2	4.8	2.0
Total manufacturing	100.0	100.0	1.0	1.0	1.0	1.0	1.0	1.0	1.0	1.0

1. Distribution of OECD R&D manufacturing expenditure.
2. The national distribution divided by the OECD distribution. The weighed average per country is equal to 1.0. If it is more than 1, the country is considered to be "R&D specialised" in the given industry and vice versa if less than 1. The most recent available OECD-14 distribution is 1995.
3. Not all sub-groups are shown.
Source: OECD (2000c), Appendix Table 28.

Innovation

As illustrated in Table 7, the available evidence on Danish ICT patenting to a certain extent follows the R&D pattern. The share, as well as the average annual growth rate 1992-98, is very low indeed. However, a comparison of the 1992-98 and 1992-99 growth rates points to some data problems that may significantly disturb the pattern.

Finland has experienced the highest growth among the countries shown in terms of the number of patents, with average annual growth of more than 30% in the 1990s. ICT patents made up almost one-third of total patents in 1999. The Swedish pattern is similar, but not quite as dramatic. The evident explanation is the absence of large "domestic" ICT multinationals in Denmark. In particular, Nokia and Ericsson have been exceptionally active in patenting mobile-communications-related technologies, whereas the Danish ICT sector is dominated by small R&D-oriented firms with a low propensity to patent. The country may be characterised as a development "hub", in particular of terminals for mobile communications and cordless phones. These are typically developed by R&D units of large multinationals, which may mean that patents are taken out by the firm's headquarters outside Denmark. In another scenario, the innovations may come from small firms hired on contract by large multinationals; in this case, the process does not usually result in patent applications.[19]

Table 7. **Patents granted in the United States**

	Share of ICT patents in total			Growth rate 1992-98	Growth rate 1992-99
	1992	1998	1999		
Finland	6.0	29.0	30.4	41.7	31.8
Sweden	7.3	16.8	16.9	28.2	23.2
United States	8.8	18.4	17.5	21.5	16.6
Ireland	14.2	24.4	16.5	17.3	11.7
Netherlands	10.2	16.6	14.4	15.2	10.3
Japan	14.1	21.0	18.5	13.2	8.9
Denmark	6.4	3.1	6.3	0.0	13.0
European Union	6.2	11.0	10.4	15.7	12.5
Total OECD	**9.5**	**17.6**	**16.4**	**18.6**	**14.1**

Source: OECD (1999b), Appendix Table 11.3.1 for 1992 and 1998; OECD (2000c), Appendix Table 36 for 1999.

Innovation data based on a Danish questionnaire carried out by the DISKO project contains a data set from which it has been possible to construct Table 8. As a rough illustration of variations in propensities to innovate across different clusters, firms reporting product innovations in the DISKO panel data have been aggregated to five of the EFS MCs.

Table 8. **Relative number of product innovative firms in five EFS mega-clusters, 1993-95**

N=1 910	Product innovation	No product innovation	Total	N
ICT MC	75.1	24.9	100	249
Medico/health MC	57.8	42.2	100	93
Agro-food MC	52.4	47.6	100	248
Construction MC	38.3	61.7	100	517
Total economy	**50.4%**	**49.6%**	**100%**	**1 910**

Note: The DISKO data set contains questionnaire data from a sample of 1 910 firms representative of the private sector of the Danish economy.
Source: The DISKO panel data, 1993-95.

According to this source, the ICT-MC has by far the highest rate of product innovation, with more than 75% of the ICT-MC firms in the sample having completed a product innovation. On average, only half of the DISKO firms completed a product innovation. It appears that a high proportion of the reported product innovations in the Danish ICT-MC might, therefore, not be patented.

Regional development patterns

In terms of the geographical distribution of the ICT industries, it would be expected that among the major location factors is: *i)* a "metropolis" effect, simply because many of the ICT service activities, such as publishing, advertising, broadcasting, computer software development and services, and telecommunication services, are typically concentrated in cities. Another, related, factor is: *ii)* the supply of skilled labour (engineers, computer scientists, business economists, etc.), which again is expected to be a function of the location of universities and business schools, often determined by government decisions. Finally, comes *iii)* the somewhat "random" location of manufacturing firms due to the personal preferences of their founders. This jargon is mainly introduced through the efforts by the economics profession to re-enter the discipline of economic geography (see, for example, Krugman, 1991, 1995). Krugman emphasises the combination of "random" initial triggering events (such as individual founder preferences) and subsequent cumulative causation effects (mainly through external economy effects on the labour market), leading to concentrations of related industries, *i.e.* geographical clusters.

As an indicator of relative specialisation, employment in a given activity in a region may be compared with the national average. A value above 1 indicates an above-average employment share – *i.e.* the region is "specialised" in that activity (and vice versa). Thus measured, only two regions are specialised in the entire ICT-MC; these are the regions around the two largest cities, Copenhagen and Aarhus.[20] These two regions are shown in grey in Figure 3.

Figure 3 also shows the regional employment specialisation pattern for the four horizontal segments described above. It clearly illustrates that ICT is mainly, although not exclusively, a "city" activity. It also shows that the Copenhagen region is specialised in the four segments, followed by the Aarhus region specialised in three segments (Broadcasting is missing). The two other observations of high specialisation among the four main segments are Telecommunications in North Jutland and Broadcasting in Ringkoebing County.

Figure 3. **Regional ICT employment specialisation, 1998**

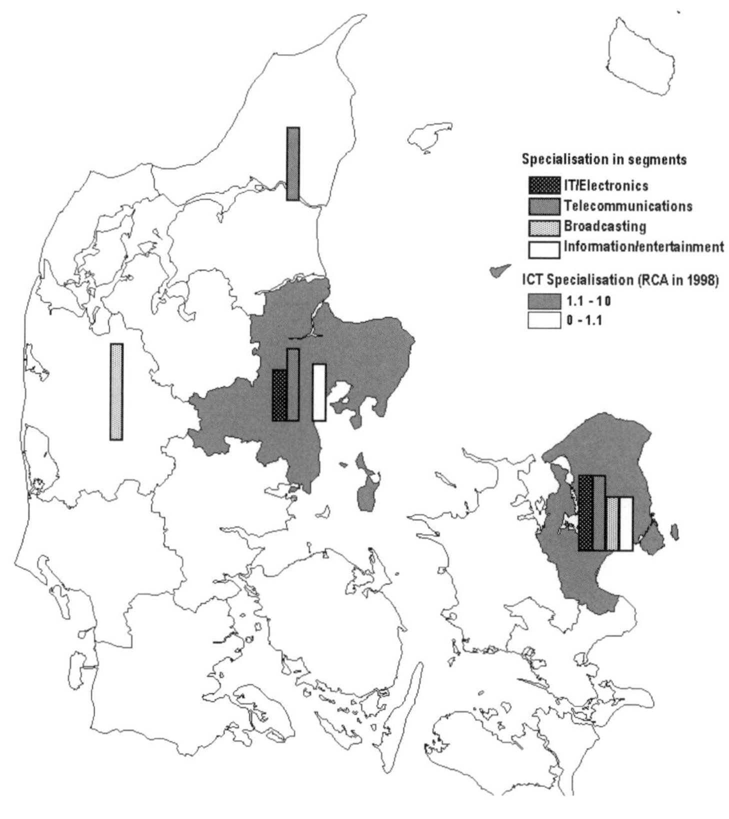

Source: Statistics Denmark, based on total employment data.

Further analysis of the regional employment data, in terms of specialisation for each of the four-digit NACE industries, points to a somewhat more differentiated geographical specialisation pattern than that indicated in Figure 3.[21] For ICT/Electronics, the strength of the Copenhagen region is located in the area north of the inner core of the city (the so-called Copenhagen County), with specialisation in all 22 four-digit NACE codes. The Aarhus region is specialised in seven industries, followed by North Jutland with four.

The ICT consultancy and processing segment (the various four-digit NACE codes belonging to 72) is concentrated in Copenhagen and, to some extent, in Aarhus. The four-digit data point to the existence of two concentrations of electronic component firms, in the county of Vejle (around the town of Horsens) and in North Jutland. In both regions, some of these firms can be traced back to spin-offs from established firms in the region, such as Arena and Kirk in Horsens and SP Radio in Aalborg, the latter being the "mother" firm of the North Jutland wireless communications cluster.

Very few of the Danish counties are specialised in one of the two four-digit NACE Telecommunication industries, hardware (3220) and telecommunication services (6420). Only one of the Copenhagen counties and North Jutland is specialised in both. The "inner core" of Copenhagen (the Copenhagen Municipality) and Aarhus are specialised in telecommunication services, while Ringkoebing County is specialised in telecommunication hardware.[22]

Telecommunication service provision (NACE 6420) is concentrated in Copenhagen, Aarhus and Aalborg. This is a direct effect of the location of the major service providers (cable as well as wireless), Tele Danmark, Sonofon, Telia and Mobilix, all of which are headquartered in Copenhagen, although the bulk of Sonofon's employment is located in North Jutland. A coherent telecommunications cluster has emerged in North Jutland, focused on wireless communications equipment. This strong specialisation emerged in the mid-1960s in maritime communications and diversified into mobile communications at the very beginning of the Nordic NMT boom in 1981.[23]

The Broadcasting segment is a mixture of production, wholesale, retail and repair of consumer electronics hardware and the so-called "radio and television" industry (NACE 9220). At one end of the spectrum, hardware production is concentrated in the only remaining radio and TV producer, Bang & Olufsen in the small town of Struer in western Jutland. At the other end, radio and TV broadcasting is a rather city-dominated industry. As shown in Figure 3, only two of the Danish regions are specialised in the entire segment, Ringkoebing County (Bang & Olufsen) and the Copenhagen region ("metropolis" effect).

The Information/Entertainment segment is to a large extent concentrated in the Copenhagen and Aarhus regions, as shown in Figure 3. The detailed pattern in Annex Table A3 shows that in such industries as newspapers (publishing and printing) and book printing, the manufacturing activities of the segment are to some extent spread throughout the country.

As can be seen, a "metropolis" effect is prevalent in the regional specialisation pattern of ICT activities. At the more detailed industry level there is, however, a certain degree of geographical diversification, which may be related to the rather decentralised nature of the public education system.

There is, thus, a rather close correlation between the distribution of government-financed R&D and higher education institutions in ICT and the regional distribution of private employment in Denmark. Engineers and computer scientists typically choose jobs close to these institutions. More specialised small-scale clusters usually emerge around these. This pattern has major implications for policy. To further encourage the development of ICT activities, a co-ordinated policy approach to such fields as higher research and education, specialised venture and seed capital and regional development of the necessary infrastructure facilities (science parks, telecommunications networks, general transport facilities, etc.) are of key importance.

Policy perspectives

Given the broad definition of the Danish ICT-MC, a wide range of policy issues can be raised in a cluster context, such as regulation of the telecommunications sector, broadcasting, publishing, production and distribution of books, newspapers and various electronic media, support schemes for film production, etc. Most of these policies clearly overlap with competition policy.

The most obvious restraint for the future development of ICT in most Western countries is a fundamental lack of skilled labour. During the early 1990s, the intake of new students in engineering schools and business schools decreased in Denmark. However, during 1995-2000, the intake of students to IT-related university programmes increased from 1 750 to 5 500. As of 1999, two new institutions, the IT High School in Copenhagen and IT-West, started R&D and education, with an intake in 2000 of more than 800 students. A dedicated science park and buildings for the IT High School are planned as part of the new Oerestad project in Copenhagen.[24]

At the national level, a major initiative has recently been launched by the Ministry of Research and Information Technology under the label "Digital Denmark". The project was first presented in a 1999 report which had five ambitious goals (Ministry of Research and Information Technology, 1999). Among these goals are the establishment of "IT lighthouses" in two of Denmark's regions. These initiatives are inspired by the strong investments in regional growth centres undertaken in other countries.[25] Ambitious investments in these projects in other countries have resulted in the establishment of high-growth regions specialising in ICT businesses; these regional growth centres provide powerful magnets in which high-technology businesses agglomerate and form clusters. A strong physical concentration of innovative environments of knowledge and education institutions has determined this development.

The goal of the lighthouse projects is to stimulate the formation of ICT businesses with links to knowledge and education institutions. The two Danish ICT lighthouses are located in North Jutland and in the Copenhagen Oerestad region. The presence of the small-scale cluster in wireless communications in North Jutland explains the location of one of the projects. In the case of the other, the major concentration of ICT activities in the Oeresund region and the deliberate, and high-profile efforts to ensure that this region becomes more visible on the international scene, played a decisive role.[26]

Another striking feature is the dominance of foreign ownership in a significant part of the Danish ICT-MC, not least in those sub-industries with very fast technological development. This is apparently an indication of a lack of capitalists willing to take the necessary high risks that are inevitably associated with ICT. The amount of seed and venture capital and the structure of its supply are issues which need to be influenced through policy, through such measures as government guarantees for risky investments in new high-technology start-ups.[27]

However, a new fashion among governments in most of the contemporary highly developed countries is to herald rhetoric intentions of becoming "*the* IT power house of the new Millennium". In the Danish context, the *Network Report* (Ministry of Research, 2000) contains a rather extensive arsenal of activities which have already been implemented as well as a series of proposals. However, the present chapter has indicated some of the serious handicaps that Denmark will face in fulfilling such a goal, particularly on the supply side. The lack of a large "domestic" multinational ICT company appears to be pose a major problem for a small country. This may be compensated by a swarm of small and medium-sized R&D-intensive firms, but too few of these have emerged during the 1990s despite the favourable economic conditions.[28] The challenges facing Denmark appear significant given the current structure of the ICT mega-cluster.

The cluster perspective, then, leads to a focus on the structure and quality of the demand side. The introduction of the Nordic Mobile Telephone system in 1981 resulted in Nordic leadership in mobile communications equipment in the 1980s. This lead continued during the diffusion of the pan-European GSM system in the 1990s, when the Nordic countries preserved their role as major users. It did not result in the formation of large Danish players at the company level, but the country held its leading-edge position in development and consumption. The shift to the third-generation system (UMTS) during the first decade of the Millennium poses yet another challenge; it would appear that Denmark has lost some ground in this process. The new system will be introduced rather late and, on the supply side, doubts have recently been raised as to whether Denmark will be capable of staying at the world frontier and preserving its position as a development "hub". It should be noted that both of these aspects can be influenced through telecommunications and R&D policy, respectively.

Some concerns have been expressed about the rather weak Danish infrastructure in terms of broadband Internet access. More could be learned especially from neighbouring Sweden in this field. A deliberate and more focused Danish telecommunications policy – *i.e.* with quantitative targets concerning broadband diffusion – appears to be necessary. The cluster perspective focuses on the necessity of being at the leading edge in terms of consumption characteristics – not least when such is not the case on the supply side.

Annex

Table A1. **Comparison of ICT definitions**

EFS	Description	Nordic Statistical Institutes	Cluster Focus Group
2221	Printing of newspapers	Not included	Not included
2222	Other printing	Not included	Not included
2223	Bookbinding and finishing	Not included	Not included
2224	Composition and plate-making	Not included	Not included
2225	Other activities related to printing	Not included	Not included
2231+2232+2233	Reproduction of sound recording, video recording and computer media	Not included	Not included
3001	Manufacture of office machinery	3001	Not included
3002	Manufacture of computers and other information processing equipm.	3002	3002
3220	Manufacture of RTV transmitters & apparatus for line telephony/telegr.	3220	3220
3230	Manufacture of RTV receivers, sound/video recording or reproducing app.	3230	3230
3320	Manufacture of instruments and appliances for control, testing and etc.	3320	Not included
3330	Manufacture of industrial process control equipment	3330	Not included
3130	Manufacture of insulated wire and cable	3130	3130
3210	Manufacture of electronic valves and tubes and other electr. comp.	3210	3210
3340	Manufacture of optical instruments and photographic equipment	Not included	Not included
2211-2215	Publishing	Not included	2211-2215
5143+5164+5165	Wholesale of ICT equipment	5143+5164+5165	Not included
5245+5247	Retail of ICT equipment	Not included	Not included
5272	Maintenance and repair of consumer electronics	Not included	Not included
6420	Telecommunications	6420	6420
7133	Rental and leasing of ICT equipment	7133	Not included
7210	Hardware consultancy	7210	7210
7220	Software consultancy and supply	7220	7220
7230	Data processing.	7230	7230
7240	Database activities	7240	7240
7250	Maintenance and repair of office, accounting and computing machinery	7250	7250
7260	Other computer related activities	7260	7260
7440	Advertising	Not included	7440
9211-9213	Motion picture activities	Not included	9211-9213
9220	Radio and television activities	Not included	9220
9240	Press agencies	Not included	9240
9251	Libraries	Not included	9251
Not included	Photographic activities	Not included	Not included
Not included	Videotape rental	Not included	Not included

Note: Differences between EFS and CFG in grey.
Source: EFS (2000), Nordic Statistical Institutes (1998) and Laura Paija (Co-ordinator of the OECD Cluster Focus Group for ICT Studies).

Table A2. **The EFS segmentation of the ICT-MC in NACE codes**

Segment	Type	NACE codes	
IT/Electronics	Content/Application	-	
	Transport/Mediation	7220	Software consultancy and supply
		7230	Data processing
		7240	Database activities
		7260	Other computer related activities
	Equipment/Infrastructure	2233	Reproduction of computer media
		3001	Manufacture of office machinery
		3002	Manufacture of computers and other information processing equipment
		3320	Manufacture of instruments and appliances for control, testing and etc.
		3330	Manufacture of industrial process control equipment
		3130	Manufacture of insulated wire and cable
		3210	Manufacture of electronic valves and tubes and other electronic components
		3340	Manufacture of optical instruments and photographic equipment
		5164	Wholesale of office machinery and equipment
		5165	Whole sale of other machinery for use in industry, trade and navigation
		7133	Renting of office machinery and equipment, incl. computers
		7210	Hardware consultancy
		7250	Maintenance and repair of office, accounting and computing machinery
Telecommunication	Content/Application	-	
	Transport/Mediation	6420	Telecommunications
	Equipment/Infrastructure	3220	Manufacture of RTV transmitters & apparatus for line telephony, etc.
Broadcasting	Content/Application	9220	Radio and television activities
	Transport/Mediation		
	Equipment/Infrastructure	3230	Manufacture of RTV receivers, sound/video recording or reproducing apparatus
		5143	Wholesale of electrical household appliances and radio and television goods
		5245	Retail sale of electrical household appliances and radio and television goods
		5272	Maintenance and repair of consumer electronics

Table A2. **The EFS segmentation of the ICT-MC in NACE codes** (cont'd.)

Segment	Type	NACE codes	
Information/ Entertainment	Content/Application	2211	Publishing of books
		2212	Publishing of newspapers
		2213	Publishing of journals and periodicals
		2214	Publishing of sound recordings
		2215	Other publishing
		7440	Advertising
		9211	Motion picture and video production
		9240	Press agencies
	Transport/Mediation	5247	Retail sale of books, newspapers and stationary
		9212	Motion picture and video distribution
		9213	Motion picture projection
		9251	Libraries
	Equipment/Infrastructure	2221	Printing of newspapers
		2222	Other printing
		2223	Bookbinding and finishing
		2224	Composition and plate-making
		2225	Other activities related to printing
		2231	Reproduction of sound recording
		2232	Reproduction of video recording

Source: EFS.

Table A3. **Regional employment specialisation in ICT segments and NACE codes**

NACE	Copenhage Municipality	Frederiksberg Municipality	Copenhagen County	Frederiksborg County	Roskilde County	Western Zealand County	Storstroem County	Bornholm County	Fyn County	Southern Jutland County	Ribe County	Vejle County	Ringkoebing County	Aarhus County	Viborg County	North Jutland County
IT/Electronics																
7220	1.39	1.29	2.49	1.66	0.51	0.26	0.23	0.06	0.51	0.32	0.11	0.59	0.40	1.25	0.21	0.82
7230	1.47	0.81	2.72	1.34	2.02	0.12	0.04	0.00	0.44	0.11	0.13	0.99	0.36	1.23	0.10	0.19
7240	1.46	3.33	5.21	0.33	0.41	0.04	0.16	0.00	0.07	0.00	0.09	0.34	0.00	0.02	0.23	0.00
7260	2.95	0.30	3.95	0.26	0.18	0.11	0.02	0.83	0.32	0.17	0.24	0.24	0.00	0.17	0.14	0.05
2233	0.53	0.00	6.42	0.12	0.00	0.00	0.17	0.00	0.00	0.00	0.00	0.00	0.38	0.29	0.31	0.00
3001	0.00	0.00	2.40	0.00	0.00	0.00	0.00	0.00	0.91	0.00	0.00	0.03	0.00	0.16	0.00	6.50
3002	0.55	0.08	2.75	2.03	0.53	0.13	1.80	0.00	1.37	0.47	0.04	0.23	0.05	1.46	0.03	0.32
3130	0.01	0.00	1.76	0.43	0.15	8.52	0.79	0.00	0.65	0.00	0.00	2.12	0.03	0.86	0.01	0.02
3320	1.50	0.01	1.62	2.38	0.22	0.66	0.08	0.00	0.27	0.99	0.03	0.31	0.15	2.07	0.86	0.56
3330	0.18	0.18	1.11	1.37	0.74	4.26	1.11	0.00	0.82	2.53	0.06	0.55	0.10	0.70	0.06	1.84
3340	0.15	2.04	1.39	0.31	2.00	4.70	0.03	0.00	0.65	0.00	0.00	0.73	0.21	2.02	0.22	1.05
3210	0.06	0.00	1.75	1.46	2.54	0.72	1.68	0.04	0.26	0.01	0.03	4.40	0.05	0.12	0.00	1.72
5164	0.76	0.79	3.12	2.42	0.55	0.24	0.26	0.16	0.32	0.27	0.31	0.55	0.48	1.14	0.22	0.50
5165	0.81	0.23	2.55	1.24	1.25	0.50	0.48	0.38	0.70	0.44	1.00	0.91	0.35	1.07	0.57	0.55
7133	2.39	1.05	2.56	1.99	0.44	0.33	0.29	1.56	0.28	0.50	0.79	0.53	0.43	0.07	0.17	0.47
7210	3.16	0.33	1.40	0.98	0.46	0.32	0.31	0.00	0.29	0.47	0.15	2.18	0.16	0.76	0.20	0.39
7250	0.26	0.62	4.30	1.02	0.41	0.11	0.07	0.18	0.59	0.28	0.53	0.98	0.29	0.85	0.18	0.36
Telecommunications																
6420	1.94	0.96	1.56	0.42	0.72	0.26	0.37	0.59	0.78	0.67	0.19	0.53	0.28	1.84	0.14	1.05
3220	0.45	0.00	1.36	0.69	0.53	0.00	0.00	0.00	0.08	0.69	0.00	0.92	2.44	0.23	0.04	4.98
Broadcasting																
3230	0.03	0.04	0.23	0.83	0.65	0.15	0.37	0.00	0.01	0.94	0.56	2.14	6.96	0.52	2.69	1.17
5143	1.50	0.56	3.02	1.04	0.76	0.36	0.40	0.03	0.28	1.22	0.09	0.36	0.34	0.70	0.39	0.78
5272	0.38	0.70	1.84	0.74	0.66	1.14	1.11	0.74	1.18	1.37	0.46	1.06	0.70	0.97	0.63	1.11
9220	1.71	8.93	2.51	0.08	0.09	0.11	0.60	1.91	1.17	0.17	0.18	0.42	0.46	0.64	0.21	0.38
5245	0.90	1.56	1.03	1.10	0.86	0.91	1.21	0.99	1.14	0.99	0.78	0.94	0.97	1.07	0.75	1.07

Table A3. **Regional employment specialisation in ICT segments and NACE codes** (cont'd.)

NACE	Copenhagen Municipality	Frederiksberg Municipality	Copenhagen County	Frederiksborg County	Roskilde County	Western Zealand County	Storstroem County	Bornholm County	Fyn County	Southern Jutland County	Ribe County	Vejle County	Ringkoebing County	Aarhus County	Viborg County	North Jutland County
Information/Entertainment																
2211	3.38	0.78	1.62	1.02	0.54	0.42	0.22	0.00	0.54	0.19	0.04	0.33	0.23	0.80	1.48	0.23
2212	1.93	0.34	0.08	0.16	0.41	1.60	1.57	1.92	0.27	0.25	0.34	0.99	0.69	2.04	1.52	1.35
2213	2.43	1.00	1.61	0.95	0.37	0.39	1.18	0.12	1.19	1.29	0.51	0.55	0.79	0.42	0.15	0.43
2214	2.10	1.42	3.30	0.42	0.26	0.19	0.15	0.40	0.29	0.06	0.20	0.32	0.06	0.43	0.20	1.37
2215	1.87	1.62	1.74	0.62	0.50	0.24	0.26	0.00	0.28	0.49	0.31	0.12	1.19	2.06	0.08	0.80
2221	0.04	0.00	0.82	0.00	0.07	0.10	0.00	0.00	1.90	0.00	0.00	5.76	0.00	1.67	0.10	1.53
2222	0.76	0.94	1.34	0.97	0.47	1.05	0.74	0.28	0.80	0.56	0.81	1.05	1.29	1.47	0.70	1.06
2223	0.87	1.53	1.13	0.08	0.79	0.26	1.78	0.43	1.02	0.55	0.23	0.27	0.32	1.53	4.00	1.03
2224	1.66	3.68	2.24	0.22	0.43	0.25	0.23	0.00	0.83	0.62	0.50	1.31	0.10	1.45	0.15	0.16
2225	1.95	2.78	1.06	0.89	0.65	0.19	0.58	0.00	0.67	0.30	0.00	3.23	0.25	1.27	0.51	0.15
2231	0.67	0.50	0.17	0.20	0.00	0.78	17.37	0.00	1.17	0.08	0.08	0.00	0.00	0.00	0.00	0.58
2232	0.43	2.00	0.70	0.00	0.00	0.00	0.00	0.00	0.00	0.00	0.38	0.00	0.00	4.93	0.00	2.53
5247	1.54	1.46	0.73	1.05	0.98	1.12	1.03	1.69	0.85	0.97	0.83	0.85	0.93	1.07	0.79	0.89
7440	2.46	2.27	0.65	0.49	0.33	0.32	0.23	0.09	2.47	0.32	0.35	0.50	0.59	0.97	0.51	1.03
9211	4.93	1.77	1.32	0.28	0.99	0.06	0.20	0.06	0.26	0.06	0.08	0.27	0.26	0.45	0.15	0.47
9212	7.50	2.20	0.17	0.06	0.00	0.00	0.08	0.00	0.35	0.14	0.00	0.05	0.24	0.05	0.07	0.00
9213	2.80	0.00	0.74	1.03	0.71	0.20	1.03	0.48	0.96	0.76	0.47	0.74	0.61	1.19	0.59	0.58
9240	5.81	1.69	0.76	0.44	0.34	0.25	0.26	0.40	0.30	0.10	0.23	0.29	0.14	0.45	0.10	0.10
9251	1.40	1.24	1.17	1.18	1.25	0.90	1.05	0.85	0.95	0.79	0.77	0.69	0.74	1.01	0.89	0.75

Note: The specialisation indicator is the share of ICT employment of total employment of a given county compared with the national average. A value above 1 indicates an above-average employment share, *i.e.* the county is "specialised", and vice versa.
Source: Statistics Denmark.

NOTES

1. The EFS definition of the ICT-MC is the point of departure of the present chapter. Annex Table A1 contains the definition in terms of a list of four-digit NACE codes. This table also contains a comparison with the cluster focus group (CFG) definition as well as the definition used in a statistical comparison of the Nordic ICT sectors by the Nordic Statistical Institutes (1998). Further, Annex Table A2 contains a division into four major segments.

2. Surveyed by and well represented in Edquist (1997).

3. In this perspective, the major EFS-sponsored DISKO project on the Danish national system of innovation (Lundvall, 1999), is strongly related to the ongoing cluster studies by EFS.

4. For more detailed information on the mega-clusters and the Danish cluster studies, see Drejer *et al.* (1999) and Holm Dalsgaard (Chapter 18 of this volume).

5. Danish Agency for Trade and Industry (2000), referred to as EFS (2000).

6. See also Holm Dalsgaard (Chapter 18 of this volume).

7. At present, no data are available for an international comparison of the relative size and segmentation of the ICT-MC based on the specific and detailed four-digit NACE codes shown in Annex Tables A1 and A2. Studies of ICT activities are hampered by a lack of comparable data on, for example, value added, employment and exports at the industry level, especially concerning the software and service activities. At the commodity level, international trade data are fairly detailed for electronics hardware but are not very developed in the case of software and services. OECD (2000b, Annex 2) contains a summary of the state of the art concerning efforts to develop an international "standard" for the definition of the ICT sector. A separate statistical publication is announced, but not yet available, drawing upon the OECD Working Party on Indicators for the Information Society (WPIIS). Until then, the OECD publication represents a pragmatic approach, with a series of international comparisons based on several definitions, depending on data availability. The more detailed studies by the OECD are mainly based on national sources, which is also the case for the present chapter. The definition presented in OECD (2000b) is more narrow, focusing mainly on three of the four horizontal segments, excluding Information/Entertainment (30% of ICT-MC employment).

8. ICT has been defined as ISIC Rev.2 classes 3825 (computers), 3832 (radio, TV and communications equipment) and 72 (communications services).

9. The proxy for ICT in Figure 2 is basically manufacturing output from the electronics industry and the paper and printing industry, which is only a subset of the ICT-MC (for example, no service exports are included).

10. Among the entire group of OECD countries, only the three countries shown in Table 3 and the United Kingdom were specialised in high-tech industries in 1990. For a definition of the categories of technology intensity, see OECD (1999b, Annex 1).

11. In 1996, Mexico and Korea joined the group of high-tech specialised exporters.

12. See also Green *et al.* (Chapter 3 of this volume).

13. At the detailed commodity level, Jensen and Jessen (2000) have calculated Danish specialisation for 1992-98 for telecommunications (proxied by the SITC Rev.3 groups 746.1, 764.3, 764.9 and 50% of 764.2). For 1992-97, the specialisation measure was between 0.7-0.9; in 1998, it increased to 1.2. Another part of the electronics industry in which Denmark has systematically specialised, is (electro)medical equipment (included in the Medical-Health mega-cluster in the EFS classification).

14. Storno was founded in 1948 by GN Great Northern and, by 1960, had become the third largest producer of then closed mobile communications systems, surpassed only by the US firms, General Electric and Motorola. During the 1970s, General Electric acquired Storno, which it sold to Motorola in 1986.

15. For further details, see for example Dalum *et al.* (1988) and Dalum *et al.* (1999b).

16 . *Financial Times*, "Survey on Nordic Information Technology", 11 May 2000.

17. See Ministry of Research (2000), Addendum to the Network Report, p. 12.

18. The relation between the pattern of R&D *vs.* export specialisation (Tables 6 and 3) is striking, although no strict statistical correlation has been performed in the present context.

19. A recent survey of "decisive" patents in telecommunications by the EU standardisation body ETSI reveals only limited Danish contributions (www.etsi.org/ipr/SR_314_V151.pdf).

20. Denmark is divided into 16 counties. Employment data (full-time and part-time) for 1998 are available from Statistics Denmark at the four-digit NACE level. The "Copenhagen Region" is interpreted here as comprising the counties of Copenhagen, Frederiksborg and Roskilde as well as the municipalities of Copenhagen and Frederiksberg.

21. Annex Table A4 contains regional specialisation for the 16 Danish counties for all 43 four-digit NACE codes of the ICT-MC for 1998.

22. The latter as a result of spin-off activities from the big (in the Danish context) consumer electronics firm, Bang & Olufsen (B&O Technology and the now Ericsson-owned DIAX).

23. For detailed studies of the North Jutland small-scale cluster in radiocommunications, see Dalum (1995) and Dalum *et al.* (1999b). See also www.norcom.dk., the homepage of this regional cluster.

24. For further details, refer to Ministry of Research (2000).

25. Silicon Valley (United States), Information Age Town (Ennis, Ireland), Oulu Technopolis (Northern Finland) and Kista Science Park (near Stockholm, Sweden).

26. An indication of the success of these efforts is illustrated in the ranking of the Oeresund region among high-tech valleys throughout the world in *Wired*, July 2000 (pp. 259-271).

27. Two new major seed funds (organised as private companies) have recently been announced, financed jointly by government funds and private finance. One will be based in the Copenhagen region and oriented towards biotechnology and pharmaceuticals. The other will be based at the Aalborg University Science Park, NOVI, with an emphasis on ICT projects.

28. In *Danmark.com*, PLS Ramboel and Boersens Nyhedsmagasin (2000b) have mapped a large number of new Internet start-ups. The field appears to be flourishing; nonetheless, few, if any, major international players have appeared on the scene.

REFERENCES

Begg, I., B. Dalum, P. Guerrieri and M. Pianta (1999), "The Impact of Specialisation in Europe", in J. Fagerberg, P. Guerrieri and B. Verspagen (eds.), *The Economic Challenge for Europe: Adapting to Innovation Based Growth*, Edward Elgar Publishing, Olso/Rome/Maastricht.

Braczyk, H.-J., P. Cooke and M. Heidenreich (eds.) (1997), *Regional Innovation Systems*, University College London Press, London.

Breschi, S. and F. Malerba (1997), "Sectoral Innovation Systems", in C. Edquist (ed.), *Systems of Innovation: Technologies, Institutions and Organizations*, Pinter Publishers, London.

Carlsson, B. and S. Jacobsson (1997), "Diversity Creation and Technological Systems: A Technology Policy Perspective", in C. Edquist (ed.), *Systems of Innovation: Technologies, Institutions and Organizations*, Pinter Publishers, London.

Clark, G.L., M.P. Feldman and M. Gertler (2000), *The Oxford Handbook of Economic Geography*, Oxford University Press, Oxford.

Dalum, B., J. Fagerberg and U. Joergensen (1988), "Small Open Economies in the World Market for Electronics: The Case of the Nordic Countries", C. Freeman and B.-Å. Lundvall (eds.), *Small Countries Facing the Technological Revolution*, Pinter Publishers, London.

Dalum, B. (1995), "Local and Global Linkages: The Radiocommunications Cluster in Northern Denmark", *Journal of Industry Studies* 2(2), pp. 89-109.

Dalum, B., C. Freeman, R. Simonetti, N. van Tunzelmann and B. Verspagen (1999a), "Europe and the Information and Communication Technologies Revolution", in J. Fagerberg, P. Guerrieri and B. Verspagen (eds.), *The Economic Challenge for Europe: Adapting to Innovation Based Growth*, Edward Elgar Publishing, Oslo/Rome/Maastricht.

Dalum, B., M. Holmén, S. Jacobsson, M. Praest, A. Rickne and G. Villumsen (1999b), "Changing Regional Systems of Innovation", in J. Fagerberg, P. Guerrieri and B. Verspagen (eds.), *The Economic Challenge for Europe: Adapting to Innovation Based Growth*, Edward Elgar Publishing, Oslo/Rome/Maastricht.

Dalum, B., K. Moeller, U. Joergensen and F. Valentin (1991), *Internationalisering Og Erhvervsudvikling*, Ministry of Trade and Industry, Copenhagen.

Drejer, I., F. S. Kristensen and K. Laursen (1999), "Studies of Clusters as the Basis for Industrial and Technological Policy in the Danish Economy", in OECD, *Boosting Innovation: The Cluster Approach*, OECD, Paris.

Edquist, C. (ed.) (1997), *Systems of Innovation: Technology, Institutions and Organisation*, Pinter, London.

EFS (2000), *Benchmarking Af Dansk Erhvervsliv* [Benchmarking Danish Businesses], Danish Agency for Trade and Industry (EFS), Copenhagen.

Enright, M.J. (1999), "The Globalization of Competition and the Localization of Competitive Advantage: Policies Towards Regional Clustering", in N. Hood and S. Young (eds.), T*he Globalization of Multinational Enterprise Activity and Economic Development*, pp. 303-331, Macmillan Press Ltd., Hampshire (UK).

Financial Times (2000), "Survey on Nordic Information Technology", 11 May.

Freeman, C. (1987), *Technology Policy and Economic Performance: Lessons from Japan*, Pinter, London.

Freeman, C. (1995), "The National Innovation Systems in Historical Perspective", *Cambridge Journal of Economics* 19(1).

Freeman, C. (2000), "A Hard Landing for the 'New Economy'? Information Technology and the United States National System of Innovation", Inst. of Economics, Federal University of Rio de Janeiro.

Green, R., J. Cunningham, I. Duggan, M. Giblin, M. Moroney and L. Smyth (2001), "The Boundaryless Cluster: Information and Communications Technology in Ireland", Chapter 3 of this volume.

Holm Dalsgaard, M. (2001), "Danish Cluster Policy: Improving Specific Framework Conditions", Chapter 18 of this volume.

Jensen, B.J. and J. Jessen (2000), "Sammenhaeng mellem geografisk branchekoncentration og eksportspecialisering" [The Connection between Geographic Concentration and Export Specialisation], Unpublished Master Dissertation, Department of Business Studies, Aalborg University.

Krugman, P. (1991), *Geography and Trade*, MIT Press, Cambridge, Mass.

Krugman, P. (1995), *Development, Geography, and Economic Theory*, MIT Press, Cambridge, Mass.

Lotz, P. (1992), "Demand-side Effects on Product Innovation", PhD dissertation, Series 9.92, Copenhagen Business School.

Lundvall, B.-Å. (1985), "Product Innovation and User-Producer Interaction", *Serie om industriel udvikling* (31).

Lundvall, B.-Å. (ed.) (1992), *National Systems of Innovation: Towards a Theory of Innovation and Interactive Learning*, Pinter Publishers, London.

Lundvall, B.-Å. (1999), *Det Danske Innovationssystem* [The Danish Innovation System], Erhvervsfremmestyrelsen, Aalborg.

Lundvall, B.-Å., N.M. Olesen and I. Aaen (1984), *Det Landbrugsindustrielle Kompleks* [The Agro-Industrial Complex], Aalborg University Press, Aalborg.

Malerba, F. (2001, forthcoming), "Sectoral Systems of Innovation and Production", *Industrial and Corporate Change*.

Maskell, P. and A. Malmberg (1999), "Localised Learning and Industrial Competitiveness", *Cambridge Journal of Economics* 99.

Ministry of Research (1999), *Det Digitale Danmark* [Digital Denmark], Ministry of Research, Copenhagen.

Ministry of Research (2000), *Netvaerksredegoerelse* [Network Report], Ministry of Research, Copenhagen.

Moeller, K. and H. Pade (eds.) (1988), *Industriel Succes* [Industrial Success], Samfundslitteratur, Copenhagen.

Nelson, R.R. (ed.) (1993), *National Innovation Systems: A Comparative Analysis*, Oxford University Press, New York.

Nordic Statistical Institutes (1998), "The Information and Communication Technology in the Nordic Countries – A First Statistical Description", *TemaNord* (587).

OECD (1999a), *OECD Communications Outlook*, OECD, Paris.

OECD (1999b), *Science, Technology and Industry Scoreboard – Benchmarking Knowledge-based Economies*, OECD, Paris.

OECD (2000a), *A New Economy? The Changing Role of Innovation and Information Technology in Growth*, OECD, Paris.

OECD (2000b), *OECD Information Technology Outlook – ICTs, E-Commerce and the Information Economy*, OECD, Paris.

OECD (2000c), *Science, Technology and Industry Outlook,* OECD, Paris.

OECD (2000d), *Education at a Glance 2000*, OECD, Paris.

PLS Ramboell Management and Boersens Nyhedsmagasin (2000a), *Frontloeberlande i den nye oekonomi – Benchmarking af Danmark.com* [Frontier Countries in the New Economy – Benchmarking Denmark.com], PLS Ramboell Management and Boersens Nyhedsmagasin, Copenhagen.

PLS Ramboell Management and Boersens Nyhedsmagasin (2000b), *Danmark.com – Kortlaegning af de danske internetpionerer* [Denmark.com – Mapping the Danish Internet Pioneers], PLS Ramboell Management and Boersens Nyhedsmagasin, Copenhagen.

Porter, M.E. (1990), *The Competitive Advantage of Nations*, Free Press, New York.

Porter, M.E. (1998), *On Competition*, Harvard Business School Press, Boston.

Saxenian, A. (1994), *Regional Advantage: Culture and Competition in Silicon Valley and Route 128*, Harvard University Press, Cambridge, Mass.

Storper, M. (1992), "The Limits to Globalization: Technology Districts and International Trade", *Economic Geography* 68(1), pp. 273-305.

Vernon, R. (1966), "International Investment and International Trade in the Product Cycle", *Quarterly Journal of Economics* 80(2), pp. 190-207.

Chapter 5

INNOVATION DYNAMICS IN THE SPANISH TELECOMMUNICATION CLUSTER: POLICY IMPLICATIONS

by

Cristina Chaminade[*]
Autonomous University of Madrid, Spain

Introduction

Today, there is no doubt that innovation is one of the key factors underlying growth and hence the manner in which innovation takes place has been a major concern since the mid-1980s. It is generally accepted that firms do not innovate in isolation but in continuous interaction with other sources of knowledge. This has been the argument of the evolutionary economists, from Schumpeter (1939) to Nelson and Winter (1982), and later in the systemic theories of technical change, at national level (Lundvall, 1992; Nelson, 1993), regional level (Saxenian, 1994), sectoral level (Carlsson and Stankiewick, 1991) and firm level (Kline and Rosenberg, 1986).

This general conception of the innovation process supports the relevance of networking and clustering of resources. The stronger the linkages with other sources of knowledge, the better the firm will perform in terms of innovation and finally in terms of growth. In this setting, the cluster approach emerged as useful framework within which to analyse the networks linking different agents. A cluster can be therefore considered as a "reduced-scale national innovation system" (Roelandt and den Hertog, 1999).

Identifying clusters of innovative activity is important for investigating both where the growth dynamics take place and where the learning capabilities of the economy are located (DeBresson and Hu, 1999). Identifying such clusters allows policy makers to place the emphasis on the innovation pools of the economy, adapting their innovation policies to the innovation dynamics.

The purpose of this chapter is to discuss whether there is a cluster of innovative activity among the ICT industries in Spain rather than a collection of unrelated firms and, if so, to analyse their innovation dynamics in order to be able to discuss the policy implications. One can argue that

* The author would like to thank the COTEC Foundation for their support in providing the statistical data for this work. The COTEC Foundation is an entrepreneurial organisation whose main purpose is to support technological innovation within Spanish business and society. She is also extremely grateful for comments by Prof. Paloma Sánchez of the Autonomous University of Madrid and Prof. Hanno Roberts of the Norwegian Business School. Special thanks to Prof. David Charles of the University of Newcastle and Prof. Pim den Hertog of Dialogic for their revisions to this text.

information and communication technologies (ICTs) are important as they are the drivers of the so-called "New Economy" (US Department of Commerce, 2000). Furthermore, they are the "enabling" technologies that support the rapid access to new knowledge and the interaction between different agents throughout the world. However, in the case of Spain, there is another reason that supports the need to deepen the analysis of the innovation dynamics in this cluster: the Spanish innovation survey shows that ICT-related industries can be considered as an innovation pole of the Spanish economy (Sánchez and Chaminade, 1999a). The innovative firms in the cluster are responsible for nearly 20% of the total innovation expenditure of the Spanish business sector. Furthermore, the proportion of innovative firms in the cluster is clearly above the Spanish average.[1] Innovative firms in the ICT industries are responsible for the lion's share of turnover and exports (INE, 1999). Compared to the rest of the Spanish economy, firms in these industries are introducing more new products, processes and services into the market.

So, ICT-related industries play an important role in the innovation dynamics of the Spanish economy but can they be considered a cluster? This chapter attempts to answer the following questions:

- *Is there an ICT innovation cluster?* This analysis will be based on the Spanish R&D survey.
- *What are the innovation dynamics of the Spanish ICT cluster?* That is, can different patterns of innovation be identified within the cluster in terms of inputs of the innovation process, results of innovation, collaboration with other actors in the system, etc. This discussion will be based on the Spanish Innovation Survey.
- *In the light of the previous analysis, what policy responses are necessary?* Would it be necessary to add any additional measures to current Spanish ICT cluster policy?

The following section is devoted to a discussion of the existence of a Spanish ICT cluster. Next, the dynamics of the cluster will be analysed and the Spanish ICT cluster policy presented. Finally, some conclusions will be drawn.

Is there a Spanish ICT innovation cluster?

The identification of what we consider a cluster is a tricky issue; it depends on the specific dimensions taken as reference point (spatial, industrial, etc.) and the criteria used to determine what falls inside and outside of the cluster (trade flows, knowledge flows, R&D flows, etc.). There are many definitions of what can be considered a cluster (based on input/output techniques, the value-chain approach, etc.) which are mainly based on trade flows among different industries. Despite the fact that this type of analysis has proven useful for the identification of industrial clusters, an alternative tool is needed when the purpose is to identify innovation clusters. In this sense, our analysis is based on Chris DeBresson's definition of a "cluster of innovative activity" (DeBresson and Hu, 1999) which uses as its point of departure the *innovative interaction matrixes* compiled from Innovation Survey Data.[2]

An innovative matrix is defined by DeBresson and Hu as "a square matrix, with the suppliers of the innovative output (for example, innovation sales or investments) classified by (...) industry in the rows, with the most frequent user industry of innovative outputs identified in the columns" (DeBresson and Hu, 1999, p. 29). In other words, their analysis is based on innovation flows.

Thus, the main requirement in identifying innovation clusters is to obtain information on the users and suppliers of the innovative output (either innovation sales or innovation investments). In the case of Spain, the Spanish R&D survey offers information on the suppliers of R&D and the users of the R&D by industry. Innovation investment is, of course, a broader concept than just R&D but, taking into account that R&D is responsible for 73% of the innovation expenditure in the ICT-related industries (INE, 1999), it can be assumed that it is a good proxy. In this sense, the results presented below are an adaptation of DeBresson and Hu's methodology, using R&D flows instead of the suggested innovation flows.

Taking into consideration the data on the users of the R&D developed by each ICT industry,[3] the Spanish innovative matrix can be developed. Each cell shows the percentage of the R&D produced by industry i (short axis) which is used by industry j (long axis).[4] That is, if the computer industry spends EUR 100 on R&D activities, approximately 60% is used by the computer industry itself, 10% is used by the trade services and mechanical machinery industry, and 5% is used by consumer electronics and telecommunication services.

To simplify the chart, only the rows related to the ICT-related industries are plotted. As Figure 1 shows, almost all R&D transactions are performed among the ICT industries. They are the main users of the R&D produced by other industries within the cluster. Furthermore, some non-ICT industry users emerge, such as financial services and public administration.

Figure 1. **The Spanish innovative matrix, 1997**

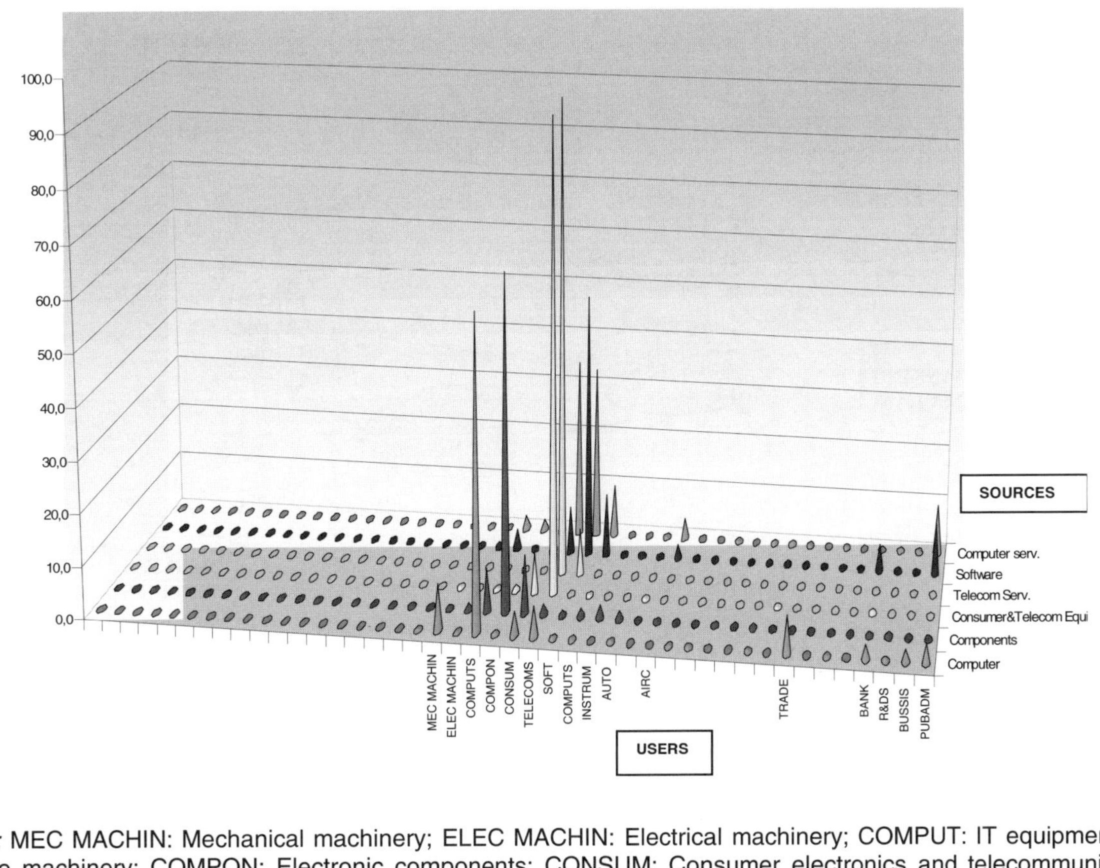

Key: MEC MACHIN: Mechanical machinery; ELEC MACHIN: Electrical machinery; COMPUT: IT equipment and office machinery; COMPON: Electronic components; CONSUM: Consumer electronics and telecommunication equipment; TELECOMS: Telecommunication services; SOFT: Software; COMPUTS: Computer services; INSTRUM: Instruments; AUTO: Automobile; AIRC: Aircraft; TRADE: Trade; BANK: Banking and insurance; R&DS: R&D services; BUSS: Business services; PUBADM: Public administration.

In principle, focusing on the previous analysis we might argue that there is an Spanish ICT cluster, in that the central actors of the cluster are the ICT industries – consumer electronics and telecommunication equipment, electronic components, IT equipment, telecommunication services, IT services and software. However, we need to go a step further by identifying the members and the shape of the innovation clusters.[5] This will be done using information collected in the survey on the users and suppliers of the R&D developed by each cluster. In doing so, we obtain information on the degree of integration of the different industries in the cluster, the type of actors (users, suppliers), and their relevance to the whole cluster.

The results of the analysis are plotted in Figure 2. Two types of arrows are used to refer to one-way relationships (unidirectional, discontinuous line) and two-way interaction (symmetrical, black line). Only those flows representing more than 5% of total R&D are plotted in the graph.

More than 88% of the R&D developed by the electronic components industry is assigned to other ICT industries, and this percentage is 99% for consumer electronics and telecommunication equipment, 73% for IT equipment, 76% for software, 100% for telecommunication services and 83% for IT services.

Figure 2. **The innovative cluster in ICT, 1997**

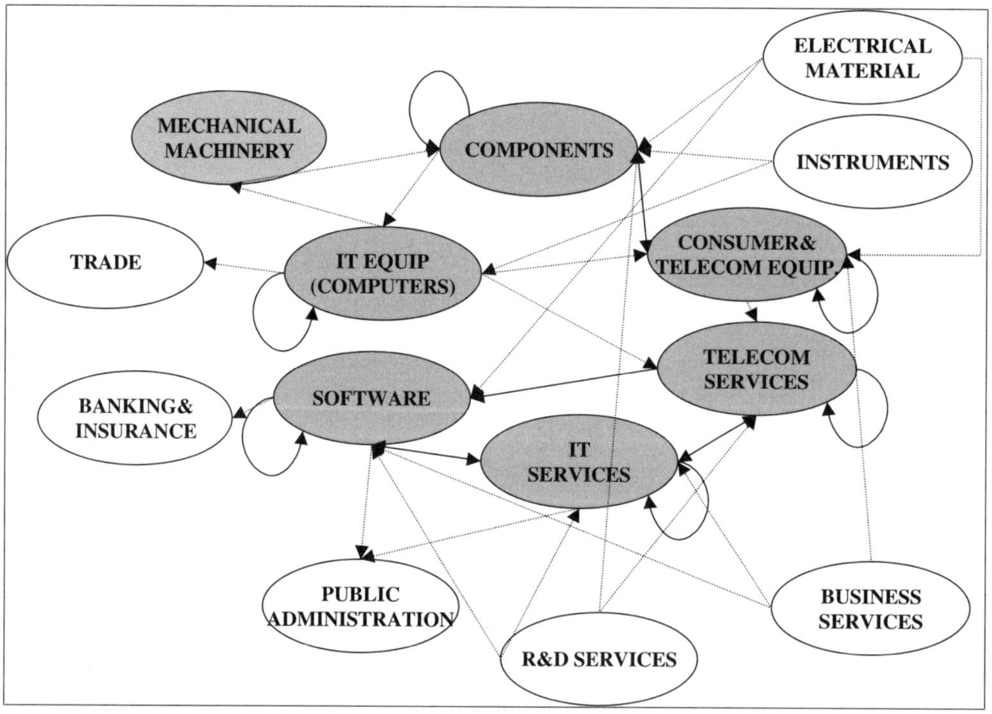

With this simple graph, it is possible to distinguish between the users and suppliers of the cluster: electrical machinery, instruments, business services and R&D services are supplier industries in the innovation cluster, while trade, banking and insurance and the public administration are exclusively users of the innovative clusters. The rest of the industries, including mechanical machinery, are both suppliers and users of the cluster. The inclusion of mechanical machinery can be explained as this industry contains part of so-called industrial electronics, which is traditionally considered as part of the ICT industries. Some key features of the expected ICT cluster emerge:

- First, we can clearly see that there is a strong relationship between telecommunication services, IT services and software. This suggests the existence of what DeBresson and Hu (1999) call a *clique*; that is, a cluster where all participant industries supply each other with innovative inputs.

- Additionally, we find that the main suppliers of that clique are electronic components and consumer electronics/telecommunication equipment, which can be considered as an innovative couple, following the DeBresson and Hu taxonomy; that is, the two industries are both suppliers and users of the R&D generated by the other industry. Furthermore, as the next section will demonstrate, the telecommunication equipment industry is the main technology supplier for ICT services.

- Finally, Figure 2 indicates a clear absence of links between IT equipment, software and IT services. The IT equipment industry can be considered, in the words of DeBresson, as a development point; that is, an industry with innovative activities totally enclosed within itself and having weak links with other industries. More than 80% of the R&D output developed by the IT equipment industry is devoted to own firms, leaving just 20% which goes outside the industry. Only 1% of the R&D from the IT equipment industry goes to software and IT services. In this sense, using this innovation matrix approach, it can be stated that the IT equipment industry is not part of the cluster.

We can conclude that we do not have an integrated ICT cluster in Spain, but rather a telecommunication cluster, involving some equipment and components supplier firms and ICT services, such as telecommunication services, software and IT services.

The next question to be addressed is whether we can talk about a broad telecommunication cluster in Spain involving the components industry, consumer electronics and telecommunication equipment, telecommunication services, IT services and software or, on the contrary, if we should talk about two regional clusters: *i)* the narrow telecommunication cluster, comprising telecommunication equipment suppliers and ICT services; and *ii)* a consumer electronics cluster, comprising the components industry and the consumer electronics industry. This issue will be further discussed using the innovation expenditure data obtained from the Spanish Innovation Survey (INE, 1999).

One broad national telecommunication cluster or two regional ICT-related clusters?

We have seen that there are two different cluster structures: *i)* an innovation couple between electronic components and consumer electronics/telecommunication equipment; and *ii)* a clique between the ICT services and telecommunication equipment. The first question is whether those two different structures correspond to two different clusters, and if those clusters are located in different regions.

The analysis of the distribution of the innovation expenditure of the industries considered in the broad cluster shows clearly that the innovation expenditure of the consumer industry is highly concentrated in the Catalonian region, as is most components expenditure. On the other hand, the majority of telecommunication equipment and telecommunication services expenditure is located in the Madrid region.

Table 1. **Percentage of innovation expenditure by region and ICT industry, 1998**

	Average	Components	Consumer	Telecom. equipment	Telecom. services
Andalusia	5.63	3.27	0.06	3.05	6.96
Castilla-La Mancha	2.83	0.00	0.01	3.41	2.90
Catalonia	9.80	40.98	71.75	1.26	8.50
Madrid	62.48	25.13	6.08	86.94	58.05
Basque Country	2.74	13.55	2.70	1.96	2.39

Source: Based on INE (1999).

We might then expect to find two different innovation dynamics, corresponding to two different ICT-related clusters: one a telecommunication cluster located in Madrid, and comprising telecommunication equipment and ICT services; the other a consumer cluster which might be located in the Catalonian region. In the following, we will discuss the innovation dynamics of these two clusters and their constituent industries.

Innovation dynamics in the cluster

All the results presented in this section are based on the Spanish Innovation Survey (SIS) which was conducted in 1998, covering 31 NACE manufacturing industries plus telecommunication services.[6] The SIS covers all earlier identified industries within the cluster except software and IT services.

What can we expect from the analysis?

The previous analysis points to the fact that there seem to be two different ICT-related clusters in Spain, a telecommunication cluster and a consumer cluster. The purpose of this section is to assess the relative strength of the two clusters through an analysis of innovation expenditure, collaboration agreements, innovation output and the role of multinational firms in each constituent industry. Our hypothesis is that we will find different patterns of innovation as we are talking about two clusters and industries that fall under different groups following the classic Pavitt taxonomy (1984):

- Components are *specialised suppliers* and, as such, to be efficient they need to develop their innovation activities in close collaboration with the users of their innovative output. They may develop their own knowledge through internal R&D and innovation activities. Design may also be an important element in their innovation input. The predominant size of the firm is small.

- Consumer and telecommunication equipment can be viewed as *science-based* industries. R&D is considered to be their main source of technology together with their suppliers, in this case, the components manufacturers. They are mainly large firms. Linkages with external sources of knowledge are also very important.

- ICT services might be seen as *supplier-dominated sectors*. They are expected to innovate through the acquisition of new equipment developed by their suppliers, in this case, telecommunication equipment and, to a lesser extent, consumer electronics.

The inputs of innovation expenditure

As expected, innovation in components, consumer and telecommunication equipment (supplier industries of the clusters) is based on R&D. In terms of innovation expenditure, R&D can be considered to be the main input for innovation in those industries. As Figure 3 shows, R&D accounts for 73% of total innovation expenditure on the supplier industries, while the average for Spanish businesses is 43%. In contrast, most innovation expenditure of telecommunication services is related to the acquisition of new equipment.

Figure 3. **Composition of innovation expenditure by individual industries in the telecommunication cluster, 1998**

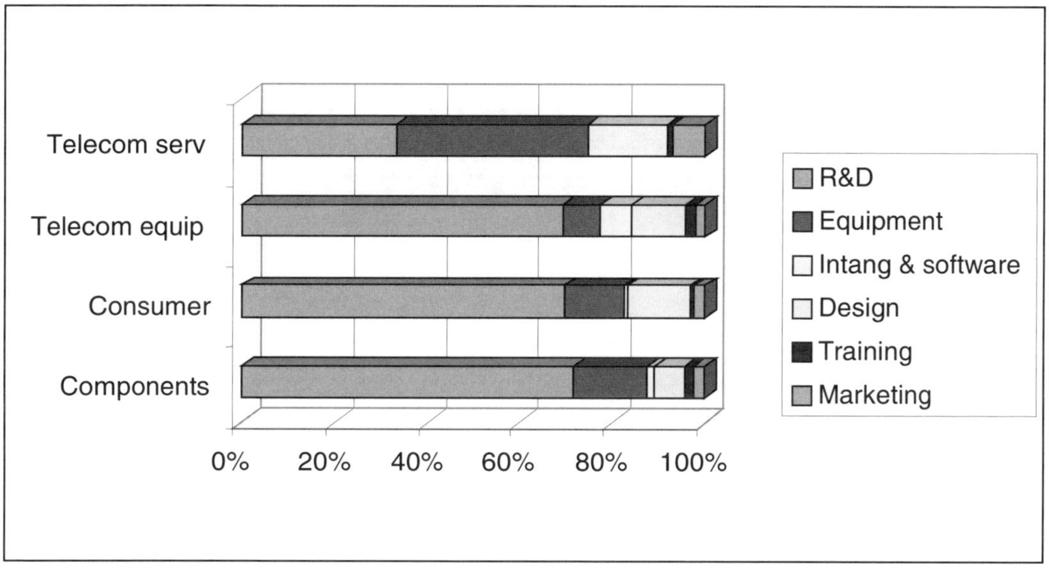

Source: Based on INE (1999).

The way in which the firms innovate differs according to the size of the firm. Large manufacturing firms focus their innovation expenditure on R&D activities, while SMEs usually do not have the required human and capital resources and innovate mainly through the acquisition of new equipment. For large firms, the acquisition of disembodied technology, such as patents, weighs more heavily in total innovation expenditure than is the case for small firms.

In telecommunication services, differences between firms according to their size are even more evident. Small firms with less than 50 employees devote 67% of their innovation expenditures to the acquisition of new equipment, the share drops to 41% in firms with more than 200 employees.

In the case of SMEs, it is evident that there is a shortage of human and technological resources. In this sense, the clustering of resources through collaboration with other firms or research institutions in R&D projects could present an alternative source, as some Spanish studies suggest (COTEC, 2000).

It would be expected that those firms which are well integrated in the cluster spend a significant amount of money on innovation and more specifically on R&D activities to provide the ICT service sector with new equipment. However, analysis of innovation expenditure by firm reveals that despite the innovation expenditure of the firms being significantly above the average for Spanish businesses, they are still low. While the average innovative firm in the Spanish economy spends EUR 365 000 on innovative activities, expenditures rise to EUR 661 000 for consumer, EUR 642 000 for telecommunication equipment and, finally, EUR 8 215 000 for firms in telecommunication services, on average. Only electronic components firms lag behind the Spanish average, with total expenditure per firm of only EUR 282 000. This might suggest that not all of the firms considered in the cluster are well integrated with each other. The results seem to support the hypothesis of the existence of a telecommunication cluster but do not support the hypothesis of the existence of a consumer cluster. Nevertheless, further research is needed.

Figure 4. **Innovation expenditure per firm and individual ICT industry, 1998**

In EUR millions

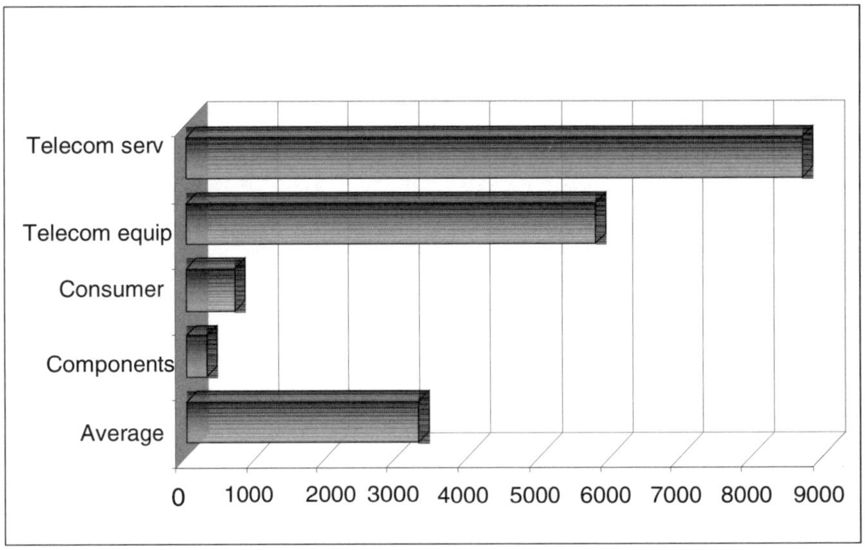

Source: Based on INE (1999).

As in many other European countries (Sweden with Ericsson, Finland with Nokia, etc.), the Spanish telecommunication cluster pivots around one large firm, Telefónica, and this is also reflected in the statistics. Telefónica is responsible for the lion's share of innovation expenditure in telecommunication services.

The high proportion of telecommunication services innovation expenditure is primarily due to the acquisition of highly sophisticated telecommunication and IT equipment with a high market value. In this sense, Telefónica might be acting as a technological demand pull for the rest of the cluster. Not only is 41% of their innovation expenditure due to the acquisition of new equipment from their suppliers; in addition, more than 40% of their R&D expenditure is extra-mural. This issue will be further developed in the next section.

R&D collaboration and other sources of innovative activity

An analysis of the innovation dynamics should also take into account the external relationships of firms with other research institutions in the innovation system such as universities and public research institutes. It can be postulated that the stronger the relationships with other institutions in the national innovation system, the better the cluster will perform in terms of innovation; this postulate is especially crucial for SMEs which usually have fewer resources.

The analysis of external relationships may also serve to detect the weaknesses of the system and obstacles to innovation which need to be overcome in order for firms to become more competitive. Furthermore, it will be useful to verify whether there is a consumer cluster.

The SIS contains information for 1996-98 on collaborative innovation agreements by partner and by geographical dimension. Unfortunately, the information focuses on the number of firms that have collaborative agreements with different institutions, while no information has been collected on the value of those agreements. For this reason, the results presented below should be used with caution.

One-fifth of Spanish innovative firms had some collaborative innovation agreement during 1996-98; this percentage doubles for telecommunication services. In general terms, the proportion of firms with collaborative agreements increases with the size of the firm. SMEs collaborate less than large firms as they probably encounter more difficulties in finding a suitable partner for their projects.

Table 2. **Percentage of innovative firms with collaborative agreements by geographical area, 1998**

	Spain	Other EU countries	Rest of Europe	United States	Japan	Other countries
Spanish average	90	24	2	4	1	2
ICT	79	40	7	8	3	3
Components	60	69	7	5	2	0
Consumer	100	23	0	0	8	0
Telecom. equipment	87	27	13	7	0	7
Telecom. services	90	20	8	15	3	6

Note: Totals across do not equal 100 as firms have agreements in more than one region.
Source: Based on INE (1999).

As Table 2 shows, most R&D collaborative agreements take place within the Spanish national boundaries. The sole exception is the components industry, where 68% of the firms had collaborative agreements with other institutions in the European Union. Table 3 illustrates the main partners by industry. For each sector, the main partners are shown in bold.

Table 3. **Percentage of innovative firms with collaborative agreements by partner, 1998**

	% of firms which collaborated in 1998	Same group	Cus-tomers	Suppliers	Compet-itors	Joint-ventures	Experts and consul-tants	Other firms	Public research institutes	Univer-sities	Research assoc-iations	Other
Spanish average	21	15	**28**	**33**	4	3	12	**27**	16	19	7	2
Telecom. services	41	**34**	11	20	20	3	5	**62**	15	**28**	3	0
Telecom. equipment	29	27	**47**	**20**	0	7	7	**33**	13	**47**	7	0
Consumer	27	**15**	8	**15**	0	8	**15**	**23**	23	**69**	8	0
Components	25	**49**	**19**	15	2	2	7	12	7	**35**	2	0

Source: Based on INE (1999).

The main partners vary significantly across the different industries, showing once more the different innovation dynamics that come into play within the cluster. As would be expected, given that telecommunication equipment and components are suppliers, one of the most important partners is the customer. The consumer equipment industry does have some collaborative agreements with its suppliers, mainly components, but they are not their main partners. The linkages between this "innovative couple" are not sufficiently strong for them to be considered as a cluster.

Universities seem to be a relevant partner for all the industries considered. However, experience leads one to think that the individual value of those agreements is relatively low in comparison to the total innovation activities of the cluster. Universities seem to be regular partners, but the collaboration agreement usually focuses on less relevant innovation projects. As a general trend, the Spanish innovation system displays weak linkages between industry and university (COTEC, 1999a), and this also holds true for the ICT industries (Chaminade, 1999).

This observation gains strength when we consider the main sources of information for innovation. For this purpose, the data collected in the 1996 Innovation Survey is used.[7] As is shown in Figure 5, universities are not considered to be an important source of information for innovation. The main sources, as would be expected, are the clients, followed by internal activities such as R&D, production or marketing, and competitors.

These results can be explained using qualitative data from other studies (Chaminade, 1999; COTEC, 1999b). Although firms are aware of the significant improvements made by the Spanish universities in terms of their internationalisation, there are still significant obstacles to true collaboration. In the entrepreneurs' opinion, there is a lack of confidence over whether the universities and research centres can provide solutions to the technological problems faced by firms. Firms argue that there are significant differences in the ways the two institutions work (cultural problems) and suggest that the knowledge generated at universities is of a scientific or academic nature and does not respond to their technological needs. In addition, experience induces us to think that, from the universities' perspective, there is a lack of information on what firms need.

Figure 5. **Sources of information for innovation in the telecommunication cluster, 1996**

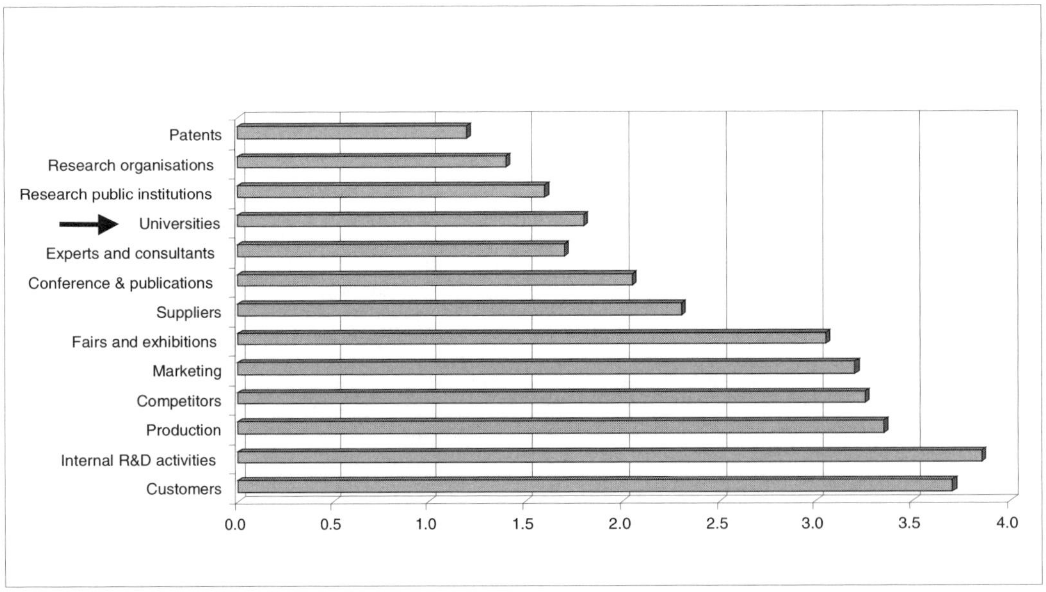

Source: Based on INE (1997).

In summary, although many firms collaborate with universities on R&D projects, it seems reasonable to think that we would arrive to different conclusions if we were to consider the volume of those agreements. Universities are not considered to be an important source of information for innovation. COTEC (1999b) suggest that there are few instruments within Spanish innovation policy that focus on the relationship between firms and R&D institutions. In their opinion, the role of policy makers should be: to foster the environmental conditions necessary for the emergence of institutions such as technology brokers; to stimulate spin-offs; and to improve information flows.

Innovation outputs

There are differences in the inputs of the innovation processes, as well as differences in the patterns of collaboration in the individual industries in the cluster. The question is: do these detected differences in inputs lead to different results in terms of innovation outputs?

Regarding the output of the innovative process, the CIS collects information on two different indicators: *i)* percentage of turnover due to new products or services; and *ii)* percentage of exports due to new products or services.

Considering the first indicator, as can be observed in Figure 6, more than 90% of the total turnover of the consumer electronic firms is due to new or improved products. In electronic components and telecommunication equipment, the proportion of brand-new products is higher than in the rest of the cluster. This is important as they are the main technological suppliers of telecommunication services.

In telecommunication services, as would be expected, the share of new services accounts for less than 20%. Traditional telecommunication services continue to represent the most important share of the industry's total turnover. These industries do not produce as much innovation as the rest of industries in the cluster but they innovate through the acquisition of technology generated elsewhere.

Figure 6. **Breakdown of turnover due to new and improved products and services, 1998**

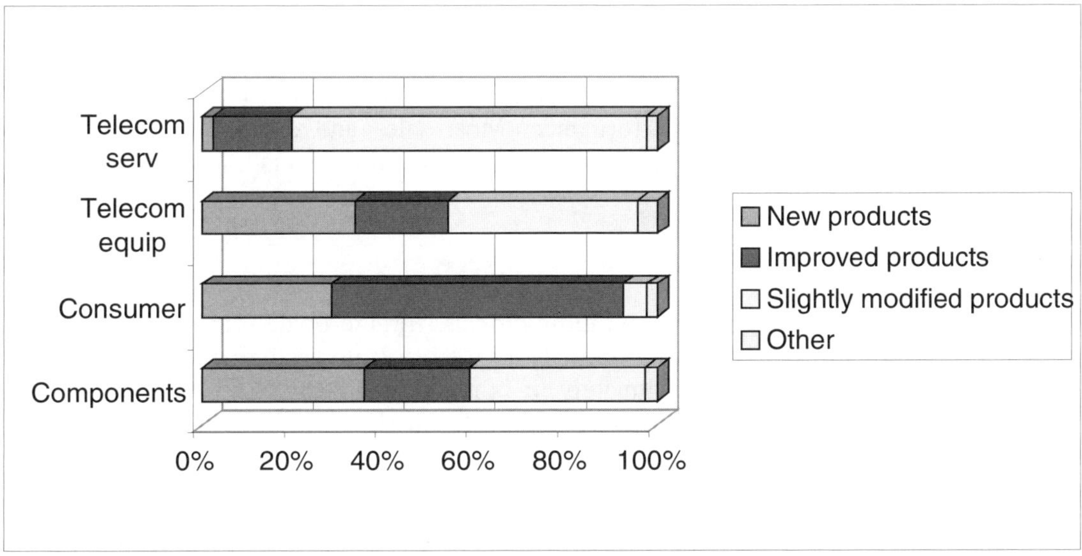

Source: Based on COTEC (2000).

The majority of the new products introduced into the market are new for the firm but not to the market. That is clearly the case in consumer electronics, telecommunication services and components. Only in the case of telecommunication equipment is most innovative output new to the market. That seems to confirm what has been previously stated, *i.e.* not all of the firms in the cluster are well integrated and it would appear that the linkages between telecommunication equipment as supplier and telecommunication services are stronger than those between components and consumer manufacturers.

Regarding the impact of innovation on the volume of exports, we can observe that consumer electronics show the highest proportion of exports of new and improved products while, in the telecommunication equipment and the electronic component industries, the highest proportion of exports is due to brand new products, as shown in Figure 7.

Figure 7. **Breakdown of exports due to new and improved products and services, 1998**

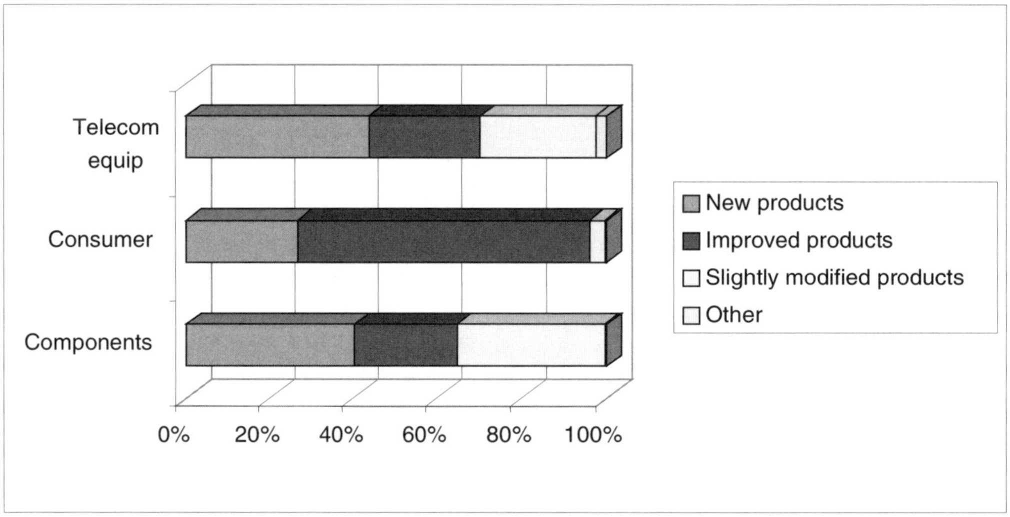

Source: Based on COTEC (2000).

These results are similar to those obtained from analysing the weight of new and improved products as a share of the turnover of the cluster, thus leading to the conclusion that there is a strong relationship between innovation and performance. Most sales and exports of the ICT industries considered in the cluster are due to new or significantly improved products.

Multinational vs. domestic firms

Differences are to be expected between multinationals (foreign-owned firms) and domestic firms, notably when dealing with innovative activities. Next we break down the data already presented by domestic and multinational firms. Unfortunately, it is not possible to break down the data between consumer electronics and telecommunication equipment, which might introduce some bias into the analysis as the consumer electronic industry is mainly dominated by foreign-owned firms while in the telecommunication equipment industry the proportion is more balanced.

As Table 4 shows, in the components industry one in four firms is multinational, accounting for 45% of the total turnover of that industry. In the case of consumer and telecommunication equipment, the situation is even more dramatic, with one in every eight firms a multinational, accounting for nearly 80% of the total turnover of the industry. In telecommunication services, the opposite situation prevails: only eight firms are multinationals and these firms account for less than 1% of the total industry turnover.

Table 4. **Key data on the telecommunication cluster**

Domestic *vs.* multinational firms, 1998

	Total turnover (EUR millions)		Number of innovative firms		Innovation expenditure by firm (EUR thousands)		% of turnover due to new and improved products	
	Multinational	Domestic	Multinational	Domestic	Multinational	Domestic	Multinational	Domestic
Total industry	86 384	134 268	921	15 718	2 450	243	43.9	22.9
Components	477	573	32	134	438	245	61.1	57.1
Consumer and telecom. equip	3 670	903	11	89	25 016	506	72.0	53.2
Telecom. services	102	14 958	8	87	794	8 898	21.2	19.8

Source: Based on COTEC (2000).

In the case of innovation expenditure, multinational firms from the components, consumer and telecommunication equipment industries account for 79% of total innovation expenditure. In telecommunication services, domestic firms account for nearly 99% of the total innovation expenditure.

When average expenditure by firm is considered, it can be observed that multinationals spend 20 times more than domestic firms; this figure rises to 50 times more in the case of the telecommunication equipment and consumer industries. This can be explained by the fact that the Spanish consumer electronics industry is dominated by large foreign-owned firms, such as Sony, Sanyo and Philips, who use Spain as a platform to address foreign markets, mainly in Europe. This sustains the hypothesis that most of the firms in the consumer electronics industry in Spain do not have strong linkages with the rest of the cluster and, consequently, we should not talk about a formal consumer cluster in Spain. Certainly, further research using more detailed data is needed to confirm this hypothesis.

Once again, the exception is the telecommunication services industry in which domestic firms spend, on average, eleven times more than multinationals.

Finally, regarding the outcome of innovation efforts, the percentage of turnover due to new and improved products is somewhat higher for the multinationals than for domestic firms, although the differences are not as significant as on the input side. Could this mean that although domestic firms invest less than multinational firms, they are better able to translate innovation investments into sales and exports of new products? Are they more efficient? Unfortunately, we do not have sufficient information to perform this analysis. Further research is required to examine the precise role of multinationals in this cluster.

Enabling conditions

Before dealing with the policy conclusions that can be drawn from the above analysis, it is important to be aware of the main obstacles that the firms in the cluster face in trying to innovate. This information can be extracted from the Innovation Survey conducted in 1996.[8] Figure 8 shows the importance ascribed to each of the main obstacles to innovation by the firms in the cluster.

As previous analysis has stressed (COTEC, 2000; Chaminade, 1999), lack of adequate finance seems to be the main obstacle to innovation in the telecommunication cluster. Figure 8 shows that the absence of risk capital in Spain *vis-à-vis* the increasing cost of innovation activities, is a very important barrier for firms, especially for SMEs. The long period needed to recover the innovation investment is also seen as an important obstacle. Third, the firms in the cluster claim to need a better legal environment (on matters such as telecommunication services liberalisation, standards, etc.) in order to reduce uncertainty and enable a better allocation of scarce resources.

Figure 8. **Main obstacles to innovation in the Spanish telecommunication cluster, 1996**

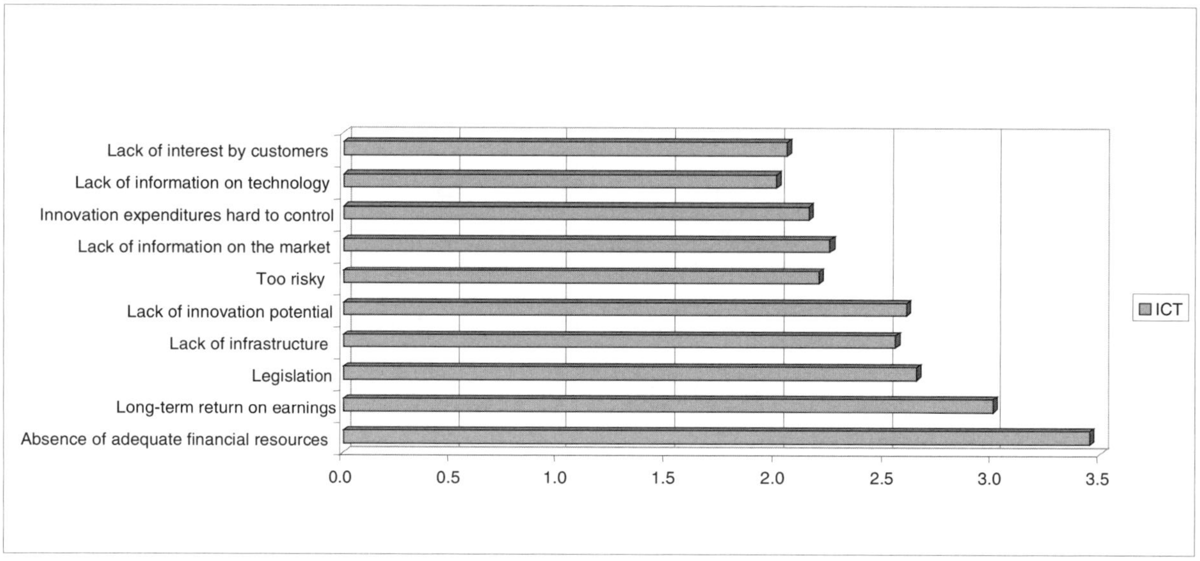

Source: Based on INE (1997).

The main differences with the Spanish economy as a whole are, first, the importance attached by the ICT cluster to legislation, followed by the lack of innovation potential and the long-term return on investments. From a policy perspective, the perceived differences between the most important obstacles for the telecommunication cluster call for specific action tailored to that cluster. We will now examine the policy implications of this analysis.

Spanish telecommunication cluster policy

All the literature on innovation systems stresses networking and clustering of resources as one of the key factors explaining growth and development. Clustering is a market-led process, but the government can support these initiatives by facilitating the networking process and ensuring a friendly institutional framework (Roelandt and den Hertog, 1999). Cluster policies aim to overcome some of the systemic and market failures that might hamper the networking process, such as the inefficient functioning of markets, informational failures, limited interactions among actors, institutional mismatches between the knowledge infrastructure and market needs, and the absence of a demanding customer (Roelandt and den Hertog, 1999).

The above analysis has shown that in Spain, rather than an ICT cluster there would appear to exist a telecommunication cluster formed by the telecommunication equipment industry, telecommunication services and probably, IT services and software (although these two latter industries could not be analysed because of data constraints). We have also demonstrated that the innovation dynamics and linkages between the consumer electronics and components sectors are not strong enough to support the hypothesis of the existence of an innovative consumer cluster. This section will therefore focus on the policy implications of the analysis of innovation dynamics in the telecommunication cluster.

In terms of innovation, some specific features of the telecommunication cluster have emerged from the above analysis:

- Innovation is one of the key factors supporting the competitiveness of the firms in the cluster. On average, one in five firms in the cluster is considered to be innovative; however, such firms are responsible for nearly 98% of the total turnover of the cluster and 70% of the total export volume (99.7% if only telecommunication equipment is considered). This points to the fact that innovation is a key driver of the telecommunication cluster's growth in Spain. Spanish ICT firms should strengthen their innovative efforts in order to reach the level of their competitors. *What policy measures could serve to encourage innovation in the cluster?*

- Research and development activities are a very important innovative input to the technology supplier side of the cluster, although the acquisition of new equipment is crucial for the users' side. *Can it be said then that policies promoting R&D activities, such as the R&D tax deduction scheme which the Spanish authorities have recently promoted, are well suited to the needs of the telecommunication cluster?*

- R&D is more important for large firms than for SMEs. The lack of human and technological resources might be an important barrier for small firms. Initiatives to encourage the mobility of qualified human resources between the research world and firms, and to promote graduate placements might be an option for these firms. *What is Spain doing in this sense? What could be recommended?*

- SMEs collaborate less than do large firms, perhaps reflecting the difficulties encountered by small firms seeking partners for collaboration. Technology transfer and collaboration, especially in SMEs, need further support. SMEs appear to have significant difficulties in networking with other agents. *Are there active bridging institutions in Spain, such as networking agencies, that can help to bridge this gap? What other instruments can be suggested?*

- Multinational firms are important in terms of innovation, sales and exports. The role of these firms in innovation in the Spanish telecommunication cluster needs to be better addressed. Further research is needed to assess the role of multinationals and the dynamics with other agents in the cluster in order to overcome the weaknesses.

- A clearer legal framework is required in order to encourage innovation. This is seen as one of the major obstacles to innovation in the cluster. Uncertainty about the future legal framework and standards in emerging services such as, for example, digital radio broadcasting, is a strong barrier to innovation when innovation costs are high. This argument is even more valid for SMEs. *What are the Spanish authorities doing in this respect?*

- The absence of risk capital institutions in the country is a serious obstacle to innovation. Although this is not exclusive to the telecommunication cluster, it seems to represent a more important barrier in high-technology clusters than in more traditional ones. It is important to promote the establishment of financial institutions that can encourage new innovative projects and enable the emergence of high-tech SMEs. In view of the fact that innovative SMEs are an important source of employment in the cluster, such support could have a significant impact.

Is there a Spanish telecommunication cluster policy? Does it provide a response to the detected weaknesses? Spain has implemented a number of general policies as well as specific ICT policies that aim at improving the framework for innovation and promoting collaboration among different agents within the system of innovation. Following Boekholt and Thuriaux's (1999) classification of cluster strategy policies, the following policy actions can be distinguished under the headings of policy programmes, framework conditions and strategic actions:

Strategic cluster policy	Generic	ICT
Policy programmes		
- Networking agencies	X	X
- Human resource mobility between research and industry	X	X
- Technology transfer programmes	X	X
- R&D collaborative programmes	X	X
- Cluster regional policies		X
Framework conditions		
- Standards		X
- Regulatory reform		X
- Policy procurement policy		X
- Privatisation		X
- Institutional reform in policy-making	X	
- Tax incentives	X	X
Strategic actions		
- Strategic market information – Info XXI		X
- Demonstration and development projects		X

Policy programmes

The national R&D programme for 2000-03 focuses on a limited group of strategic areas, with information and communication technologies being one of the main technology areas (CICYT, 1999).[9] The R&D programme is built around three main horizontal actions: human resource development; international co-operation; and technological innovation and technology transfer.

Regarding *human resource development*, the Spanish Government has implemented a Human Resource Mobility Programme whose main purpose is to promote the mobility of researchers and professional between the academic and the industrial world. ICT is considered a priority area. With this initiative, the government hopes to increase the qualification of human resources and to create and reinforce linkages between industry and the research centres. As we have seen, this is especially relevant for SMEs. In practice, this programme has encountered a number of obstacles, mainly due to the scarcity of researchers in the ICT area.

The main aims of the horizontal actions on technological innovation and technology are to: *i)* provide support to innovative firms through setting institutional framework conditions that favour innovation; *ii)* strengthen clusters that show strong potential, supporting networking and removing existing obstacles to clustering; *iii)* support the emergence of technology-based firms; *iv)* foster technology diffusion; *v)* protect the results of the innovation process; and *vi)* provide support to networking agencies. A variety of tools are being used to achieve these goals.

Since the mid-1980s, the Spanish Government has supported the creation and functioning of the OTRIs – the public system of *networking agencies* whose main function is to transfer research results from research centres to industry. During the 1980s, these networking agencies were based inside the universities and were generic in nature. Our experience leads us to believe that the impact of the networking agencies has been very low. Given that the majority of the OTRIs emerged in the universities, their linkages with industry have in most cases been very limited. In response to firms' clear demands for a radical change in the role played by the networking agencies, in 1996, the Government supported the creation of a networking agency for the ICT industries which was linked to the Spanish Electronic and Telecommunication Association (ANIEL); that is, to the firms themselves.

Support for *collaborative R&D projects* takes the form of subsidies or preferential loans at very low interest rates for concerted and co-operative actions between a firm, or a group of firms, and a research centre or university, as well as for specific technological innovation projects in single firms.

In addition to the national initiatives, some regions, such as Catalonian and the Basque Country, have been developing specific *cluster policies* since the 1990s. In Catalonia (Conejos, Duch *et al.*, 1997), the regional government identified 20 regional clusters in manufacturing and services, one of them being the consumer electronics cluster located in the region of El Vallès near Barcelona. In the Basque country (Escorsa and Camacho, 2000), three ICT-related regional clusters were initially identified as emergent clusters in 1993: professional electronics, telecommunications and computers. However, only one, telecommunications, was finally selected as a priority cluster in view of its potential for improved competitiveness. The telecommunication cluster is supported by the regional government and the firms in the cluster come from the electronics industry, software, telecommunication equipment and telecommunication services.

Framework conditions

Considering more general policy actions, in 1999 the Spanish Government created the Ministry of Science and Technology and most of the innovation-related policies were transferred to this Ministry. The main objective was to improve co-ordination among the various initiatives originating from the different ministries – identified by firms as an important factor hampering the effectiveness of the R&D programmes (Chaminade, 1999).

The Spanish Government has actively supported tax incentives for R&D activities. In this respect, the real costs of R&D activities in Spain were at the low end of the OECD range (Chaminade, 1998, based on OECD, 1996); that is, Spanish firms benefit from one of the most beneficial R&D tax incentive regimes in the OECD area.

In terms of specific ICT policy actions, one of the main tasks of the Spanish Government has been the deregulation of the telecommunication services market and the creation of a competitive environment. In 2001, the deregulation process will conclude with the introduction of competition in the market for local fixed phone calls.

The Spanish Government participates in several EU projects relating to the definition of IT standards in the public sector as well as in a number of working groups with ICT firms.[10]

Strategic actions

In 2001, the Spanish Government will launch the INFO XXI initiative, which includes a number of actions aimed at increasing the use of ICT in Spanish society. The absence of a qualified customer base is highlighted by Spanish ICT firms as representing an important obstacle to innovation. With this initiative, the government aims to increase the number of users benefiting from ICT.

Conclusions

Spanish innovation policy appears to be addressing the main weaknesses of the innovation system, both through general policies such as the national R&D programmes and through ICT-specific actions. However, the varying dynamics of the different industries within the cluster call for different actions.

Support for R&D activities in the telecommunication cluster seems to be oriented in the right direction in the telecommunication equipment industry, in view of the share of R&D in the total innovation input. However, it is less effective for those groups of firms, such as SMEs and telecommunication services firms, for which the acquisition of new equipment makes up an important portion of total innovation expenditure. Additional measures could be taken for those groups in order to encourage innovation.

General policies aimed at promoting networking agencies and fostering a favourable institutional framework also focus on some of the problems which have been identified as hampering innovation in the cluster. However, firms argue that the current bridging institutions are not as effective as they could be since they tend to be more focused on university research than on firms' needs (Chaminade, 1999). The role and linkages of these institutions with firms need to be redefined.

Finally, the different dynamics of the ICT industries should be taken into account in the design of policies. As has been seen, there are important differences between the innovation processes in the telecommunication equipment industry compared to the rest of the cluster. Some of the industries appear to be better prepared for international competition while others continue to lag behind their competitors.

NOTES

1. Approximately one in four firms in these industries can be considered innovative, compared to an average of 10% of Spanish firms that have introduced a new product, process or service during the period considered.

2. This tool allows the identification of clusters at the national level. One point for discussion is whether ICT can be considered as a national cluster, given the high level of globalisation of this group of industries, or as a European or international cluster. On the other hand, it can be claimed that most interactions take place at the regional level, so a reduced-scale cluster should be considered. However, for this analysis, only the R&D and innovative activities that take place within Spain by both national and multinational ICT firms are considered.

3. Similar cluster results are obtained when we focus on the suppliers of R&D to the ICT industries.

4. Spanish firms have to show the distribution of their total R&D expenditure by user industry, *i.e.* the industries that use the R&D developed in the source industry.

5. In doing so, we follow DeBresson and Hu (1999, pp. 42-47).

6. Innovation surveys have been largely used to map innovation processes at the national and international levels (Arundel, 1996; Calvert *et al.*, 1996; Evangelista *et al.*, 1996, Sanchez and Chaminade, 1999a, 1999b). Industry-specific analyses using Innovation Survey data are beginning to emerge, usually at national level.

7. The most recent Spanish CIS does not provide information on the sources of information for innovation or the main obstacles hampering innovation. The Spanish Statistical Office considers that this kind of information does not change significantly over time and that the results obtained in the 1996 survey continue to hold true.

8. See note 7.

9. Other areas include: biomedicals, biotechnology, materials, chemistry, industrial design, natural resources, agro-food and socio-economy.

10. Based on the information provided to the OECD by the Spanish Government. Available at http://www.oecd.org/dsti/sti/it/prod/it-out2000profiles/spain.htm.

REFERENCES

Arundel, A. (1996), "Enterprise Strategies and Barriers to Innovation", paper presented at the International Conference on Innovation Measurement and Policies, Luxembourg.

Boekholt, P. and B. Thuriaux (1999), "Public Policies to Facilitate Clusters: Background, Rationale and Policy Practices in International Perspective", in OECD, *Boosting Innovation: The Cluster Approach*, OECD, Paris, pp. 381-412.

Calvert, J., C. Ibarra, T. Patel and K. Pavitt (1996), "Innovation Output in European Industry: Analysis from CIS", paper presented at the International Conference on Innovation Measurement and Policies, Luxembourg.

Carlsson, B. and R. Stankiewick (1991), "On the Nature and Composition of Technological Systems", *Journal of Evolutionary Economics*, Vol. 1, No. 2, pp. 93-118.

Chaminade, C. (1998), "Innovación y Flujos de Conocimiento en las Industrias de Nuevas Tecnologías de la Información y las Comunicaciones en España", PhD dissertation, Autonomous University of Madrid.

Chaminade, C. (1999), "Innovation Processes and Knowledge Flows in the Information and Communications Technologies (ICT) Cluster in Spain", in OECD, *Boosting Innovation: The Cluster Approach*, OECD, Paris, pp. 219-242.

CICYT (1999), *Plan Nacional de Investigación Científica y Desarrollo Tecnológico 2000-2003*, Madrid.

Conejos, J., E. Duch *et al.* (1997), *Canvi estratègic i clusters a Catalunya*, Dep. Indústria. Generalitat de Catalunya, Barcelona.

COTEC (1999a), *El Sistema Español de Innovación. Diagnósticos y Recomendaciones*, COTEC, Madrid.

COTEC (1999b), *Relaciones de la empresa con el sistema público de I+D*, COTEC, Madrid.

COTEC (2000), *La Innovación en las Tecnologías de la Información y las Comunicaciones*, COTEC, Madrid.

DeBresson, C. and X. Hu (1999), "Identifying Clusters of Innovative Activity: A New Approach and a Toolbox", in OECD, *Boosting Innovation: The Cluster Approach*, OECD, Paris, pp. 27-59.

Escorsa, P. and J.A. Camacho (2000), "Industrial Policy of Providing Support to Clusters: The Case of the Basque Country", in R. López-Martinez and A. Piccaluga (eds.), *Knowledge Flows in National Systems of Innovation*, Edward Elgar, Cheltenham.

Evangelista, R., T. Sandven, G. Sirilli and K. Smith (1996), "Measuring the Cost of Innovation in European Industry", paper presented at the International Conference on Innovation Measurement and Policies, Luxembourg.

INE (1997), *Encuesta sobre innovación tecnológica en las empresas. 1996*, INE, Madrid.

INE (1999), *Encuesta sobre innovación tecnológica en las empresas. 1998*, INE, Madrid.

Kline, S. and N. Rosenberg (1986), "An Overview of Innovation", in Landau and Rosenberg (eds.), *The Positive Sum Strategy*, National Academy of Sciences, Washington, DC.

Lundvall, B.Å. (1992), *National Systems of Innovation*, Pinter, London.

Nelson, R. (1993), *National Innovation Systems. A Comparative Analysis*, OUP, New York.

Nelson, R. and S. Winter (1982), *An Evolutionary Theory of Economic Change*, HUP, Cambridge, Ma.

OECD (1996), "Fiscal Measures to Promote R&D and Innovation", OCDE/GD(96)165, OECD, Paris.

OECD (1997), *OECD Proposed Guidelines for Collecting and Interpreting Technological Innovation Data – The Oslo Manual*, OECD, Paris.

OECD (1999), *Boosting Innovation: The Cluster Approach*, OECD, Paris.

OECD (2000), *OECD Information Technology Outlook 2000. ICTs, E-commerce and the Information Economy*, OECD, Paris.

Pavitt, K. (1984), "Sectoral Patterns of Technical Change: Towards a Taxonomy and a Theory", *Research Policy* 13(6), pp. 343-373.

Porter, M. (1990), *The Competitive Advantage of Nations*, MacMillan, London.

Roelandt, T. and P. den Hertog (1999) "Cluster Analysis and Cluster-based Policy Making: The State of the Art", in OECD, *Boosting Innovation: The Cluster Approach*, OECD, Paris, pp. 413-427.

Sanchez, M.P. and C. Chaminade (1999a), "El Proceso de Innovación en las Empresas Españolas. Análisis de las Encuestas de Innovación", *Colección Estudios*, No. 14, Fundación Cotec, Madrid.

Sanchez, M.P. and C. Chaminade (1999b), "Instrumentos de Medición de la Actividad Innovadora. El Papel de las Encuestas de Innovación. Resultados del Caso Español", in COTEC (1999), *Informar sobre Innovación*, APIE-COTEC, Madrid.

Saxenian, A. (1994), *Regional Advantage: Culture and Competition in Silicon Valley and Route 128*, HUP, Cambridge, Ma.

Schumpeter, J.A. (1939), *Business Cycles*, McGraw-Hill, New York.

US Department of Commerce (2000), *Digital Economy 2000*, Washington, DC.

Chapter 6

ICT CLUSTER ORGANISATIONS IN FLANDERS: CO-OPERATION IN INNOVATION IN THE NEW NETWORKED ECONOMY

by

Jan Larosse, IWT Vlaanderen, **Patrick Slaets,** Agoria, **Jan Wauters,** Imec,
Stephan Bruninx, FMV, **Peter Simkens,** DSP Valley, and **René Wintjes,** Merit

Introduction

This chapter presents the results of a first attempt to summarise the experience of Flemish cluster policy and cluster organisation with the engagement of the actors involved (Larosse *et al.*, 2001). One of the characteristics of cluster development in Flanders is its pragmatic approach. This approach allows for a variety of initiatives that can lead to experimentation and learning-by-doing but that can also hamper policy learning through best practices on a more systematic level. The lessons and conclusions drawn from these experiences are presented with the aim of stimulating a debate that has now matured with the development of several new policy initiatives. The Flemish Government is preparing a "Digital Action Plan" to spur ICT in Flanders. Among various action lines (promoting ICT competencies among the population, improving diffusion of ICT in SMEs, becoming a leader in e-government), the stimulation of innovative clustering aimed at strengthening the ICT base will be a major concern. In addition, a new legal framework will be presented in support of "collective innovation" to the benefit of existing cluster organisations and new initiatives.

Traditionally, the Flemish economic clusters are largely based on agglomeration economies, *e.g.* the strong presence of automotive assembly, the petrochemical industry in Antwerp, or carpet manufacturing in the Kortrijk region. These are all activities in which Flanders has a strong market share. However, the economic structure of these industries remains "firm-based", draws on the "involuntary" benefits of co-location (a common infrastructure, labour market and tacit knowledge base) and is often weakly coupled to the rest of the economic tissue. This general observation also applies to the Flemish ICT cluster. Interactions among its more traditional components remain rather limited as most of the larger firms are part of international value chains with limited local clustering.

Cluster dynamics of the scale and intensity found in the well-known knowledge-driven ICT clusters such as Silicon Valley (United States), Shalom Valley (Israel), Silicon Isle (Ireland) or Silicon Fen (United Kingdom) do not exist in Flanders at present. However, the observation that a Flemish ICT cluster does not (yet) exist at the macro level does not exclude the fact that more specialised ICT clusters may be operating on a smaller scale. On the contrary, as will be illustrated below, these emerging clusters operate at a more disaggregated level of specialisation and increasingly show typical cluster dynamics. Cluster dynamics in a knowledge-driven economy are, in our view, not limited to classical agglomeration economies; they also exploit knowledge spillovers through co-operative learning.

We will illustrate how these clusters are developing into small-scale innovation systems, combining the necessary systemic factors for successful innovative performance enhanced by the conscious efforts of dedicated cluster organisations or cluster platforms. We will highlight some of the organisational paradoxes of regional clusters, and in particular the global/local paradox that capitalises on the combination of proximity for informal knowledge diffusion and international sourcing of state-of-the-art formal knowledge. Another paradox is the symbiosis between the spontaneous, market-induced character of cluster development and the significant contribution of voluntaristic and organised cluster promotion.

In the following, we substantiate these key observations on the development of Flemish ICT clusters.[1] First, the development trajectory of Flemish cluster policy is outlined. Subsequently, we undertake a qualitative and quantitative analysis of the converging ICT sectors and ICT technologies that are giving rise to a new industrial cluster dynamics in Flanders, the so-called "cluster space". This concept maps the trans-sectoral space in which the emergent economic cluster is likely to take shape and focuses policy on the strategic components and actors involved.

We then go on to identify some central points in this broader cluster space, using three case studies of small-scale ICT clusters: Flanders Digital Signal Processing Valley, Flanders Language Valley and Flanders Multimedia Valley.[2] Finally, we present some conclusions and policy lessons.

The trajectory of cluster policy in Flanders

Flanders was one of the pioneers in the adoption of Porter's cluster approach (Porter, 1990, 1998), introducing cluster policy in the early 1990s. The development of this policy has followed a three-stage trajectory that has mirrored the political business cycle.

In the first period, cluster policy started out very ambitiously, adopted a top-down approach and was designed to become the cornerstone of a general development strategy of "endogenous growth". In 1989, as a further step towards regionalisation, the Belgian State transferred most of its economic policy competencies to the regional governments. In search of a new regional identity, the Flemish Minister-President proned the cluster approach as a new development strategy designed to encourage enterprises to exploit synergies in innovation, training, export promotion, and the like. This endogenous growth policy was geared towards fostering the advance of Flemish economic strongholds so that they no longer relied on external direct investments. In 1993, the cluster policy was announced, together with indications as to the targeted clusters. However, a first "trial run" of the cluster programme to the benefit of the distressed textile machinery industry strengthened many actors in their conviction that cluster policy was simply one more instrument for discretionary policy intervention. Under pressure, mainly from the employers' organisations, the Minister-President had to rein in his ambitions. A compromise was established in November 1994, whereby the Flemish Government would provide support to clusters conforming to certain criteria, but that the clusters would be chosen through a bottom-up selection process.

Over the next few years, a dozen cluster organisations were awarded government accreditation; these organisations received annual funding of about EUR 3 million for their activities which included animation, brokerage and advice, as well as funding for infrastructure. They constitute a rather heterogeneous group, ranging from initiatives in more mature sectors (textile machinery, steel plate utilisation, electro-technical equipment installers, synthetics industry, furniture) to more high-tech platforms (digital signal processing, telematics in road transport, speech and language technology, product development techniques, aeronautics and the space industry). The accreditations came to a halt after 1996 due to the political and regulatory problems involved in the ad hoc procedure of case-by-case accreditation by the Flemish Government. The DSP Valley cluster initiative described below stemmed from this policy initiative, the origins of which date back to the 1980s with the creation (on the impulsion of the Flemish Government) of IMEC, now the largest independent microelectronics research institute in Europe.

The second period coincided with the new government which took office in 1995. Cluster policy was increasingly limited to the S&T domain and innovation policy. In 1996, the Flanders Language Valley initiative was started (see below); it became a role model for the new policy of "technology valleys" (Vlaamse regering, 1998) which aimed to create networks in emerging, technology-driven sectors (as opposed to "clusters" in the mature sectors). At the end of 1998, ten such technology valleys initiatives were in preparation. However, the new policy was not well documented and took an approach which was too close to "picking winners" to win sufficiently broad political support. Some new initiatives (Flanders Graphic Valley in the graphics industry and Flanders Drive in the automotive suppliers industry) were established in the framework of the European structural funds (EFRO subsidies to development zones).

The third period started with the taking of power of a new coalition government in 1999. An "Innovation Decree" provided the legal base for innovation policy as a whole, including for the cluster programmes that were under regulatory scrutiny. The latter thus became part of a general scheme to support "collective innovation" in a Regulatory Decision on "Flemish Innovation Co-operation", which should be approved by the Flemish Government in 2001. The draft Decision defines these organisations as "structured co-operation of – mainly – Flemish enterprises, possibly including one or more organisations or institutions, in view of exercising activities of collective research, technology advice and/or innovation promotion".

The general idea is to open up the "market" of collective innovation services to new entrants in order to enhance quality and provide a transparent framework of support measures. Therefore, the "recognition", or "labelling", of organisations, will be replaced by competitive funding by IWT (the technology agency of the Flemish Government) of three types of innovation activities: collective research (up to 50%); technology advice; and stimulation of innovation (up to 80%). Innovation stimulation covers a set of activities, ranging from awareness creation to first-line assistance and assistance in applications for government support. Start-up finance (feasibility and business plan) will be given special attention. The criteria for funding are the "innovation potential" and "quality" of the projects. A third criterion is that of "additionality" and "positive externalities". Funding will be "quasi-structural" (2 x 2 years) aimed at enhancing continuity but also allowing adaptation in order to improve complementarities among actors. This Decision provides a framework for collective action in innovation and will be a cornerstone in the "bottom-up" promotion of networking and clusters. Networks and cluster organisations can obtain substantial quasi-structural funding for their organisational activities. IWT will be responsible for monitoring the initiatives.

Having described the policy background in which the Flemish ICT clusters are developing, we will now deal with the so-called ICT cluster space (macro level), before going on to provide three examples of specialised ICT clusters (micro level).

The Flemish ICT cluster space: the macro perspective

The identification of "an" ICT cluster in Flanders is rather tricky. First, ICT is an international business and its value chains cross borders. National markets have been deregulated and the preferential user-supplier relations which existed in the past have been dissolved. Overall, in the ICT field, Flanders is currently specialised in telecommunication equipment (Siemens, Alcatel), consumer electronics (Philips, Barco), IT system integration and installation (IBM, Telindus), IT services and software (EDS, Dolmen, Origin), speech and translation technology (Lernout & Hauspie Speech Products) and digital photography and printing systems (Agfa-Gevaert, Xeikon). Second, geographical concentration is a necessary, but not sufficient, condition for the constitution of a cluster. Real cluster dynamics need a certain threshold level of interconnections to produce synergies which go beyond the static agglomeration effects. Flanders is too small to host a complete ICT value chain, but it is big enough to develop industrial competitiveness in ICT that can benefit from the locational advantages, the local knowledge specialisations and the network synergies that exploit the available tacit knowledge.

This process of restructuring and cluster formation is one of constant flux. The concept of a "cluster space" attempts to grasp this dynamic and increasingly complex reality of fragmentation, convergence and co-operation/competition ("co-opitition") in an emerging industry in a regional setting. Within the cluster space, some nodes of specialisation will develop into "real" clusters (as will be shown below). Cluster organisations can help to accompany cluster formation by consciously delimiting their cluster space into new value-creation models on the basis of shared objectives, and by organising interconnections between core components on the basis of shared knowledge activities.

In this section, we briefly present the main changes taking place in the ICT value chain and, more generally, the new industrial organisation of the ICT cluster space. We then discuss how this is contributing to changing innovation patterns. Finally, we provide some basic quantitative indicators on the size and economic impact of the Flemish ICT cluster, including the content or multimedia cluster.

Changing value chains

The ICT sector has been radically transformed in the last decade and the process of rapid structural change has yet to stabilise. On the one hand, the simple vertical value chain has been broken (*i.e.* the chain stretching from telecommunication suppliers to telecommunication operators in national markets). On the other hand, sector borders are becoming increasingly blurred (*e.g.* between telecommunications, consumer electronics, content producers). The result is that in the last decade the industrial landscape has become nearly unrecognisable, mainly because of the massive entrance of new players in the market, but also because of the dramatic restructuring of the traditional players. This process of change is also evident in Flanders and is clearly encouraged by the export orientation of the Flemish economy.

This type of industrial dynamics is probably most visible in the telecommunications market, although this market was opened to new operators at a rather late stage. A strategically important moment was the 1994 decision of the Flemish Government to promote the formation of a second telecommunication operator that would offer telephone and broadband services by integrating the local (coax) networks of the cable service providers, controlled by the local authorities. This was a strategic move to provide mass access to new broadband services since more than 90% of households already had cable connections. Telenet, the new fixed wire operator, started operations in late 1997 and is now gaining significant market share in terms of telephone and Internet connections. The liberalisation of the mobile telephony market in 1996 introduced two newcomers and, at the end of 2000, more than 55% of Belgians owned a GSM, with growth of the market running above average. In 2000, over

40 companies had acquired a licence to offer public network services (of which 30 in voice telephony), involving huge investments in new infrastructure.

The new operators, Telenet and Mobistar, offered Flemish industrial and financial players the opportunity to enter this new market, although in both cases they had to hand over the initiative to a strong foreign player in this global business (Calahan for Telenet, France Telecom for Mobistar). Telindus is one of the few Flemish players that could use this strong dynamics to launch itself as a major international player in the field of network integration. Other Flemish companies, being latecomers and lacking financial support, are at present mostly engaged as niche players in very different segments of the ICT market. Some, like ICOS Vision with its inspection systems, are connected to the global ICT industry. Others have successfully entered specialised niches in speech technology (LHSP) or security (Ubizen). Most often, there is a link with local strengths in research (IMEC) or with mature sectors such as the banking and graphics industry. In banking, for example, Belgium is a pioneer in information technology, in particular in electronic terminals, with the establishment in 1989 of Banksys, a joint venture among the Belgian banks to implement a common network for electronic payment. In 1995, Proton, the "electronic purse", was introduced. Proton is a smart card which allows payment from stored value at any terminal. The system has become a world standard, with licences in 15 countries and 35 million users. In the graphics industry, firms such as Barco, Xeikon, Artworks and Agfa Gevaert are international players, building on local expertise in digital communication technology.

Hundreds of new hardware and software firms have started up in the last few years (50% of ICT firms are less than five years old), although the main players in Flanders remain the big industrial constructors of the 1980s. These firms have undergone a profound metamorphosis: a number of Philips' production facilities in Flanders have ceased to produce standard consumer articles and have become "centres of excellence" in their specialisations, pioneering new consumer products (in lighting, video screens, audio, etc.). For example, Phillips Hasselt was responsible for the worldwide development of the CD-ROM. The telecommunication equipment producers have succeeded in transforming themselves from hardware companies to software companies. Siemens ATEA is Siemens' "excellence centre" for UMTS, while Alcatel Bell is the "excellence centre" for IP protocol and leads the world in high-speed ADSL modems. The Flemish ICT industry employs many multilingual engineers who are extremely sought after for international management positions in multinational firms.

Finally, the integration of the traditional media companies in the ICT cluster space – which has only just begun – will further modify the organisation of industry. Media companies have started to diversify into the ICT realm, and vice versa, and to participate in joint ventures aimed at providing Internet-based services. However, the most important platform for the convergence of network technologies, broadband services and content industry in Flanders might be the advance of digital TV – and Flanders ranks among the world leaders in cable TV infrastructure.[3]

New innovation patterns

The disruption of vertically integrated production chains in the large manufacturing companies calls for new methods of strategic management and innovation. In order to better master the complexity of globalisation and technological change, a kind of "semi-integration" in networks through alliances and partnerships of a different kind is replacing the integrated large-scale organisations. Proximity (in both the cultural and the geographical sense) could provide a competitive advantage in this respect, thus promoting local clusters.

"Technology-push" innovation, practised in the vertical value chain by the incumbents (the linear innovation model), is being challenged by the new interactive model of innovation. In the past, R&D in telecommunication depended to an important degree on public funding: the R&D budget of the national operator was fully "outsourced" to the "national champions" in equipment construction; these firms presented their solutions to the national operator, acting as a reference for later export contracts. The Flemish telecommunication constructors developed very successful export products thanks to public R&D. Today, the former national operator is independent, but the inheritance of R&D-dependence lies at the origin of the weak R&D effort in telecommunication services. R&D by Flemish constructors represents a high proportion of turnover (more than 15%) and has become far less government-dependent. Global market developments are driving changes and proprietary systems have been turned into more open systems. The need for new technologies is leading Flemish R&D departments into international alliances, steered by their headquarters. In addition, the role of (local) knowledge-intensive business services in the re-engineering of market and innovation strategies is growing. R&D in ICT companies accounts for about 25% of total in-house R&D expenditures in the business sector (total ICT-related research in all sectors probably accounts for nearly 50%); in 1997, this represented some EUR 500 million. Outsourcing of R&D remains limited (a mere 12% of total R&D). A small number of large firms perform the lion's share of this R&D.

Start-ups are important sources of innovation. This is bringing about a new pattern of innovation in the form of learning-by-doing (supported by own-R&D), immediately sanctioned by the market. Seed capital for start-ups is no longer a bottleneck in Flanders.[4] In particular, ICT projects dominate VC investment. This share – 54% in 1998-99 – represents an estimated EUR 300 million a year, invested in innovative ICT ventures (taking into account capital exports). However, successful start-ups are often taken over. One reason for this is that venture capital firms are not yet big enough for major growth investments. Few newcomers grow to an international dimension. Sometimes, large corporations try to emulate the success conditions of the new innovators: the rapid development of Alcatel Bell as a world leader in ADSL can be largely attributed to the special status of "internal start-up" that gave the fledgling department full operational and financial independence to react to the demands of the new market.

Another recent development in the ICT innovation pattern – linked to the expansion of the venture capital market – is the role of spin-offs from local knowledge institutions. Flemish universities give birth to about a dozen ICT spin-offs a year. The transfer of knowledge from universities to business through education and research contracts is now being paralleled by this new channel of technology development in which the microelectronic research institute, IMEC, plays an important role.

Basic economic indicators on the Flemish ICT cluster space

The evolution of the Flemish ICT sector during the 1990s can be divided in two main periods (Tables 1 and 2). In the beginning of the 1990s, the number of firms was fairly stable and employment decreased. From 1994 onwards, the *number of firms* began to grow more rapidly. Of the present ICT firms in Flanders, 50% are less than seven years old and 25% were founded in the last three years. Likewise, *employment* started to increase from 1995 onwards. Within five years, employment in the ICT sector had increased by almost 15 000 people, or 3 000 jobs per year, and the growth rate of employment in ICT is increasing over time. Most of these jobs are situated in the IT services and software sectors. The decrease in telecommunication equipment was compensated by the uptake in telecommunication services (operators and service providers). As in most other European countries, the incumbent telecommunication operator lost part of its market share to the new operators entering the market due to the gradual liberalisation of telecommunication markets. At the end of 1999, the estimated 60 000 workers in the ICT sector represented about 3% of total employment in Flanders.[5]

Value added in the Flemish ICT sector amounted to an estimated BEF 200 billion in 1999 (EUR 5 billion or 4% of Flemish GDP – a mid-ranking position in the international context). Most importantly, the Flemish ICT sector is directly responsible for almost a quarter of total R&D investments in the region. The R&D intensity of the total ICT sector was 6%. IT hardware and telecommunication equipment and the IT services and software showed almost identical R&D intensities. R&D expenses per employee rank among the highest in Europe. In telecommunication services, R&D intensity is very low. Software development is relatively strong in the international perspective.

Table 1. **Number of enterprises in the ICT sector in the 1990s**

Sector	1992	1993	1994	1995	1996	1997	1998	1999
30	22	24	15	12	10	10	10	11
32	36	32	47	54	59	61	60	55
51.64	523	526	543	605	657	726	744	784
64.2	5	7	14	23	40	63	81	88
72	789	796	833	870	942	1 020	1 141	1 279
Total	**1 375**	**1 385**	**1 452**	**1 564**	**1 708**	**1 880**	**2 036**	**2 217**

Source: Social Security statistics.

Table 2. **Employment in the ICT sector in the 1990s**

Sector	1992	1993	1994	1995	1996	1997	1998	1999
30	511	361	441	457	335	350	341	352
32	22 339	21 011	19 997	19 958	19 868	19 006	19 150	19 112
51.64	6 212	6 179	6 191	6 426	7 128	8 158	9 475	9 879
64.2 (37% Belgacom)	9 809	9 496	9 385	9 350	9 713	9 281	8 267	7 445
64.2 (other operators)	261	265	362	716	1 002	1 465	2 042	2 599
72	8 136	8 733	8 960	9 850	11 270	13 698	16 250	19 590
Total	**47 268**	**46 045**	**45 336**	**46 757**	**49 316**	**51 958**	**55 525**	**58 977**

Source: Social Security statistics.

If we extend our analysis to include the *multimedia* or content sub-cluster (Table 3) in the wider ICT cluster space, it is noteworthy that the industries in this sub-cluster are integrating ICT in their products and/or are using ICT intensively in their internal and external business processes. The Flemish content cluster is specialised in five fields: editing and printing; broadcasting (radio and television); publicity and marketing agencies; photography; and film and video. An estimated 35 000 employees (in FTEs, 2% of total Flemish employment) together created a value added of about BEF 100 billion (EUR 2.5 billion or 2% of Flemish GDP). The largest groups in the Flemish multimedia cluster are clearly located in the press world (newspaper, radio, television, etc.). Almost one-third of employment in the content cluster is concentrated in the ten largest companies.[6] The regional content cluster is ready to adopt the local world-class photography and printing technology and speech and translation technology.

Table 3. **The content sub-cluster in Flanders, 1999**
Estimations

Sector	Enterprises	Employment	Value added (in BEF billions)
Editing & Publishing	493	5 000	20
Printing	1 574	17 000	45
Advertising	1 646	4 000	15
Photo & film laboratories	323	2 000	4
Movies	470	1 000	4
Broadcasting (radio & TV)	104	4 000	8
Other sectors	104	2 000	4
Total	**4 862**	**35 000**	**100**

Source: Own estimations, based on Annual Reports (National Bank of Belgium).

Clusters as small-scale innovation systems: the micro perspective

DSP Valley

DSP Valley was founded as a private initiative by the University of Leuven (K.U.Leuven), IMEC and Philips Leuven in 1993-94, to add value to the research performed by IMEC and K.U.Leuven, and to stimulate industrial activities based on research in the domain of Digital Signal Processing (DSP). In 1998, the DSP Valley board and its members decided to accelerate the operations and activities of DSP Valley by a full deployment. Therefore, DSP Valley vzw (a non-profit organisation) was set up as the legal base and a full-time management was appointed. Since its inception seven years ago, DSP Valley has grown into an archetypal network or cluster organisation in the Flemish context.

DSP Valley is a network organisation with a strong technology focus on the *design of digital signal processing systems*. Within this network, DSP groups members, selected on the criterion of activities in the targeted domain in a region of about 200 km around the centre of gravity of Leuven, a small university town near Brussels. Today, DSP Valley counts 19 members, ranging from small and medium-sized enterprises and young start-up companies, through the local R&D branches of large multinational groups, to independent R&D organisations, universities and higher polytechnic schools. Even more impressive is the network's pool of 1 500 highly educated design engineers.

DSP Valley is recognised as an official cluster by the Flemish Government. Many cluster conditions are met within the DSP Valley environment:

- Dedicated, world-class *research for knowledge creation*, available from the university partners and IMEC. Indeed, the university members are very well ranked among European universities. IMEC is the largest independent research centre in microelectronics and ICT in Europe and attracts a great deal of interest from industry based on the available know-how.

- *Industrial leadership:* as a technology network, DSP Valley focuses on a specific technology. Today, a very strong leader is available for each of the sub-domains of digital sound (Philips Leuven), digital imaging (Agfa-Gevaert) and telecommunications (Alcatel Microelectronics). IMEC leads development activities in the domain of design methods and tools. The availability of several strong leaders is clearly an advantage, since they can provide an early insight into their roadmap for the smaller DSP Valley members, who can adapt their own strategy and technology development to the needs of the leaders. In addition, the presence of several leaders ensures a very stable environment, independent of the success or failure of a single player.

- *Pool of trained labour:* the availability of 1 500 highly trained DSP design engineers represents a very high concentration of expertise, even at world level.

- *Training and education:* are provided by top institutes like K.U.Leuven, RU Gent and VU Brussels. Applied and specialised training is available from IMEC. Several companies provide training on the utilisation of their products, often incorporated in a programme that includes the principles of the methodologies.

- *Networking*, with a flow of informal information on common business. DSP Valley acts as the information channel, collecting and distributing interesting and relevant news for the selected business domain, while respecting confidentiality where necessary.

- *Venture capital*: over the last few years, the availability of venture capital (VC) in Flanders and the Leuven area has increased significantly. Several specialised VC Funds are active, each in their own specific target market. VC Funds such as IT Partners and Capricorn Ventures are headquartered in the DSP Valley region and specifically target IT, of which DSP is a sub-domain.

- *Hosting infrastructure*: DSP Valley provides access to research parks in the area around Brussels and Leuven, with excellent access roads. New areas are under construction. In addition to the industrial research parks, hosting infrastructure is available in Innovation and Incubation Centres. These centres rent out offices and multi-use spaces as well as services to start-up enterprises setting up a research- or technology-oriented business. They also provide an incubation function for university or industry research spin-offs.

- *Entrepreneurship*, including a positive approach to risk. Both large and small DSP Valley companies are headed by experienced managers. Through the networking function of DSP Valley, such experienced leaders can help each other and the young entrepreneurs in the new spin-off companies to avoid the pitfalls they encountered in their rich professional career.

Transfer of know-how is realised among the various parties (see Figure 1) in many ways: publications, joint education programmes, exchanges of PhD students, recruitment from universities, academic and industrial training, joint R&D programmes and creation of spin-off firms.

IMEC plays a particular role in this and other networks. IMEC was founded in 1984 by the Flemish Government to act as a catalyst for industrial renewal in the region. From the outset, the institute has deployed a three-phase strategy.

In the first phase, IMEC built up a technological portfolio, becoming a renowned research centre with a world-class reputation. Innovative ideas and technologies can only be created by operating at the frontiers of scientific research.

In the second phase, a worldwide network of industrial partners and renowned institutions was set up, based on joint R&D projects. Research and development can only be carried out successfully by acting on a global scale – research has no borders.

In a third phase, through various technology transfer activities, the technology portfolio (IMEC's intangible assets of know-how) and its industrial network are being fully exploited to the benefit of Flanders. In addition to the creation of new companies, IMEC has attracted corporations such as

ASM-I, Heraeus and Philips to establish a branch or R&D centre near IMEC in order to engage in joint research. IMEC is at the crossroads of international co-operation networks and regional networking and thus provides a good illustration of the global-local paradox of successful clustering in small open economies.

To fulfil its mission, DSP Valley distinguishes three generic classes of activity:

- *Business support* to member companies and research organisations in addition to members' individual marketing and sales efforts, recruitment and training programmes.

- *Business development,* promoting the available knowledge and competencies on a common basis (trade missions and other collective actions).

- *Regional development* to increase economic activity in the region, not only through the growth of the DSP Valley members, but also by attracting new players in the same domain. If the skills and competencies available through the network serve to attract new players, this will automatically generate new activities through co-operation and contracts for existing DSP Valley members.

Figure 1. **Co-operation and transfer of the know-how concept in DSP Valley**

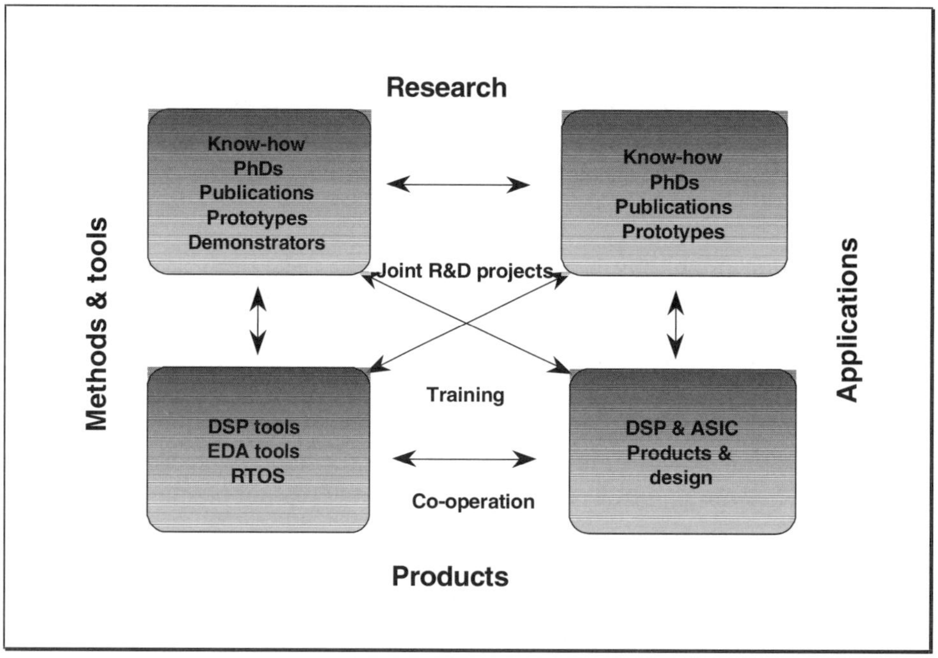

Both short-term business support and long-term regional development form part of the same mission for DSP Valley. DSP Valley members expressed their interest in three particular activity categories within business support, business development and regional development, that are closely interlinked: development of human resources, necessary for attracting new players to the region; networking with industry; and interfacing with government. The latter is relevant for discussing immigration rules for internationally recruited talent or for discussing the incentives to be made available to foreign investors and foreign start-ups in the DSP Valley region.

Flanders Language Valley

Flanders Language Valley (FLV) is a cluster of localised technological change, which had developed successfully over the last five years in Ieper, a town situated in rural Flanders. During 2000, this ICT cluster was in distress due to financial problems, mainly concerning its core firm Lernout & Hauspie Speech Products (L&H). Since clustering and networking serve to accumulate both success and problems, the virtuous cycle may turn into a vicious one, affecting the community in which the cluster is embedded. In fact, since the end of 2000, the FLV initiative has almost come to a halt. A new evaluation will be necessary once the future strategy of L&H becomes clear.

To draw conclusions of general policy relevance, we will focus on the development of the FLV cluster before the onset of the crisis. Three phases have been identified using a life-cycle approach: creation, growth and exploitation. A description of these phases will be followed by a number of lessons which can be drawn from the FLV case study. Some of these lessons refer to the inherent danger and risks of clustering of innovation since the prospects for FLV as a cluster, its core firm L&H, and its core speech and language technology, were highly uncertain from the very beginning, albeit mainly in a technological sense.

The three development phases of FLV

The first phase in the development of Flanders Language Valley concerns L&H, a pioneering ICT firm. Its founders, Jo Lernout and Paul Hauspie, made the initial choice of technology and location. In 1987, after working for several years outside Europe, they returned to their home region, met in Ieper and decided to set up a business in language and speech technology. Jo Lernout had worked with Bull and Wang computers on a technology to codify natural speech, but this work focused exclusively on English. In the multilingual context of Belgium and Europe, L&H focused on a multilingual technology. They started their company in a building set up by the regional development agency (GOM) as a regional employment project. The first years of research and development were funded by "regional", and mostly informal, venture capital, which is very much in line with the inherited business mentality for which this area is famous: "Persuading local bakers, butchers and farmers to invest was part of our daily routine for a long time in those early days, because traditional investors did not have faith in the project" (*FLV Magazine*, 1998). GIMV, the Flemish regional investment company, was the first main institutional investor.

Throughout the growth period of 1994, L&H sold licences, opened offices in the United States and began to acquire other pioneering companies. After years of internal orientation, the autarkic, closed mentality evolved into a network mentality. At the beginning of 1995, L&H received many requests from clients wanting access to their R&D core, the language laboratory in Ieper. Many small firms had signed licence agreements in order to integrate L&H technology with their own technologies. Communication had to be carried out over long distances and, despite their ICT capabilities, this led to misunderstandings and delays. In spite of the close technological proximity between L&H and their clients, a great deal of effort was put into integrating their technologies and combining separate and different, albeit complementary, technologies. This process of (re-)combining the codified knowledge of two firms is not a matter of simple exchange, adding an extra set of codes to the existing stock of codified knowledge. To facilitate the necessary intensive communicative interactions, bridging the physical distance between the two research teams was inevitable. Therefore, clients began visiting Ieper, usually for periods of several months. At first L&H was not very enthusiastic about providing access to their technological core. However, with time, L&H learned that the more the technology is used by companies to make applications, the better for L&H since this would help them in making their technology the industry standard.

Once L&H had adopted a network mentality, a second phase started with the FLV initiative in 1995. As soon as the idea of creating FLV was launched, a group of institutional and private investors were attracted to what they saw as good investment opportunities. A venture capital fund (the FLV Fund), focusing on speech- and language-related technology applications, was established in 1995. Some 50 companies from Asia, the United States and Belgium contacted the FLV Fund. From among these requests, six to seven concrete opportunities were selected. This is an important point, since the screening procedure typically canvasses the opinion of L&H and the other firms in order to validate the technological and commercial capabilities of potential candidates to investigate any common interest or complementarity. In other words, part of the strategy is to look for possible technological spillovers between the firms. After only three years of dynamic technological clustering through licensing, and financial clustering through participations, several language- and speech-related technology firms had clustered geographically. In 1999, the first group of firms were located at a campus on a 65-acre business park in Ieper. The favourable communication conditions thus created induced localised innovative linkages. Companies "find" each other at FLV to their mutual advantage. They learn from each other and benefit from using and developing common pools of resources in proximity, *e.g.* companies hire employees from the "collective pool of labour" created by the many education and training programmes. In 2000, more than 40 young, growing, innovative FLV firms were listed in the FLV Fund portfolio, of which a dozen are supported by the incubation services of the FLV Foundation. The two most important tasks for the foundation in those days were to create the infrastructure and to increase the availability of human resources for language and speech technology in the region. The Foundation was sub-divided into 12 organisations, of which the most important are the following:

- *FLV Business Development* operates promotion platforms in Flanders, Boston, Singapore and San Francisco, offering FLV companies a springboard for global expansion.

- *FLV Education* proposes a range of training and education programmes in co-operation with other training institutions.

- *FLV Human Resources* works with a Flemish employment agency (VDAB) and several selection and recruitment agencies and serves to acquire highly educated employees for FLV companies.

- *FLV Finance* manages a local start-up fund and is responsible for contacts with the FLV Fund. It is a local investment organisation that co-ordinates the activities of a number of local business angels.

- *FLV Telecom* provides the necessary telecommunication infrastructure (*e.g.* thousands of high-bandwidth internal access points were installed in private households in the vicinity of FLV, in exchange for active participation in testing the products of FLV companies). As a result, interactive user-producer relationships are localised within the regional tissue of the Ieper area.

In November 1999, the FLV Foundation was renamed "S.AI.L. Trust". With SAIL (Speech, Artificial Intelligence and Language), the cluster entered a third phase. From a focus in the second phase on local development, investments and interaction, which created localised, intra-cluster interdependencies, the third phase focuses on exploitation in inter-regional or inter-cluster networks through external linkages. With FLV as the first "S.AI.L. Port", the Foundation planned to replicate and localise the FLV concept in nine international centres of excellence, aiming for a worldwide network of centres for the development of S.AI.L technologies. This third phase has been labelled "exploitation" since it concerns local wealth creation based on local multipliers and spillovers to other sectors and other parts of the Flemish economy and on "exporting" the FLV concept under the name of SAIL. With this new structure, support from the FLV Fund and S.AI.L Trust was adjusted to the development stage of the firms. The crisis which occurred in 2000 has brought a temporary halt – and for some features perhaps indefinite postponement – to the implementation of this blueprint.

Three key questions

Why has FLV developed in such a small, relatively isolated rural area and not in a region with a much stronger knowledge base, such as Leuven?[7]

It would seem that the technological idea and assistance for L&H was destined to come from foreign sources. Therefore, from a technological point of view, it did not really matter whether or not the company was located in a large centre (apart from the fact that an airport might have been convenient). Either way, it would be necessary to travel and "shop" outside the country (and, as it turned out, initially outside Europe) to obtain the required "brains". However, the entrepreneurial spirit, the availability of venture capital in the region and the multilingual aspect can all be conceived as a sort of "first nature" (Krugman, 1993) of the region-specific advantages on which the success of L&H and its technology was partly based. In fact, the strong family-oriented business ethic seemed to be crucial with regards to the faith shown by local private investors in putting money into the vision, *i.e.* would Lernout and Hauspie have been as successful in persuading local butchers and bakers in a community with which they were not familiar, such as, say, Brussels or Leuven? Social links drew and "tied" them to Ieper. Another point is the fact that the human resources needed by the cluster were scarce in other regions as well, so that the cluster firms had in any case to invest in training and education and the presence of a large unused labour reserve in the Ieper area may have helped in this respect. This "answer" has an important policy implication: despite the present crisis, the FLV case shows that even regions that do not have a strong and preferably diversified knowledge base may have an inherent potential to generate innovative clustering in a modern industry. Moreover, very little public money was needed to get the clustering process started.

Why is it that geographical proximity remains important in the case of FLV?

When explaining innovation clusters, reference is often made to the importance of tacit knowledge and non-tradeable technological interdependencies. In the case of FLV, one would expect – since we are dealing with experts in communication and codification here – that the knowledge at hand could be codified and sold through licences? However, the FLV case shows that even the exchange of codified knowledge in the form of licence agreements can be a highly localised learning process. Just as "we can know more than we can tell" (Polanyi, 1966, p. 4), L&H knew more than it could sell and their clients wanted more than they could licence. The need for face-to-face contact in interacting and exchanging the tacit knowledge which is used to combine the codified knowledge began to attract other firms with similar and complementary technologies to Ieper. Over time, later in the life cycle (whether the life of a young firm, a cluster or a technology), more and more tacit aspects become codified and standardised, thus reducing market imperfections. The notion that firms and clusters evolve and mature suggests that private and public policy needs to evolve and mature accordingly. This is an argument for providing learning opportunities early in the firm's life, *e.g.* by investing in infrastructure and education and support for "start-ups". At a later stage, the reduction in market failures weakens the justification for policy intervention.

What then is the role of public policy?

Vigorous policy intervention, aimed at creating an innovation cluster from "scratch", such as that created in Sofia Antipolis (Longhi and Quéré, 1999) could provide one answer. Although FLV received public support and has served as a flagship for Flemish cluster policy, FLV was not born from policy invention. Although the present crisis which FLV is going through does not seem to be based on problems related to technological or locational choices, and although it may even turn out to be a blessing in disguise for the regional economy, it does call into question the notion of providing public support for this very risky kind of clustering. One lesson for policy is that the clustering may have been too much, too fast or too specific in the technological, organisational or geographical sense. The FLV case shows that the networked economy offers opportunities for economic entities (firms, clusters or regions), which lack a strong and diversified knowledge base to overcome the problem of "critical mass" by opting for a highly specific "niche" strategy. However, in accordance with Arthur's law of the New Network Economy: "of networks there will be few" (often translated as "the winner takes all"), there is a high risk of supporting a cluster that is not a winner. Another policy option would be not to frustrate private peripheral "experiments", and to provide more generic temporary support to start-up phases, focusing on (hard and soft) infrastructure and education, since such assistance may still be useful in the event that the clustering firms fail to survive or leave the cluster.

Flanders Multimedia Valley

Flanders Multimedia Valley (FMV) is a government-supported membership organisation, clustering the Limburg (Flanders) ICT and multimedia sector. Limburg ICT and multimedia represents over 110 companies, totalling 4 500 skilled workers and a global turnover of USD 720 million.

Until 1985, the Limburg region was dominated by the coal-mining industry. When the mines were shut down in the early 1980s, Limburg policy makers decided to draw up plans for stable economic growth. The focus was on three core competencies: automotive, logistics and ICT/multimedia. The automotive and logistics sectors both had longer histories in the region, while the growth of the ICT sector in Limburg was a fairly new phenomenon. Philips had a long history in Hasselt with its R&D facility where the first audiocassette, audio CD and CD-I were developed, but during the 1980s and 1990s, other ICT and multimedia firms also experienced explosive growth. To strengthen this sector and ensure its future growth, a separate cluster organisation was created: Flanders Multimedia Valley.

FMV is a fairly young organisation, existing only since July 1999, yet it already groups nearly all of the ICT, and especially the multimedia companies, in Limburg. Limburg's ICT sector is very diverse – with companies active in sectors such as content creation, computer design, laser optics and multi-platform content distribution – and this is why a deliberately vague name was chosen for the cluster: all the companies in the region could relate to "Flanders Multimedia Valley". Today, FMV focuses on four core activities:

- Optimising the influx of skilled IT personnel to the member companies by setting up educational initiatives and by bringing education providers and member companies together to work on new training programmes. The recent educational E-lab project is an excellent example of this co-operation.

- Stimulating knowledge exchange and networking between local firms through the organisation of events where members can present new products or services, with an opportunity for building up contacts. These FMV events – such as the monthly "Whats'n Hows" evening sessions – are appreciated by the members since many of the companies are active in small niche markets and these gatherings offer the possibility to promote a product and get together with other FMV members active in a different part of the market. FMV also communicates through newsletters and its Web site.

- Joint actions are undertaken to obtain cost reductions and service level agreements between member companies and providers of hardware, bandwidth, etc. FMV will act as the representative of all the member companies in negotiations with large multinationals that do not normally allow cost reductions to smaller companies, but are willing to close deals with larger organisations like FMV. In autumn 2000, FMV announced a deal with a telecommunications provider that could result in a 40% reduction in the cost of bandwidth for local multimedia companies.

- FMV puts Limburg on the world map. Its role is to promote Limburg as an active participant in the knowledge economy and to assist member companies in building up their international contacts. In particular, FMV focuses on international trade fairs by organising participation or joint visits. Furthermore, FMV is in close contact with the Flanders Foreign Investment Office and the other Valleys in Flanders, and frequently receives business delegations from all over the world.

In short, Flanders Multimedia Valley strengthens the local multimedia sector and firms places Limburg on the world map as a flourishing participant in the new economy.

Conclusions and policy lessons

Conclusions

Continuous restructuring and industrial renewal are constituting a "cluster space" in which new small-scale ICT clusters are emerging. What are their characteristics?

A first observation is the idea that successful clusters constitute *small-scale innovation systems*. Basic components for "systemic" development at the regional level include, for example, the presence of an (international) leading actor or a concentration of competitive SMEs, internationally outstanding specialisations of the knowledge base, proximity relations and networking, venture capital, a strong education and research infrastructure, and public/private partnerships. The case studies presented in this chapter present a combination of these ingredients, although the exact recipes differ according to local conditions.

The new cluster dynamics in Flanders emerges at the edge of traditional economic and technological strongholds. In research, this relates to microelectronics, digital signal processing, artificial intelligence and language technology; while in industry, it is imaging in the graphics industry, materials, banking. Start-ups are bridging the gap between the old and new economies in certain fields of application. New hybrid organisations, such as the incubators and cluster organisations exemplified in the various Flanders Valley initiatives, are emulating these conditions and fostering these processes. An essential element is the growing involvement of the leaders of the "old economy". The Valleys constitute small-scale innovation systems that combine the essential elements that breed innovative success.

A second observation is the idea that *ICT clusters are mixed local and global phenomena (global-local paradox)*. These regional clusters are parts of the international ICT value chain. In a small economy the cluster orientation of its major players is of strategic importance. Most of the big ICT companies have to align their alliance strategies with headquarters abroad, thus limiting the scope of local cluster strategies. Regional clustering in the regional value chain is weak. Nevertheless conditions are there to develop the virtuous dynamics of clustering, because of the presence of technologically strong players in industry and in research that are open to the new challenges of the network economy, and because of the appearance of many new entrants. In particular the development of the research institute IMEC, with a balance of international and local partnerships and spin-offs, is a showcase for the successful operation of the global-local paradox. Conditions in Flanders are good for emerging ICT-based industrial clusters that capitalise on the openness of the economy.

The case studies presented above show that small-scale clusters are typically *innovation systems on a human scale*. As knowledge diffusion mechanisms, they are particularly well-suited for the transfer of tacit knowledge embedded in people. Know-how is obtained through learning-by-doing or through "apprenticeship" and is often very context-specific. Because of the difficulty of written transmission, interpersonal contact is the main vehicle for the transfer of practical knowledge. Mobility of personnel, exchanges of experiences, informal networking, mediation of local experience via local consultants or specialist suppliers, joint ventures *in situ*, joint seminars, etc., are all channels of knowledge transfer that are very much dependent on proximity. The organisation of knowledge diffusion on a regional level is therefore of crucial importance. As the case studies show, the strengthening of the "absorption capacity", or learning ability, is a precondition for innovation. The Flanders Valleys have demonstrated "good practices" for cluster and network behaviour.

A fourth observation relates to the fact that, in practice, successful cluster growth is often the result of *market-induced clustering* and clustering that involves a *planned organisational effort*. To obtain a better understanding and a more effective operation of the cluster approach, one has to clearly distinguish between the two different contexts of clustering. The first context is that of "spontaneous" clustering in the periphery of markets, based on the quality of infrastructure, economies of agglomeration. The second context is that of "voluntaristic" clustering, involving a planned organisational effort to exploit external economies through learning efforts and collaborative interaction. In reality, we observe a mixture of factual market-based cluster dynamics due to self-reinforcing economic mechanisms, and emerging network-based cluster dynamics arising from a communication-enhancing cluster organisation. One of the main policy lessons in the Flemish context is that optimal cluster growth is achieved by combining both types of dynamics. In most cases, an important role is played by various intermediary organisations, such as development agencies and enterprise associations, often combined in new cluster organisations. These cluster organisations are the drivers of the hybrid interaction mechanisms that constitute the new cluster dynamics.

Policy lessons

The case study of ICT in Flanders and of the small-scale clusters that are important drivers of its development, offers opportunities for policy learning.

In the first place, it can be observed that different kinds of market failures in the Flemish economy prevent firms from taking full advantage of local cluster dynamics. Information asymmetries, short-sightedness, unwillingness to co-operate and lack of trust hinder the emergence of networks and cluster ties. The overall advantages are not well understood when firms only narrowly maximise their private benefits. Co-operation is a collective good that is often not easily generated in a competitive context. The omnipresence of spillovers in the knowledge-based economy indicates a *new role for public policy in handling systemic failures such as lack of systemic connectivity or clustering*. There is a new role for government and intermediary organisations as information brokers and catalysts of networking among potential partners, especially in the early stage of cluster formation.

Second, the experiences of the existing ICT cluster organisations are very *important role models for co-operative learning and innovating* in the economy as a whole. They create new co-ordination mechanisms to manage knowledge spillovers to the benefit of the participants. Clusters enable common interests to be defined in terms of strategic goals, end-product markets and resources among participants. The interaction within and between these organisations is a learning process that is requires a great deal of pragmatism while appealing to a high degree of ambition. The ICT-related platforms such as DSP Valley, Flanders Language Valley and Flanders Multimedia Valley are already networking with each other. This may result in the emergence of a "virtual" cluster park in Flanders, making it a backbone of the networked economy!

Third, the start-up of cluster policies during the early and mid-1990s will be remembered as a period of experimentation and policy learning. Most of the cluster initiatives of the past were largely based on an "intuitive" approach. However, this may limit their potential. Specialised assistance for analysis and process management in the "definition stage" of cluster formation by specialised intermediaries (process managers, cluster analysts) can be of strategic importance. Public support in this stage may have strong additionality.[8] The provision of foresight studies can be an enabling condition for cluster platforms in which "neutral" public agencies can have some input. However, the ability to draw lessons for policy making is still hampered by a *shortage of absorption capacity or intelligence within the public services*. Policy development departments that can rise to the new role of government as a strong catalyst of knowledge interactions in the innovation system (knowledge on knowledge) do not yet exist.

Finally, the cluster cases demonstrate the need for a *comprehensive innovation policy* that is not limited to technology creation. The need for such policy is now widely acknowledged but the institutional setting in which integral and integrated policy can be implemented does not exist to date. New horizontal linkages among policy departments and feedback mechanisms with the various actors need to be established. Cluster organisations and cluster policy can contribute to this new setting.

New instruments are now being implemented which will fully support "bottom-up" clustering. Policy makers have a key role to play here. The approach and purpose reflected in the priority concern of promoting the platform and networking initiatives that enhance clustering will have a much bigger impact if they are championed by policy makers themselves.

Less obvious and brand new for the Flemish situation is the potential interaction between cluster platforms and policy makers in various policy areas (the environment, energy, public works, education, for example) in the case of innovation. Innovation can make a direct contribution to the bid to achieve policy aims, taking advantage of local know-how. What's more, policy decisions can help stimulate or act as a brake on potential innovation (regulations, government procurement, etc.). The *public/private partnership* option may form part of the official discourse, but a meaningful implementation of it still calls for a great deal of creativity. This implies that technological innovation policy is to be addressed and developed to some extent by authorities that are not directly responsible for it. After all, "innovation" is of key importance not only for the business sector but also for the *modus operandi* of government itself.

NOTES

1. The concepts of clusters and networks are often interchangeable. In this chapter, we define clusters as more implicit and informal organisation structures at the meso level emanating from market behaviour, while networks are more explicit (both formal or informal) co-ordination mechanisms at the micro level.

2. Other initiatives exist. One of the most important, Flanders Graphics Valley, organises networking among the Flemish graphics industry which boasts world leaders in digital communication, such as Agfa Gevaert, Barco Graphics, Artworks.

3. The Flemish Government recently conferred VRT, the public broadcaster, with the mission of establishing a digital platform "e-VRT" that will provide all Flemish citizens with access to a rich information base and a portal to the wider Internet.

4. The risk-capital market is growing at a rapid pace. With EUR 673 million invested in 1999, Belgium ranked fourth in terms of venture capital/GDP at the European level (EVCA). In terms of seed capital, Belgium came first on the basis of the ratio of seed capital to GNP (Van Seebroeck, 2001). On both measures, Flanders scores a more than proportional share.

5. In a global context, the ICT sector represents 4% of world GDP. This means that the share of ICT in Flanders is average, compared to the clearly above-average shares of the ICT sectors in the United States (8%), Finland (6%) and the Netherlands (5%).

6. VRT, Roularta, V.U.M., de groep Hoste, Mediaxis, Uitgeversbedrijf Tijd en Het Volk, Wolters-Kluwer (publishing), Promedia (publishing) and Brepols (printing).

7. Theory holds that innovation performance will be stronger if innovative activity is geographically agglomerated. Many studies, from different perspectives, supply empirical support for this view (*e.g.* Antonelli, 2000).

8. This lesson has been usefully applied in the new cluster programme in the Walloon region of Belgium: the experimental cluster programme is geared to the central role of a thorough cluster analysis and definition of the action programme from the outset.

REFERENCES

Antonelli, C. (2000), "Collective Knowledge Communication and Innovation: The Evidence of Technological Districts", *Regional Studies*, Vol. 34, No. 6, pp. 535-547.

Bergman, E. and E. Feser (1998), "Industrial and Regional Clusters: Concepts and Comparative Applications", www.rri.edu/Webbook.

Jacobs, D. (1997), "Knowledge-intensive Innovation: The Potential of the Cluster Approach", *The IPTS Review*, No. 16, Seville.

Krugman, P. (1993), "First Nature, Second Nature and Metropolitan Location", *Journal of Regional Science*, Vol. 33, No. 1.

Larosse, J. *et al.* (2001), "ICT Clusters in Flanders", *IWT Studies* No. 35, Brussels.

Longhi, C. and M. Quéré (1999), "Sohia-Antipolis" in *INLOCO – Industrial Districts and Localized Technological Knowledge,* Final Report to the European Union, DG XII, Targeted Socio-Economic Research – TSER Programme, Contract No. SOE1-CT97-1058.

OECD (1999b), *Boosting Innovation: The Cluster Approach*, OECD, Paris.

Polanyi, M. (1966), *The Tacit Dimension*, Routledge, London.

Porter, M. (1990), *The Competitive Advantage of Nations*, Free Press, Harvard.

Porter, M. (1998), "Clusters and the New Economics of Competition", *Harvard Business Review*, November-December, pp. 77-90.

Van Seebroeck, H. (2000), "Financiering van innovatie met risicokapitaal", Working Paper 9-00, Federaal Planbureau, Brussels.

Vlaamse regering (1998), "Valleien aan het werk – Werk in de vallei", De Vlaamse regering: katalysator voor technologievalleien, Brussels.

Chapter 7

INNOVATION IN AN ADOLESCENT CLUSTER: THE DUTCH MULTIMEDIA CLUSTER

by

Pim den Hertog, Sven Maltha and Erik Brouwer[*]
Dialogic Innovation & Interaction and Department of Innovation Studies, Utrecht University

Introduction

The use of terms such as e-commerce, e-banking, e-marketing, e-learning and e-government signifies that electronic infrastructures are being set up for the purpose of shaping interactive information, communication and transaction services in a wide range of activities. An increasing number of businesses and organisations in the Netherlands are giving shape and form to such services and the associated products and infrastructures. In the world of ICT, such lines of business include consumer electronics, telecommunications, content (AV/broadcasting, publishers, entertainment), advertising and marketing communications.

Nevertheless, there is a growing group of often small and medium-sized enterprises (SMEs) operating at the crossroads of multimedia enabling, content distribution, content provision and e-marketing, thereby combining all four worlds. These are occasionally businesses that originated from one of the aforementioned traditional sectors; sometimes they are existing offline multimedia businesses and, surprisingly often, they are dynamic start-ups that to a large extent are shaping the "new economy". These actors produce an ever-increasing variety of products and services and together form the Dutch multimedia cluster.[1]

In this chapter, we present some of the results of a recent study into the Dutch multimedia cluster. Basically, this study delves into the innovation dynamics and future prospects of this cluster. Issues include:

- How is the assumed multimedia cluster structured?
- Are the conditions that relate to the development of the multimedia cluster favourable or unfavourable?

---

* The authors can be contacted at Dialogic, Wilhelminapark 20, 3581 ND, Utrecht, The Netherlands (e-mail: denhertog@dialogic.nl). The research reported here is largely based on the ClusterMonitor Multimedia performed for the Dutch Ministry of Economic Affairs.

- To what extent is the Dutch multimedia cluster a part of international production networks, and might the cluster possibly be driven from abroad?

- To what extent are the relevant parties able to collaborate constructively in a positive spiral of knowledge generation and knowledge transfer (without hindering competition)?

- To what extent do demanding clients stimulate the multimedia cluster to innovate on a constant basis?

- How do the players in the multimedia cluster innovate? Is there a specific style of innovation involved?

- How do the players in the multimedia cluster perform in terms of business economics and innovation?

It will be shown that innovations in this cluster require cross-sectoral combinations of knowledge calling for various types of interaction between cluster actors. The type and shape of these interactions is highly specific to – in this case – the Dutch multimedia cluster. In other words, each cluster has its own innovation style or innovation culture and the barriers to innovation encountered by the cluster members are also cluster-specific. Likewise, the set of options for improvements identified are cluster-specific. Therefore, the required actions need to be taken by the actors involved in the cluster, including (innovation) policy makers.

In contrast to most of the contributions presented in the preceding chapters, this chapter will focus on the more narrowly defined multimedia cluster. The client of the study on which our contribution is based[2] was most interested in the characteristics, innovation dynamics and performance of the Dutch multimedia cluster, which we perceive as a sub-cluster of the wider ICT cluster. This multimedia cluster (see also Box 1) is defined as "those businesses and organisations that actively shape offline and online[3] information services, communication services, and transaction services for intermediaries and end users, whereby several media are combined and interactivity is one of the main features". The businesses and organisations selected were mainly those that:

- Realise a substantial part of their turnover (>50%) from multimedia products and/or services.

- Or play, or will play in the near future, a major role in the cluster on the basis of their strategic positioning.

- Or are recognised and accepted by market parties as belonging to the cluster.

Box 1. **Some definitions of multimedia and multimedia clusters**

"Multimedia is usually defined in technical terms. In this sense it signifies a combination of a variety of media available in a digitised format that can be accessed interactively. We consider multimedia to be a networking technology. Our definition is geared towards the end product of a potential multimedia chain or value network in which a variety of actors can be present. These actors include, among others, network operators (for online multimedia), hardware producers (servers, computers, etc.), the media industry, service providers, software producers and the entertainment industry". (Pavlik, 1999, p. 81)

"... the merging of traditional audio, visual and print media through digitisation in an interactive format. Our study of multimedia firms includes established corporations that have moved into the field of multimedia such as graphic design companies, computer software developers, and book publishers. It also comprises newly developed 'multimedia-only' firms, whose core business offering is the development and production of multimedia products based on CD-ROM, Internet/Intranet, Kiosks and so on". (Brail and Gertler, 1999, p. 98)

The next three sections describe the characteristics, functioning and performance of the multimedia cluster, with an emphasis on innovation. The final section formulates recommendations on how to improve the innovativeness and competitiveness of the Dutch multimedia cluster.

Basic characteristics of the multimedia cluster

With regard to multimedia, a distinction can be made between four types of activity, namely:

- *Multimedia-enabling activities:* the production of ICT hardware and software, consumer electronics, design and business consulting.
- *Content distribution:* distribution via a variety of electronic infrastructures.
- *Content provision:* broadcasting, entertainment, publishing and associated business activities such as AV productions and printing.
- *E-marketing:* advertising, direct marketing, media acquisition and marketing communication.

These four types of activities are further explained in Box 2. They also form the pie slices shown in Figure 2, in which a further distinction is made between the core and ring players in the multimedia cluster.

Figure 2. **Four segments of the multimedia cluster**

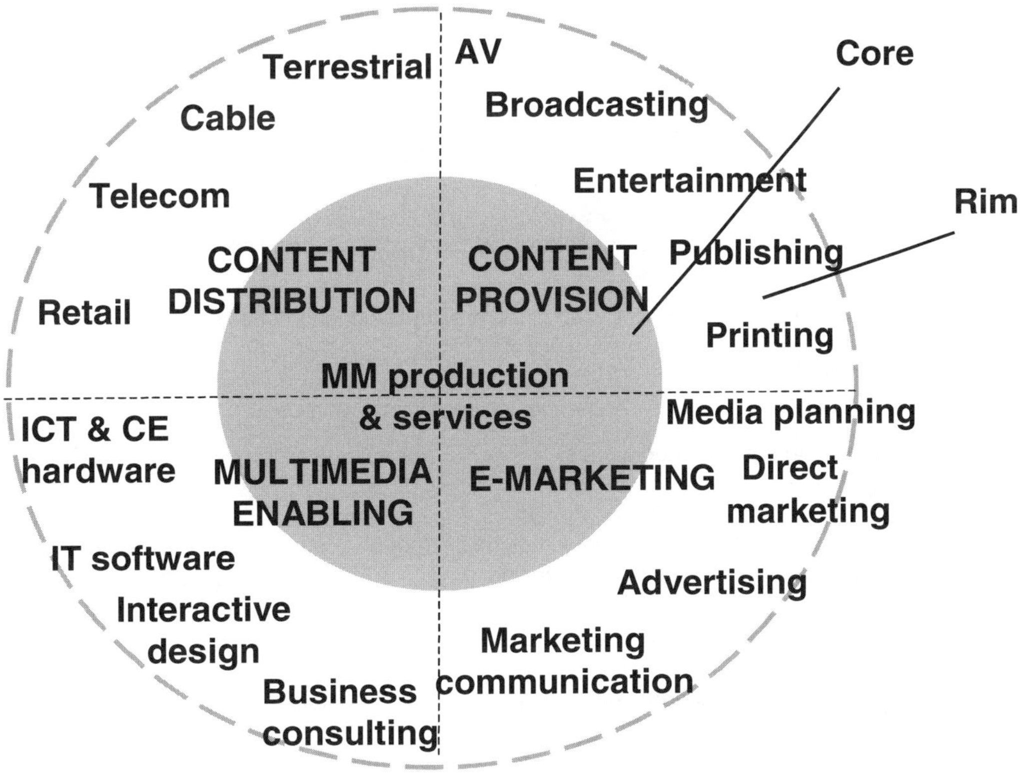

Box 2. **Multimedia products and services: a continuum**

Below is a list of the activities (which is by no means exhaustive) performed in those businesses active in the multimedia cluster on the basis of the four segments of the multimedia cluster identified above. It can be regarded as a continuum with pure hardware at one end and pure services at the other, with many possible combinations in between.

Multimedia-enabling

- Development and production of multimedia hardware, such as computer hardware, CD-ROM players, DVDs, modems, decoders, network components, household appliances (consumer electronics), mobile telephones suitable for multimedia use, etc.
- Development of packaged software for the creation of multimedia applications, such as multimedia authoring systems, intelligent agents, search engines, content management systems, measurement of Web site usage (Netstatistics), browser-based banking, etc.
- Creation of a development environment for certain kinds of application, *e.g.* e-commerce applications, a virtual learning environment, etc. Making existing back office systems Web-enabled.
- Interface design.
- Web hosting.
- Integration (Web enabling) of existing back office systems with online applications.
- Advising on e-commerce and Internet strategies.

Content distribution

- Provision of access to the Internet (sometimes in combination with telephony and the distribution of radio and TV signals) through a variety of infrastructures.
- Access services, generally combined with forms of content aggregation (own portals) and the creation of communities.
- Distribution of multimedia devices and software.

Content provision

- Development of new formats, concept development and content creation (*e.g.* for combined TV and Internet).
- Production and maintenance of content for Web sites, *e.g.* the production of e-magazines.
- Web site quality assessment and comparison.
- Production of instructive CD-ROMs and DVDs, *e.g.* for educational purposes.
- Compilation and publication of multimedia titles.
- Electronic publishing.
- Development of new service concepts/new business models, *e.g.* virtual marketplaces, purchasing platforms, specialised portals, creation of communities, etc.

E-marketing

- "Webvertising": exploiting advertisement space on Web sites and portals.
- Online brand development/online promotion.
- User profiling and user management (overlap with multimedia enabling).

The *core* of the cluster consists of an estimated 500 to 1 000 businesses (freelancers and sole traders excluded), which are to a large extent dependent on offering advice on, contriving, designing, building and maintaining multimedia applications.[4] These are the pure multimedia businesses such as Lost Boys, Bitmagic, Razorfish, Jamby, NOB Interactive, Virtual Affairs, Netlinq, Nedstat, Agency.com, Inforay, Motion Container, SQR, ACSi, ICATT, etc. Their role consists of manipulating content, refining it and making it suitable for platform-independent distribution. The innovative players active in this core are able to link the world of multimedia-enabling with those of electronic distribution, content provision and E-marketing (the shaded area in Figure 2). These firms are often SMEs which do not always possess the necessary in-house knowledge to undertake full multimedia applications. In addition to a few product developers that (wish to) market new products, and have the world market as their frame of reference, these firms are for the greater part service providers who on the basis of available technology realise multimedia applications, usually for business customers. Among the service providers, several – sizeable – businesses are now presenting themselves as full-service providers, with the aim of concluding long-term contracts with major clients. This core is augmented by firms that aggregate content, package it and are able to make it suitable for a wide public (end consumers) by, among other things, creating communities (*i.e.* World Online, Planet Internet, @Home, Chello, Veronica Digitaal, Ilse, BOL, Macropolis). These businesses must be included among those operating at the core of the cluster.

A group of businesses operating in a *ring* immediately surrounding these core businesses focuses to a greater or lesser extent on multimedia and thus contribute to the production and application thereof. The main difference with the core businesses is that these firms have a clearly established position in one or several of the supplying disciplines and sectors. As yet, they have not always recognised the omnipresence of the new multimedia products and services and are moving towards the core at varying speeds. It is noteworthy that the established "ring players" have a valuable asset at their disposal (unique content, extensive distribution network, thorough knowledge of an area of application) and often act as both supplier and customer for the core businesses. Nevertheless, by no means all the businesses in this ring have developed into full providers of, or eminent clients for, multimedia products and services. These firms are often the large to extremely large parties who are considering entering online media services but who see "mass" as a prerequisite for doing so. For instance, they can develop their online applications into massive services using their existing client base, distribution potential, public profile or capital. The players at the core of the cluster often depend on these "consolidators" for their economic survival.

Figure 3 sets out the most important players in the multimedia cluster in the form of a value chain or value network. The figure shows that very different businesses contribute to the creation of multimedia products and services.

The multimedia cluster is a highly dynamic cluster which is taking form "along the way"; it continues to expand and its borders are fuzzy.[5] In a highly stylised way, clusters can be depicted as moving through various stages in a kind of life cycle. For example, Peneder (Chapter 15 of this volume) indicates that at each of these stages, the relative importance of variation, cumulation and competitive selection differs. Although rather crude, the terms "emerging" and "mature" or "established" are commonly used to differentiate between the stages of development of a cluster.

We have labelled the Dutch multimedia cluster an "adolescent cluster" to indicate that is at an in-between stage: there is a clear value chain, there are a number of more or less specialised players originating from various industries that compete and collaborate using various technologies and standards, and the cluster is increasingly "recognised" as such by the cluster actors themselves. We therefore see the Dutch multimedia cluster as a dynamic, rapidly growing cluster in its puberty (an "adolescent cluster"). We use this notion to indicate that some of the early-phase start-ups typically associated with emerging clusters are growing up; some of the firms have grown to a considerable size while others have managed to form alliances with established larger firms.

Figure 3. **Players in the multimedia cluster**

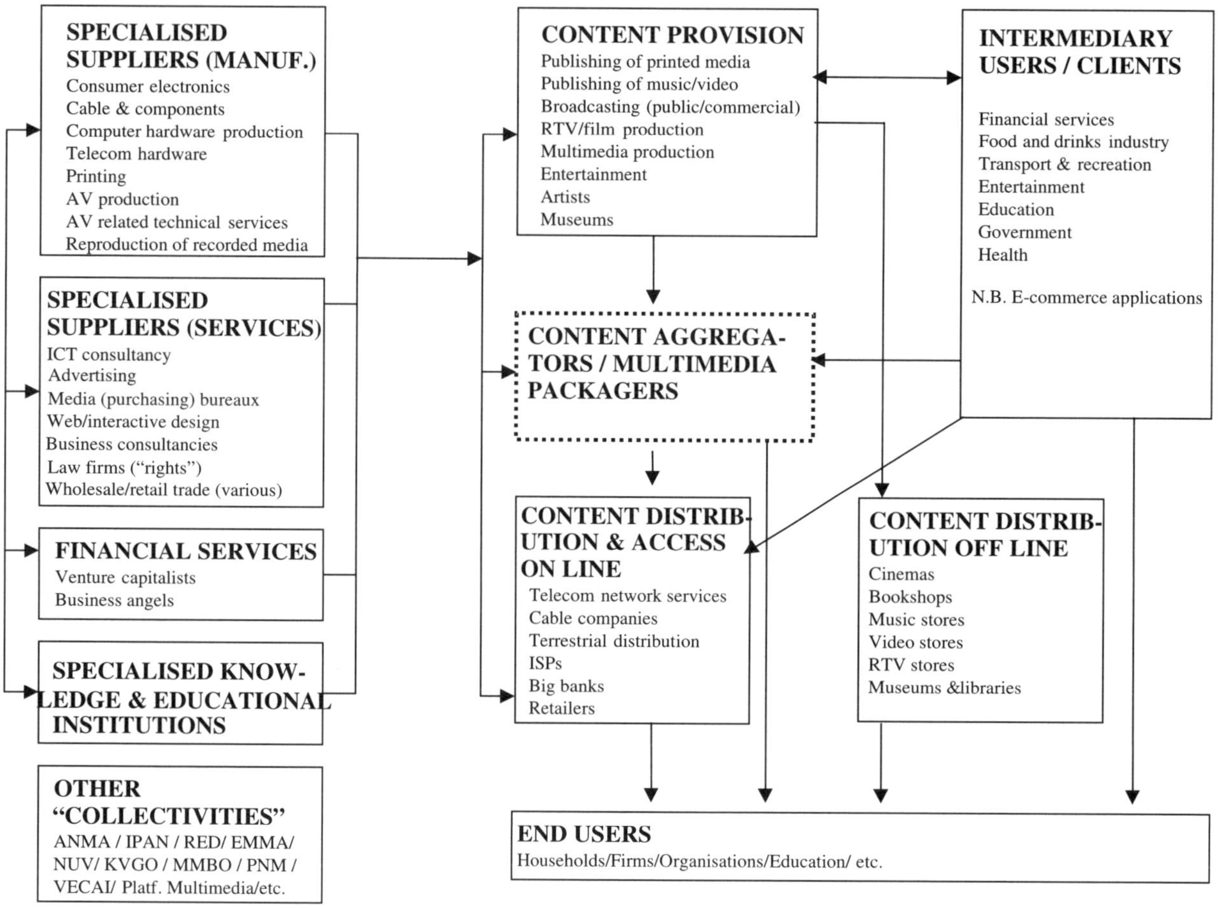

However, at the same time the multimedia cluster continues to witness a substantial number of start-ups and new entrants. These entrepreneurs typically have different backgrounds: opportunistic fortune-seekers, technical and creative "whiz kids", employees from universities and research organisations, as well as established "old economy" firms and major consultancies. In addition, entrepreneurs that have "sold up" can continue to play a role by reinvesting in new start-ups. However, as noted by Feldman (2001), it is important to differentiate between the initial development of new businesses by entrepreneurs and the subsequent creation of the institutions and the conditions that allow them to thrive. Equally, conditions that are supportive of entrepreneurship should not be confused with conditions that are supportive of innovation (Feldman, 2001, p. 22).

The aspect of internationalisation is emerging. On the one hand, major foreign players see the Netherlands as a good stepping stone to Europe. Companies like UPC and, even closer to the centre of the cluster, Agency.com (which took over one of the relatively large Dutch interactive firm) are illustrative of this trend. It is also apparent that, alongside the Dutch informal investors, foreign parties – Scandinavian businesses for instance – have shown an interest in taking over and participating in the Dutch multimedia businesses operating at the core of the multimedia cluster. At the hardware end, the application development for broadband services, Gigaport, and therein Gigaworks[6], exercise a technology-pull effect on foreign firms wanting to experiment (such as Lucent, IBM, Ericsson, Nokia and Cisco).[7] On the other hand, Dutch companies at the multimedia core are starting to eye up foreign markets; these include the rare product businesses that (wish to) launch a standard for the world market, such as Nedstat (Webstatistics software), Oratrix (multimedia presentation authoring tool) and Tryllian (intelligent agent technology). These businesses have the world market as their frame of reference. Service providers remain primarily locally and nationally oriented, although making use of internationally available technologies, knowledge and tools. However, a number of smaller service providers have indicated that they are thinking about internationalisation, either in order to serve multinational customers abroad or to collaborate with like-minded businesses in foreign countries in opening up foreign markets.

The comparative advantages of the Dutch multimedia cluster include: strengths in content provision; reasonably advanced telecom facilities (there is, however, a need for rapid availability of low-cost broadband networks); the availability of capital; the entrepreneurial spirit; the suitability of the Netherlands as a test-bed and an experimental market; and the high quality of education and research. A more complete list of the comparative advantages and disadvantages is set out in Table 1.

Table 1. **Relative strengths and weaknesses of the Netherlands with regard to multimedia activity**

Relative strengths	Relative weaknesses
• Geographical location/international accessibility	• Small domestic market
• Reasonable study programmes	• Lack of human resources
• International orientation/advantage of multilingualism	• The need for better co-ordinated training courses
• Availability and quality of the telecom infrastructure	• Adaptation of education to the online media is only moderate
• Availability of sufficient capital	• Interface between the knowledge infrastructure and the commercial world is moderate
• Labour is relatively cheap, certainly when compared with the United States	• Lack of ambition
• Strength of and tradition in consumer electronics, publishing, AV production, graphic design and, to a certain extent, film (documentaries)	• Lack of a tradition of product ideas (traditionally, the Dutch are traders)
• Specialisation in constituent areas, *e.g.* educational CD-ROMs	• Limited knowledge of the top companies of the Internet
• Completeness of the cluster and the presence of powerful players (Endemol, NOB, publishers, KPN, UPC)	• Obstacles in legislation and regulations (copyright, privacy)
• Adequate room for innovation (ideas)	• Too much regulation (*e.g.* cable)
• The Dutch market has the characteristics of an experimental market/testing ground	• Negative self-image (thinking small)
• High-quality knowledge institutes and strong interdisciplinarity	
• Liberalised telecom market	
• Availability of governmental innovation instruments	

Source: Interviews reported in den Hertog *et al.* (2000).

The main bottlenecks associated with the basic characteristics of the multimedia cluster are:

- A lack of transparency and – the associated – low level of organisation in the multimedia cluster. Even the parties that are active in the cluster are not always aware of one another's knowledge and skills. The strong focus on today's rapidly-growing market means that less attention is paid to jointly improving factorial conditions (supply of qualified labour, exchange with education and research) and limiting conditions (government policy, export focus, "intelligence"). This is amplified by the lack of a useful cluster index (activities, products and services), outdated statistics and the resulting limited insight into the cluster's economic significance.

- The parallel operation of regional multimedia crystallisation nodes with different specialisations without sufficient attention being given to consolidating these different cores. Various regions view themselves as "*the* ICT", or more specifically, "*the* multimedia" region: Greater Amsterdam (communications and content), Twente (manufacturing), Hilversum (AV), Eindhoven (the cluster surrounding Philips) and Rotterdam. This creates a risk of optimising cluster conditions at too low a scale level.

- The widely experienced lack of suitably qualified manpower (for the time being, this is concentrated in the multimedia-enabling side and is related to the general lack of ICT specialists), and the fact that many training courses appear to be inadequately attuned to the requirements of business practice.

Functioning of the multimedia cluster

The dynamics of the multimedia cluster are to a large extent linked to the ease with which the core and ring players – the (in practice) difficult-to-achieve collaboration between relatively small, young and relatively large, established businesses – are able to find one another. It goes without saying that the internationalisation of firms – both incoming and outgoing – provides new impulses. Aggressive foreign firms do not automatically conform to the existing structures on the Dutch market and are conspicuous, for example, for their speedy reaction times. At the very least, national players must become familiar with new technologies and developments on foreign markets in order to remain competitive. Therefore, to embed a cluster such as the multimedia cluster in a region like the Netherlands, it is essential not to focus merely on optimising the innovative links between players at the regional level. Any attempt to create the much praised new multimedia industrial district at the small scale of the Netherlands might even be counterproductive. To make the Netherlands a "sticky place"[8] for firms in the multimedia value chain, it is essential to embed foreign firms locally (and to learn from them) and to ensure that Dutch firms tap into the available international reservoir of knowledge (among other ways, by going abroad). It is therefore quite logical to expect that the functioning of the actors in the Dutch multimedia cluster will evolve and change over time as they find their niche in what is essentially an international cluster. The method of functioning described here is essentially a snapshot. In this section, we will briefly focus on dynamism in the cluster, the style of innovation (including the relationship between businesses and the knowledge infrastructure) and the quality of demand.

Cluster dynamism

An important part of the dynamism of the multimedia cluster results from the ability of firms to combine the four basic ingredients of multimedia (multimedia-enabling, content provision, content distribution and e-marketing), either internally or through entering into the required strategic alliances and joint ventures. For the time being, it would appear that most of the dynamism stems from the multimedia-enabling and content distribution segments (technology-driven and supply-driven innovations). However, the knowledge of the players in the content provision and e-marketing segments is required to create successful products and services for specific target groups. Below we look closer into three aspects of dynamism in the multimedia cluster: *i)* the exuberant forming of alliances; *ii)* co-operation at the project level of especially the smaller providers; *iii)* and regional networking.

The strong increase in the number of *strategic alliances* and special formations in the cluster is typical of the cluster's current dynamism. The stand-alone scenario of businesses that want to do everything themselves is becoming uncommon. Temporary joint ventures on the basis of complementary competences are typical. Well-known major players are involved or are seeking involvement (broadcasting companies, publishers, ICT businesses, the national incumbent KPN), as well as many other new businesses which are essential for the creative component of the cluster. It is clear that, large and small, the firms need one another. Good ideas no longer suffice: economies of scale and the ability to push out a service or an infrastructure means that only the very wealthy players in the cluster (venture-backed firms) can play this game. Alliances are by no means unique to the multimedia cluster but, given the enormous investments and the associated risks involved, going it alone is almost out of the question.

Joint ventures at project level are very common. Production of a multimedia application generally calls for a wide range of specialist knowledge and, although the larger businesses in particular give themselves the proverbial pat on the back, claiming that they are full-service providers, reciprocal supplies and contracting out to one another are generally accepted practices. This has become the rule for many smaller businesses and the relatively large group of freelancers. However, the larger businesses also rent components, for instance for shooting AV material, bringing in the Netstatistics component, farming out of Web hosting, etc. Considerations of capacity force the larger businesses to maintain contact with smaller firms and freelancers so that they can fulfil their obligations should their own manpower prove insufficient. Smaller businesses in particular have informal networks of businesses and freelancers that they can take on should they conclude a major contract, and vice versa. Smaller businesses sometimes come together to make joint bids and play alternatively the roles of contractor and subcontractor.

In addition to project-linked networks, parts of the multimedia cluster are characterised by social and *regional networks*. People know one another from training courses, have worked together in the past, chat together at meetings, come across one another in the various platform organisations, etc. This is apparent in Hilversum, Amsterdam and Eindhoven. In Amsterdam, for example, the Amsterdam New Media Association (similar to the New York New Media Association)[9] and Internet Society Netherlands (ISOC) contribute to regional network forming by organising meetings and events. In a similar fashion, local organisations are also active in other regions, such as Digital Region Eindhoven (RED) and Eindhoven MultiMedia Association (EMMA), the Teleport Foundation in Twente, the Platform Multi Media Hilversum, etc.[10] The emergence of the Twinning Centres[11] – whether or not associated with science park type initiatives – also contribute to the birth of local concentrations and networks of multimedia activity.

The complement of collaboration is competition. *Multimedia product businesses* compete on the world market in terms of quality, speed and participation in standardisation processes. Knowledge and innovation are essential for these businesses and they generally have a somewhat formalised innovation process at their disposition (a separate software development unit). *Multimedia service providers* compete in terms of price, quality (interactivity, ease of use, integration with existing information systems) and a complete package of services (the full-service concept). These businesses generally work with internationally available tools and knowledge. However, even ensuring that one is well-informed and remains so, calls for considerable efforts. It is noteworthy that a number of service providers indicate that their main competitors are those businesses that commission work from them, often because they implement their own Web-based systems.

Style of innovation

The multimedia cluster is developing with an unprecedented dynamism. Large numbers of firms are able to enter the market on the basis of high expectations and the immaturity of the market. New multimedia technologies and products are popping up and flooding the market. In addition, new technologies and tools – ever-more advanced versions of software – to produce these new products and accompanying new multimedia services are now coming onto the market and standards have not yet been fixed. Specialisation and one-stop shopping concepts are parallel developments. As the worlds of media, publishing, advertising, IT and telecom converge and new platforms for developing new applications and services continue to pop up, the creative process of new combinations and (literally) the gales of creative destruction that sweep the growing multimedia community will most likely stay with us for some time.

It is clear that these trends affect the style of innovation of the multimedia cluster. In general, multimedia business innovation is strongly linked to projects. Alliances between specialist firms or between larger and smaller firms for these projects are common and provide important learning environments. However, many different sources of information are used in combination. One important source of information is obviously the Web. In particular, technical specialists make frequent use of this medium to keep themselves informed about new tools, to discuss with colleagues, etc. In addition to (alternating) project partners, information which is important for innovation reaches these businesses through customers, through their own personnel,[12] through study tours, through suppliers and through competitors (what applications are they working on?). Some entrepreneurs stress that in respect of technology the aspect of timing is extremely important. Lagging behind – or getting too far ahead – is risky: the important thing is to make use of just-proven technology. Customers are not always ready for technologies or solutions. Consequently, some service providers talk not of an innovation problem but rather of application problems with their customers.

Having said this, it is necessary to differentiate between:

- Multimedia product businesses and multimedia service providers which have different strategies and needs in terms of knowledge.

- Technical multimedia knowledge and non-technical multimedia knowledge and the role the knowledge infrastructure has both currently and in the near future in jointly developing this together with industry.

- Young, recently established and fast-growing multimedia firms that are in the steep curve of their life cycle (and learning curve) and well-established and mature firms that, in terms of Figure 1, are moving from the rim towards the core of the cluster.

The available evidence suggests that at one end of the scale the "real product businesses" generally dispose of a more formal innovation process, sometimes run independently, and are engaged in "real" product development. In general, they need to develop a detailed and more fundamental understanding of multimedia technologies in order to develop new technology-based products and succeed on the world market. At the other end of the scale, relatively very little formal R&D is carried out by the multimedia service providers. Yet, there is a growing need for knowledge management and formalisation of business development processes in these firms, especially among growing businesses

(>20 employees). The larger service firms develop some form of knowledge management, and some have even started to work with competence centres similar to the type of specialisation that is visible in multinational (manufacturing) firms. However, their game is offering increasingly complex service products.

What then is the role of the knowledge infrastructure[13] in this fast developing field? What do knowledge institutions have to offer (apart from a highly educated workforce) and with whom do they interact? Here, opinions differ. Most respondents in the (technology-oriented) knowledge infrastructure highlighted the importance of a thorough knowledge of at least the technological basis or the most crucial field of technology or knowledge (*e.g.* a knowledge of server technology, content management systems, a standard) for the current, and particularly the future, competitive strength of the businesses in the multimedia cluster. These respondents stressed that the firms that will be capable of maintaining themselves in the longer term (also internationally) will be those that have a thorough understanding of the "underlying technology" and are able to exploit that knowledge to commercial ends. They feel a connection with the typical product businesses (and vice versa), sometimes because they themselves have a knowledge institute background or because – as indicated above – they rely on technological knowledge for success on the world market. Such firms share a fascination for new technology and new applications.

However, there is also a need for non-technical multimedia knowledge, which would in principle deal more with the softer type of knowledge that could contribute to multimedia service innovation. One could think of knowledge and expertise on the changes taking place in the Internet economy, changing business models, changing life styles, distribution patterns, and a knowledge of electronic markets as well as the didactic and educational side of multimedia applications. This type of non-technological multimedia knowledge is currently difficult to track down in the knowledge infrastructure. In practice, if they show any interest at all in what is happening in the knowledge infrastructure, the multimedia service providers operating at the core of the sector have neither the means, the know-how, nor the willingness to spend time tracking down this type of knowledge,. These, often extremely market-driven businesses have their hands full running their businesses, ensuring growth, and carrying out assignments. These businesses complain that they are barely aware of what knowledge can be obtained from the knowledge institutions; they do not always believe that these institutions have the type of knowledge that is relevant to their particular needs, and they devote very little time to finding out. The pressure of projects is often heavy and the planning horizon is frequently limited to three months due to rapidly changing technologies and markets. The two worlds are quite simply miles apart and in fact it is questionable whether the rather weakly developed interface between multimedia service firms and knowledge institutions seriously hampers innovation in an already fast-moving and innovative business.

The, at best partial, interface described above might also result from the fact that we are dealing with mostly young, recently established and fast-growing multimedia firms in the core. The well-established, more mature and larger firms have more experience with finding their way to the knowledge institutes.[14] Similarly, a number of institutes have good contacts and working relationships with at least the well-established firms. However, through spin-offs from academia, science parks, working with lecturers who are also engaged in the practical side of business, and the growth of business centres in the vicinity of the knowledge institutes, contacts are becoming more varied and these contacts will most likely develop further as multimedia knowledge becomes more institutionalised and the multimedia cluster moves on to the next phase of its life cycle.

Quality of demand

Opinions differ with regard to the quality of demand, particularly concerning the quality of demand among business users.[15] Producers of multimedia products indicate that they do not attach that much importance to the quality of the Dutch market: their frame of reference is first and foremost the world market, not least because the small size of the Dutch market offers insufficient support for the development of products primarily intended for sale on the domestic market.

Various opinions were expressed by the service businesses operating at the core. Several providers complained that the demand for multimedia services is badly specified (clearly indicative of a "me too" market) and that customers often need educating. Customers are said to have difficulty in articulating their demand and often fail to fully realise that the introduction of multimedia services has consequences for the organisation of their internal business processes. Establishing a presence on the Net, for example, is quite a different kettle of fish from building a Web site to enable transactions. All the same, providing appropriate advice to customers on the possibilities offered by multimedia and online business is a lucrative source of income, even for those businesses that claim to prefer developing and building.

The quality of demand is, in general, not considered exceptionally high, although a differentiation is made between those customers "that have and those that have not understood". Several customers have even invested to gain knowledge about interactive media (these are in fact the most important rivals of the multimedia providers!) and are now able to better formulate their demands. It is quite remarkable that some publishers and businesses in the financial services industry, for instance, as well as players like Heineken or KLM, are considered to be relatively advanced users. It is also noteworthy that since demand for the provision of multimedia services exceeds demand in several segments, some providers are able to choose the parties for whom they wish to work. A choice is made for repeat business among customers they want to work for (because it is pleasant work, because they hit it off together) and who make product innovation possible, and the addition of several new accounts on an annual basis.[16]

The question is to what extent the Dutch multimedia businesses will be able to accommodate the released budgets for interactive and multimedia strategies of the internationals (for instance, the major banks, publishers, the retail trade). Can the businesses in the Dutch multimedia cluster strengthen the strong players in the Dutch economy even further by guiding them into the age of interactive media and e-commerce? While the actual scale of the Dutch multimedia service providers is probably inadequate to achieve this, it still offers a major opportunity. To this end, at a minimum it is essential that the multimedia businesses support these clients when they are innovating on the basis of multimedia.

In summary, the main bottlenecks associated with the functioning of the multimedia cluster include the following:

- The fact that by no means all multimedia businesses devote systematic attention to organising the intelligence function (systematically keeping up to date on new technologies and market opportunities) and the consequent upgrading of products and services.

- The partial development of the interface between knowledge infrastructure and multimedia activity and, if present, the focus on the transfer of technological knowledge in the field of multimedia. One particular problem would appear to be the low level of interaction between multimedia service providers operating at the core of the cluster and the knowledge institutes.

- The under-utilisation of the demand for multimedia products and services as pursued by the internationals on the domestic market. Supporting this category of clients with their multimedia strategies could make a significant contribution towards reinforcing the competitive strength of the Dutch multimedia cluster.

Multimedia cluster performance

The performance of the multimedia cluster is co-determined by its character and the way it functions. For instance, the adaptation capability of the multimedia cluster reflects, among other things, the ability to achieve the following in good time:

- Establish a connection between the different types of knowledge and the associated activity.

- Select the appropriate partners to work with.

- Invest in the crucial fields of knowledge and the interface with knowledge producers.

- Anticipate the demands of potentially major clients.

The economic performance of players in the multimedia cluster is currently difficult to judge on the basis of partial and outdated statistical data. High growth figures (employment, turnover), large profit margins and large numbers of newcomers point in the right direction. In addition, among multimedia service providers, the consultancy element, particularly in the preliminary phase of creating multimedia applications, still provides a major source of income. The providers of products – of which there are few – are characterised more by the "boom or burst scenario". It takes time to mastermind and develop a new product, and there is little or no turnover during this "construction phase". These businesses are uncertain about their future success and acceptance of the product in the market (including the financial market) is more important than turnover in the initial stages. At the same time, these are the same businesses that are (or at least were until recently) highly valued in accordance with the laws of the Internet economy in the event of a take-over or flotation. Only a small number of businesses at the core have indicated that their exports are substantial; exports and a presence on the international scene are rare for the majority of firms.

The absolute innovation performance would not seem to be exceptional. It is true that new products and services are inherent in this cluster but it is apparent that they do not always go hand in hand with sufficient attention to the innovative process in services and future-oriented R&D. The latter is restricted to a few of the larger market players and players in the public (mainly technical) knowledge infrastructure. Nevertheless, in addition to innovation, the adaptation capability (*i.e.* "timely anticipation") is at least equally as important. In the multimedia cluster, this means first and foremost, speed: the willingness to invest in knowledge innovation in good time, to probe new fields of knowledge and business models, or the ability to work out an idea within the space of a few weeks and market a product (first-mover advantages).In second place comes adaptation capability, *i.e.* knowing one's client groups, and building up and obtaining a commitment from communities. The third important aspect is the fundamental willingness to continue investing in upgrading knowledge and multidisciplinarity, combined with a constant urge to experiment with new products, tools, infrastructures and applications. To this end, it will frequently be necessary to look beyond the borders of one's own business and branch of industry and to invest in teamwork with parties alien to the branch of industry, with players who have complementary knowledge.

Finally, adaptation capability in the multimedia cluster means the timely professionalisation of business processes (regular management, knowledge management) and the professionalisation of matters that transcend the level of one's own business. The latter implies that, in addition to carrying out projects, investments will also need to be made in facilitating the functioning of the multimedia cluster as a whole; for instance, drawing up an agenda of the bottlenecks which can be addressed by government (labour market, training, intellectual property rights, innovation incentives, telecommunication infrastructure specifications), and, for instance, the optimisation of knowledge transfer with the relevant knowledge infrastructure, the joining together of regional multimedia specialisations or the formulation of an export strategy. It is clear that the adaptability of the ring players (Figure 1) in the multimedia cluster in particular is currently being put to the test.

The main bottlenecks associated with the performance of the multimedia cluster are:

- The insufficiency of both professional (regular) management and knowledge management in a number of expanding multimedia businesses at the core of the cluster.

- The lack of an unambiguous position of the Dutch multimedia cluster for the clients, as well as the lack of an export strategy.

Conclusions and recommendations

Innovation in multimedia is not limited to innovation in multimedia technology. Very often, multimedia technology provides a tool for creating, packaging, distributing and marketing content in a novel way. Innovative Web sites, novel electronic marketplaces, new broadcasting formulas using the Internet or new interactive strategies to interact with customers using new media are as much about innovation in multimedia as a new intelligent agent technology or a multimedia authoring tool. Therefore, multimedia potentially affects all trades of business. For example, in the Dutch context, innovation in multimedia will most likely enable such industries as financial services, publishing, agro-food and logistic services to remain competitive in the unfolding digital economy.

The main bottlenecks that will need to be overcome to allow the Dutch multimedia cluster to grow into a flexible, proactive, recognisable and competitive cluster which constantly upgrades and adapts itself, have been discussed in the above. Their interrelation is summarised in Figure 4.

In a similar fashion, Figure 5 sets out the main options for improvement and prioritises the potential role for government in the actions to be undertaken. These are to be found first and foremost in the sense of support as far as an improvement in the understanding of the structure and economic significance of a multimedia cluster is concerned. This can be achieved, for instance, by investing in the setting up of a statistical index of the cluster and by providing greater insight into the economic significance of the cluster. The government can also support initiatives aimed at improving the incorporation of the multimedia component in training courses and promoting regional multimedia specialisations that reinforce one another and that are linked. Both could provide impetus to the cluster's dynamism and innovation capability.

The second group of improvements is connected with promoting the division of knowledge among the various actors in the cluster. In addition to measures focusing on providing greater insight into the demand for and the supply of knowledge in the field of multimedia (including the non-technological knowledge component), this can primarily be attained through joint (knowledge) projects. Wherever necessary, existing instruments of innovation policy should be brought in line with the speed and characteristics of the multimedia cluster. In a number of fields of application, the government should act as lead user and thereby spur the players in the multimedia cluster to produce innovative products and services. Electronic government (e-government) is an obvious candidate in this regard.

Figure 4. **The main bottlenecks in the multimedia cluster and their interrelations**

Items which have been underlined indicate a possible role for government

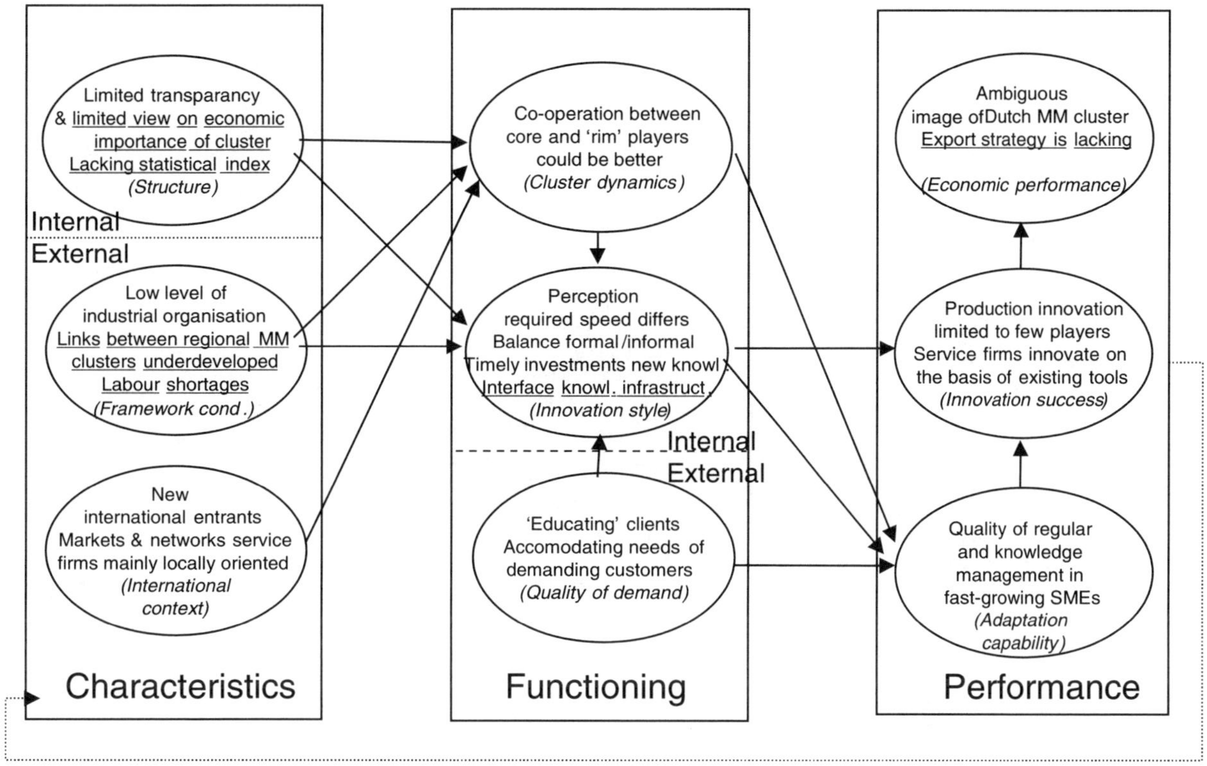

The third group of measures through which the government could help to improve the competitive strength and innovativeness of the Dutch multimedia cluster is related to improving the export orientation and supporting the formulation of an export strategy in the field of multimedia.

It is clear that businesses, educational institutes and knowledge institutes will need to develop initiatives covering all these points. Actions are required to improve the level of organisation in the multimedia cluster, to invest in joint technology and market intelligence, and the professionalisation of regular and knowledge management. The latter applies particularly to those multimedia businesses that are expanding from small to medium-sized or large companies. They must be able to operate in alternating networks and alliances and be able to stand up to international competitors and demanding clients.

Figure 5. **The main options for improvement in the multimedia cluster and their interrelations**

Items which have been underlined indicate a possible role for government

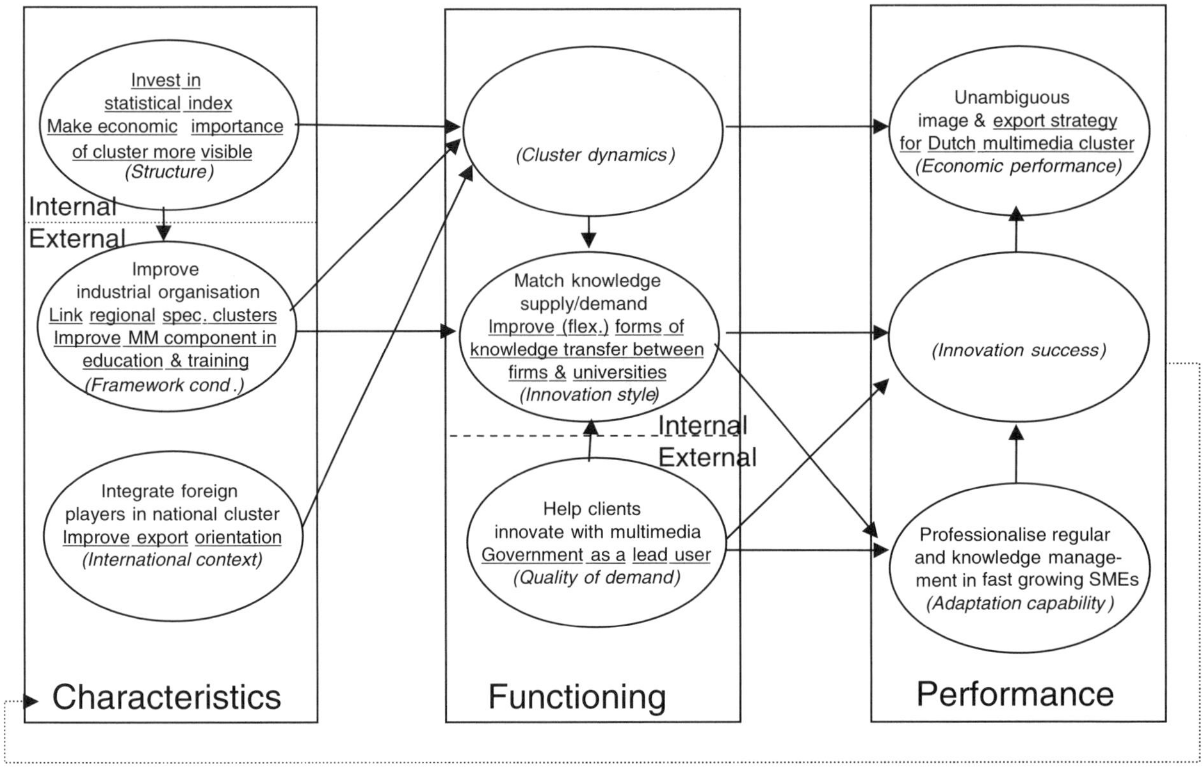

Cluster studies such as the one presented here must be perceived as a tool allowing firms, knowledge institutes and policy makers to better understand the specific innovation style of a cluster, its bottlenecks and points for improvement.[17] Undertaking such a cluster study can contribute considerably to improving dialogue and agenda setting within the cluster and can help to identify improvements to functioning of and innovation in such a cluster. The resulting agenda, as well as the facilitating role to be played by policy makers differs depending on the cluster considered as cluster structure, cluster dynamics and cluster innovation style are highly cluster-specific.

Possibly the biggest risk to be faced is that entrepreneurs, policy makers and researchers mainly focus on so-called "high tech" clusters and the obvious "success stories" that abound. This presents a major risk and it is sometimes forgotten that the rise of such clusters is the result of a combination of a unique mix of often strongly localised factor conditions and development trajectories that have evolved over decades and cannot be replicated overnight. Secondly, the economic importance, innovation dynamics and adaptation capability of established or mature – often misleadingly labelled "low" or "medium tech" – clusters tend to be easily forgotten.[18]

Annex

CLUSTERMONITOR AS AN ANALYTICAL TOOL

The ClusterMonitor developed for the Dutch Ministry of Economic Affairs is an analytical instrument. It supports the identification and analysis of clusters and makes specific recommendations relating to innovation, cluster formation and the improvement of competitive strength with regard to the actors in the cluster (businesses, knowledge organisations and other intermediary organisations, and policy makers). The core of the monitor methodology is formed by the relation model below. This model is used to analyse the various aspects of competitive strength and innovation. The relation model, and the nine dimensions set out therein, offers an initial stepping stone to better describe the basic characteristics, the functioning and the performance of the cluster. This model was used as the basis for the quantitative and qualitative collection of data (for a detailed description of the methodology developed, see Berenschot, Dialogic and Technopolis, 2000).

Figure A1. **Relation model**

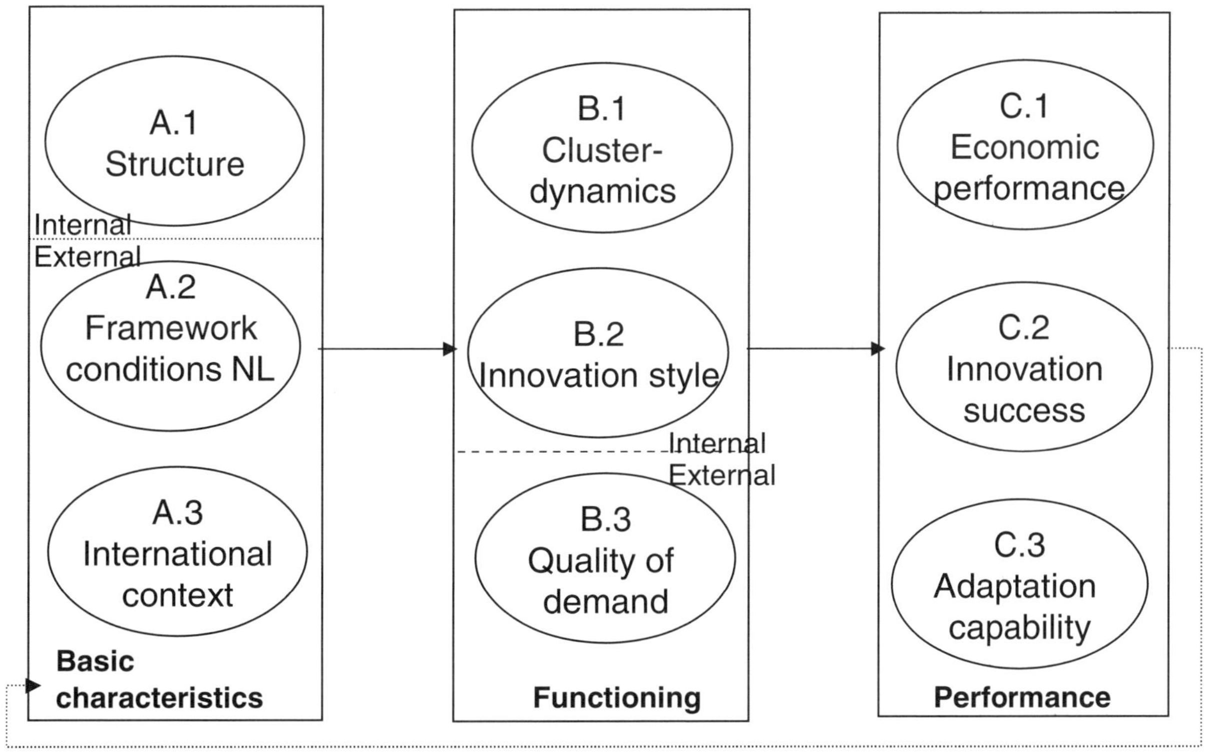

First facts and figures on the basic characteristics, functioning and performance were collected for the wider, statistically defined, ICT cluster (see also den Hertog *et al.*, 2000). This quantitative analysis was carried out in association with the Dutch Statistical Bureau (CBS). It resulted in a large database in which the various statistics (*e.g.* production and innovation statistics) at the level of individual business enterprises (the micro level) were linked and where the performance of the cluster under study can be compared with the overall score of all firms. A total of 32 indicators are now available for 46 900 businesses. These indicators were coupled to the nine dimensions contained in the relation model presented above. This ClusterMonitor database, if updated regularly, will most likely be used for further ClusterMonitor analyses in the future as well as for other purposes. The quantitative analysis is not reported here as it was, due to data constraints, only applied to the wider ICT cluster of which multimedia is a part.

Secondly, a subset of the firms which together make up the emerging multimedia cluster were analysed qualitatively in more detail and suggestions were made as to how to improve competitiveness and innovation. In addition to desk research, the qualitative analysis consisted of a series of 25 interviews with businesses and organisations active in the relevant cluster, plus extensive discussions on the subject with members of the sounding board group made up of people active in the various segments of the multimedia cluster. The provisional analysis and identification of bottlenecks and points in need of improvement were discussed at a workshop held on 7 March 2000 with some 35 participants engaged in the Dutch multimedia cluster. The outcomes of this workshop were incorporated in the final report (den Hertog *et al.*, 2000).

NOTES

1. A cluster is defined as chains of suppliers, customers and knowledge centres (universities, research institutes, knowledge-intensive services, intermediary organisations) that: *i)* have at their disposal complementary competences; *ii)* are interconnected through production chains or value chains; *iii)* improve joint industrial processes and end products; and *iv)* (possibly) participate in networks focused on innovation and technology development.

2. Within the framework of the further development of cluster policy, the Netherlands Ministry of Economic Affairs has developed an analysis instrument, the so-called ClusterMonitor (see Berenschot, Dialogic and Technopolis, 2000), one of the three pilot studies related to the Dutch multimedia cluster. Additional information on the ClusterMonitor is included in Annex.

3. The difference between online and offline multimedia is diminishing in significance. Producing a multimedia presentation is a skill that has nothing to do with the medium used. On the understanding that consideration is given to the aspect of downgrading beforehand, it is often a case of multimedia presentations being downgraded to a variety of different formats. Businesses that produce online multimedia are often engaged in the production of offline multimedia as well. In producing multimedia applications, it is important that the designers anticipate the fact that the information can crop up in many different environments for the user (a variety of different sorts of carriers, user interfaces, usage environments). For designers, this results in an "accumulation" of different requirements.

4. The overwhelming use of English names or at least names with an English-friendly sound is already an indication of the international character of cluster actors. Some of the firms mentioned are foreign (mostly American) firms, others are Dutch but all flag their international orientation or at least international image.

5. It is sometimes hard to discern where the production of multimedia ends and the (advanced) use of multimedia applications begins. Due to the enabling character of ICT as a technology, some users have built up considerable knowledge on how to put to use ICT for their particular business. Often, this results in internal business units that are as capable as specialised suppliers of ICT suppliers and in a way are part of the ICT or in this case the multimedia value chain. Sometimes these nodes of application knowledge develop into separate businesses.

6. In 1999-2002, the Dutch Government will invest NLG 142 million in GigaPort network and GigaPort applications. GigaPort is the upgrading of the present SURFnet 4 research network into one of the world's fastest research networks (a direct link to the US Internet 2 networks is already available). GigaPort applications relate to the development of applications made possible by the GigaPort network, *i.e.* a huge playing-ground for the creation of new electronic applications and services.

7. Some of these foreign companies have had major divisions located in the Netherlands for some time now (as well as development units) and are consequently already integrated in the Dutch IC cluster.

8. See, for instance, the provoking contribution of Markusen (1996) on "sticky places". She concludes that "in reality, sticky places are complex products of multiple forces: corporate strategies, industrial structures, profit cycles, state priorities, local and national policies. Their success cannot be studied by focusing only on local institutions and behaviours, because their companies (through corporate relationships, trade associations, trade, government contracts), workers (via migration and international unions), and other institutions (universities, government installations) are embedded in external relationships- both co-operative and competitive – that condition their commitment to the locality and their success there." (p. 309).

9. The ANMA organised 32 meetings for its members in 1999. The total number of member businesses is now 350. In addition to the relevant component, it is mainly the networking and informal knowledge transfer taking place at these meetings that is important for the cluster effect.

10. In their comparison of the three clusters, Silicon Valley, Silicon Alley and Route 128, Bouwman and Hulsink (2000) refer, among other things, to the importance attached to networks and platforms (communities) for the collective creation of knowledge and protection of interests, the exchange of information and the establishment of new contacts. Exactly how these contacts are moulded differs from one regional cluster to another.

11. Incubator centres for ICT start-ups sponsored through the Ministry of Economic Affairs. ICT Twinning Centres are now located in Amsterdam, Enschede and Eindhoven.

12. Mobility of scarce talent is an important source of knowledge diffusion as well.

13. We should beware not to automatically start from the notion that knowledge institutions have superior knowledge since the ICT industry in general is an industry that largely produces its own technologies and sells them either as services or products. This unique ICT aspect renders knowledge institutions in this field relatively unimportant, unless they are at the cutting edge of scientific innovations that can be converted to technologies by smart entrepreneurs.

14. Establishing relationships between knowledge institutes and firms is not always easy; it takes time to developed the needed trust (see also Gilsing, Chapter 19 of this volume).

15. When compared with the business market, the quality of the consumer market is said to be relatively higher. The market for devices (PC, modems, DVDs, etc.) and packaged software is basically a global market in which several Dutch parties play an important role (Philips, for instance). Consumers are prepared to experiment and the number of online consumers is rapidly increasing.

16. We came across examples of larger businesses that were clearly engaged in a process of upgrading and chose to clean up their current customer portfolio, to discontinue their marginal customers, and to focus on a selection of their weightier, central customers with whom they have concluded long-term full-service contracts.

17. The strategic framework presented by Gilsing (Chapter 19 of this volume) did not exist at the time the ClusterMonitor study on which this chapter is based was undertaken. However, the multimedia cluster was recognised by many players, including the Ministry of Economic Affairs, as a promising emerging cluster some time ago. Some actions had already been set in motion. Currently, one could say that, in terms of the strategic framework, the multimedia cluster is in the initiation phase (phase 2 of the framework) and, depending on how multimedia scores on the set of criteria mentioned under gate 2, it will soon enter the implementation phase. Having said this, some actions that would, under the new strategic framework, fall under the phase of implementation are currently being implemented; these include support for the Telematics Technical Top Institute, support for other multimedia research, investments in Internet-2 infrastructure and applications (GigaPort and Gigaworks), establishment of three incubators, etc. This might be an indication that it might be difficult to follow the strict discipline suggested by the strategic framework once a cluster (or a technology) is designated as promising and important for the Dutch economy.

18. See also Hauknes (this volume) and Whalley and den Hertog (2000).

REFERENCES

Aarts, E. and R. Harwig (2000), "Ambient Intelligence. A New User Experience", paper distributed at CeBIT 2000, Philips, Eindhoven.

Bakker, K., J.A. Becht, K.R. Jonkheer, A.P. Muizer and D. Snel (1999), *Multimediabedrijven in Nederland* [Multimedia Firms in the Netherlands], EIM, Zoetermeer.

Berenschot/Dialogic/Technopolis (2000), "ClusterMonitor", report prepared for the Dutch Ministry of Economic Affairs describing the ClusterMonitor methodology, Berenschot/Dialogic/ Technopolis, Utrecht.

Bouwman, H. and W. Hulsink (2000), *Silicon (V)alley in de Polder. ICT-clusters in de lage landen Silicon (V)alley in the Polder* [ICT Clusters in the Low Countries], Lemma, Utrecht.

Braczyck, H.J., G. Fuchs and H.G. Wolf (eds.) (1999), *Multimedia and Regional Economic Restructuring*, Routledge, London.

Brail, S.G. and M.S. Gertler (1999), "The Digital Regional Economy: Emergence and Evolution of Toronto's Multimedia Cluster", in Braczyck *et al.* (eds), *Multimedia and Regional Economic Restructuring*, Routledge, London, pp. 97-130.

Brouwer, Erik and Pim den Hertog (2000), "Key Figures on the Dutch Information and Communication Cluster", Centre for Science & Policy/Dialogic, Utrecht.

den Hertog, P. and S. Maltha (1999), "The Emerging Information and Communication Cluster in the Netherlands", in OECD, *Boosting Innovation: The Cluster Approach*, OECD, Paris, pp. 193-218.

den Hertog, P. *et al.* (2000), "ClusterMonitor Multimedia", report prepared for the Dutch Ministry of Economic Affairs, Berenschot/Dialogic/Technopolis, Utrecht (www.clustermonitor.nl).

Feldman, M.P. (2001), "The Entrepreneurial Event Revisited: Firm Formation in a Regional Context", paper prepared for the International Conference on Regional Economic Development: New Interpretative Paradigms and New Policy Instruments, LIUH University.

Gilsing, V.A. (2001), "Towards Second-generation Cluster Policy: The Case of the Netherlands", Chapter 19 of this volume.

Hauknes, J. (2001), "Innovation Styles in Agro-food Production in Norway", Chapter 8 of this volume.

Heydebrand, W. (1999), "Multimedia Networks, Globalization and Strategies of Innovation: The Case of Silicon Alley", in Braczyck *et al.* (eds), *Multimedia and Regional Economic Restructuring*, Routledge, London, p. 49-80.

Jacobs, Dany and Ard-Pieter de Man (1996), "Clusters, Industrial Policy and Firm Strategy: A Menu Approach", *Technology Analysis & Strategic Management*, Vol. 8, No. 4, pp. 425-437.

Markusen, A. (1996), "Sticky Places in Slippery Space: A Typology of Industrial Districts", *Economic Geography*, Vol. 72, No. 3, pp. 293-313.

OECD (1999), *Boosting Innovation: The Cluster Approach*, OECD, Paris.

Pavlik, J.V. (1999), "Content and Economics in the Multimedia Industry: The Case of New York's Silicon Alley", in Braczyck *et al.* (eds), *Multimedia and Regional Economic Restructuring*, Routledge, London, pp. 81-96.

Peneder, M. (2001), "Dynamics of Initial Cluster Formation: The Case of Multimedia and Cultural Content", Chapter 15 of this volume.

Whalley, J. and P. den Hertog (2000), *Innovation and RTOs. A Synthesis of the Findings from the RISE Cluster Studies*, University of Strathclyde/Dialogic, Glasgow/Utrecht.

PART II

LEARNING FROM EXPERIENCES IN MATURE CLUSTERS

Chapter 8

INNOVATION STYLES IN AGRO-FOOD PRODUCTION IN NORWAY

by

Johan Hauknes
STEP Group

Introduction

The purpose of this chapter is two-fold. First, it describes innovation patterns and related dynamics in agro-food production, based on a cluster approach to economic change and innovation. It is now generally recognised that innovation and wider economic development must be understood in a context of agglomeration characteristics. Innovation is substantially – perhaps even fundamentally – shaped by agglomeration characteristics, be they technological, industrial or geographical. This insight, which goes back to Alfred Marshall's description of industrial districts and their external economies, is now reshaping the debate on the objectives and priorities of industrial innovation policies (OECD, 1999).

Second, this chapter seeks to compensate for a serious bias in the current discourse on innovation research and innovation policies. It highlights two issues that should be central to these debates: *i)* the neglect of so-called low-tech and mature industries in growth policies and innovation studies; and *ii)* the narrow, over-simplistic approach adopted in outlining and mapping innovation dynamics. These two related aspects imply that innovation policies may be seriously misguided, neglecting major opportunities for growth and structural change at the cost of an undue focus on the creation of new, technology-based industries that are, and will continue to be, only a minor component of national and regional industrial systems.

Our starting point is the decomposition of Norwegian input-output tables described in Hauknes (1999). That decomposition allowed six mega-clusters to be identified in the Norwegian economy, based on the strengths of trade links at the detailed product group level. Taken together, the six clusters describe the major structural composition of the economy, accounting for some 75% of GDP and 55% of employment in the market sector of Norway. The economic dimensions of these clusters were described in Hauknes (1999), as were the structure and impacts of innovation and technical change. One of these clusters – a mega-cluster comprised of *agro-food production* and related industries – is the focus of the present chapter.

The goal of this chapter is to contribute to improving our understanding of innovation as a driver of economic growth and change. Any study of firms' innovative behaviour as a generator of industrial and economic change must be based on an understanding of the relevant industry and market dynamics and firms' perceptions of these – this forms the basis for firms' expectations and their innovation decision making. We will outline the patterns and dominant modes of change in a mature, low-tech manufacturing-based cluster – food production, a cluster which is far removed from the current notion of a "new" economy. "Old" economy clusters such as agro-food are central to the modern innovation economy and will continue to remain so.

The chapter is split into three parts. First, we outline a general conceptual framework for analysing innovation that will allow us to put technological innovation into a reflexive perspective. Second, we briefly describe the main structure of the Norwegian food industries and ongoing changes taking place in these industries. This forms the basis for an analysis of current innovation styles in the food industries along a series of dimensions, integrating a wide range of innovation indicators and related empirical data. The final section presents the general conclusions arising from this work.

Innovation and economic change[1]

It is argued that innovation is a main structural determinant of industrial production structures and economic growth. However, innovation is basically a firm-level activity, shaping firm-level competitiveness and company growth. It is through the diffusion and adoption of new forms of economic behaviour that innovation ultimately shapes industrial and wider economic production structures. Our understanding of the determinants of innovation, and which innovation modes have industry- and economy-wide repercussions, remains limited.

In the Kuznets-Schumpeter tradition, it is often argued that economic growth depends on the creation of new industries involving major new technologies and associated radical innovations, and that the latter are in some sense clustered together. According to this view, a group of allegedly "knowledge-intensive" and high-technology industries are driving economic growth in the present era. However, empirical examination of the sectoral structure of growth suggests that growth has a widely distributed sectoral basis, and that many significant growth sectors are situated in what are often referred to as "low-tech" industries. These sectors are typically large and thus contribute strongly to overall performance.

Innovation is a pervasive phenomenon. It is a core aspect of competition in all product markets, both markets for intermediate goods, and capital and consumer goods markets. The "technological" competition that provides firms with incentives for innovation and new modes of economic behaviour basically arises from the struggle for market position. Innovations may involve product and process innovations, as well as changes in organisation, and market-related and strategic choices. Any conception of innovation should, in principle, cover all these behavioural changes.

Mapping innovation patterns and their impacts requires an understanding of the core characteristics of the competition that companies face in their economic context: What are the present "hot spots" of markets and competition? These "hot spots" are reflected in firms' perceptions of the incentives for, and opportunities and potential rewards to be gained from, innovation. They reflect how firms view their competitors and how they emulate successful behaviour. This requires us to consider economic competences and capabilities beyond technical competences (Hauknes, 2000b). The present chapter attempts to describe innovation patterns in agro-food production based on these considerations. In contrast to the input-output-based cluster approach of a previous study (Hauknes, 1999), the basic approach taken here is to view the cluster as a reduced-scale innovation system around the value chain in which individual companies evolve. Hence, innovation is about changes in the company's role in the value chain and its relations with surrounding organisations and sources of competences, information and skills. This immediately puts the market dimension and the market behaviour of the firm into the central position.

The food cluster in the Norwegian economy

Some basic features of the Norwegian agro-food cluster

The Norwegian agro-food cluster is a dominant feature of the Norwegian economy. Its production represents nearly 10% of total GDP, a share that has remained fairly stable over the 1990s. Employment in the cluster accounts for more than 10% of total domestic employment. The output of the food industries (NACE 15) – and more specifically agro-food (NACE 15, excl. 15.2), the core of this cluster – comprises either products for final consumption or intermediate products used in processing and industrial activities, mainly by the food industries. Over the last few decades, the agro-food clusters of the industrialised countries have been subjected to increasing competitive pressures. This is partly a consequence of structural changes in related industries, of changes in regulatory and governance systems and of increased competition from imports. In addition, consumer preferences for food quality, food safety and nutritional content, product labelling, variety and convenience are evolving, exerting a strong pressure on the development of these industries.

In spite of increased competition, several sets of data suggest that the Norwegian cluster has a low level of innovation dynamics (Hauknes, 1999):

- Its capital intensity is fairly low.
- Less than 3% of its workers have higher education.
- Only 5% of total R&D expenditures in the business sector were spent in this sector.
- The use of knowledge-intensive business service (KIBS) inputs in the cluster is weak.
- Traditional innovation inputs are generally characterised as being relatively low.

All in all, this seems to suggest a cluster of industries with fairly weak innovation dynamics and a reliance on capital-embodied technical change as the main mode of industrial development. A main argument underpinning this chapter is that the validity of this conclusion is limited. On the contrary, we find significant innovation dynamics, with the main reason for its invisibility in the indicators we have used here being that these dynamics lie outside the traditional focus of technical and capital-embodied economic change.

Table 1. **Share of innovating firms, 1995 and 1997**

	Period	Food industries	Other manufacturing industries	All industries
Share of innovating firms	1993-95	38.0%	42.8%	24.9%
	1995-97	38.7%	38.8%	29.6%

Note: 1993-95 estimates are based on supplementary questions on innovation performance included in the 1995 R&D survey. For further explanations, see Hauknes (2000a).

In fact, the share of innovating firms in the food industries is surprisingly high. In national innovation surveys, nearly 40% of firms had introduced *technological product or process innovations* in accordance with *Oslo Manual* definitions (OECD/EUROSTAT, 1997) over the preceding three-year period. In both surveys, the share of innovating firms in the agro-food cluster is close to the overall share of innovators across all industries. There are no indications of less intensive innovation dynamics in agro-food than in the rest of the economy. To render these two observations – of weak traditional innovation inputs and high rates of innovation outputs – compatible, we need to gain a broader understanding of the industrial context and dynamics in which these innovations appear.

In the food cluster, including food production based mainly on fish and other marine products, three independent sub-clusters may be identified:

- *Agro-food* production, such as dairy production, milled products, fruits, meat and grain-based fodder.
- *Marine food* production, mainly fish products and fish-based feeds.
- Production of *beverages.*

These sub-clusters are reflected in the organisation of the institutional and wider governance system supporting or structuring these industries. In all three, there are specialised knowledge suppliers specific to each group; the regulatory system and the public support systems are specifically organised for each, even up to the ministerial level between marine-based and agriculture-based sectors. This extends also to the organisation and, until recently, the funding, of secondary and higher education. Broadly speaking, the first two sub-clusters are characterised by strong cultural and social traditions and networks, spanning industrial production, distribution and wholesale trade, policy systems, regulatory agencies, etc., around each sub-cluster.

Ownership and organisation of production and distribution are distinct for each sub-group. Agro-food production has been based on collective ownership by the agricultural sector of dairies, slaughter-houses, vegetable and fruit distribution while, until recently, grain distribution was a state monopoly. Retail trade and distribution has, on the other hand, traditionally been strongly atomised. Marine production has primarily been based on private ownership, with occasionally a strong role for collaborative marketing and brand systems.

There are significant differences in the geographical location and distribution of production. The heart of agro-food production is centred around the more fertile regions in inland Norway, while marine production takes place dominantly in coastal areas.

Industrial organisation

Industrial organisation and ownership combine aspects that are quite specific to the cluster with more general institutional characteristics. Three types of firm organisations and corporate governance structures dominate the cluster.

Multi-product corporations and concerns. Production companies are privately owned subsidiary companies of national or multinational corporations, often with domestic control of multiple daughter companies. Almost all the largest companies and corporations in the cluster are of this type. The corporations are generally vertically integrated in processing, marketing and sales activities, with strong market orientation. They were vanguards in the early development of branding and other marketing strategies. Some of these are today major multinationals, whether controlled abroad or in Norway.

Co-operatives and co-operatively owned companies. Co-operatively organised production and distribution of agro-food products is an old tradition in most European countries. These co-operatives are today generally multi-facility quasi-corporations operating at national level. They are still based on co-operative ownership, but ownership is increasingly linked to the national corporation rather than regional production and distribution facilities. The companies are generally operating as vertically integrated corporations, with strong links, even to the extent of being monopsonic, to suppliers of agricultural production. In addition to these roles, co-operatives are allotted a national role as market regulator, with obligations to supply all further processing industries. Co-operatives are increasingly also venturing into the more competitive processing industries. The process of corporatisation of co-operatives goes hand in hand with the dominant position of vertically integrated retail chains and an emergent "nordicisation" of markets for food products.

The co-operatives generally have a large market share on domestic markets. Increasingly, over the last decades, they have developed and used a series of extensive marketing strategies such as branding and trade marking systems as a strategic market positioning tool.

Independent producers. These are mostly small or medium-sized, often family-owned, autonomous companies. Generally, they are firms with weak integration into upstream and downstream value chains and in which the day-to-day management and production is carried out by the owner.

Food industries: changing innovation frameworks

Some of the noted processes of change are more or less specific to Norway, often reflecting historical contingencies of food production; others are more general, with more or less immediate parallels in other countries.

The cluster has a *strong position in public opinion.* Norwegian food production is generally seen as high quality but perhaps not sufficiently differentiated to satisfy "modern" urban tastes. The implicit rationales for policy making reflect, *inter alia,* a strong alliance between regional development and agriculture/food production interests. Arguments of "consumer interests" and multi-functionality of agricultural production afford a forceful rationale for industrial policy making and regulation, focusing on domestic production, quality and safety.

Public interest in and concern for food *safety* has grown substantially over the last 25 years. This has placed pressure on producers to supply supplementary information on content, production, additives, etc. To some extent, the link between the agro-food system and the public has become increasingly more precarious due to a series of incidents. From the industry side, the perception of an information gap has led to the implementation of various information-signalling initiatives.

At the same time, higher income provides greater scope for product differentiation and for differentiating information strategies, and an increased willingness to pay for perceived *quality.* The abundance of planned and implemented labelling strategies – AOC, green production, etc. – are examples. The strong domestic tradition in food demand implies that an obvious aspect of these market differentiation strategies is the potential role they may play as new forms of protective schemes for regional or domestic production.

Related to this is the enhanced focus of "specialised foods", supported by the increased weight placed on food quality, differentiated and income-elastic consumer demand and internationalisation of tastes. Similarly, the increased focus on food *health* is followed by an enhanced awareness and utilisation of the concept of "functional foods". The related growth of and strengthening of markets for eco-foods is a forceful means of capturing public concern for food safety, quality and health.

International trade in agro-food has traditionally been weak, supported by quota systems, tariffs and other trade regulation, sheltering domestic production from competition. As international regulation has changed, import barriers have been dismantled and tariffs reduced, thus opening up national markets to foreign competition. In addition to the EEA Treaty, the implementation of Uruguay Round and the present discussion around the WTO Millennium Round have been decisive in shaping the recent debate on national policies and strategies relating to food production. With the 1994 GATT Round, the WTO Millennium Round and its Built-in Agenda have led to wide concern over the impact of increased international trade, competitiveness, market governance and regulation of public industrial support. The Millennium Round may lead to significantly altered regimes of public policy formulation for these industries.[2]

Changes such as these are reflected in a range of recent policy initiatives concerning agro-food production, with *substantial public involvement* even in market development. Most directly, the major public agro-food restructuring programme which took place in 1994-95 grew directly out of the national debate over EU membership. Jointly organised market-signalling measures, such as the certification system Godt Norsk,[3] and, more recently, the initiation of AOC and Label Rouge certification systems for agro-food, have been launched. A large-scale ten-year public programme for "value creation" in Norwegian agro-food will be launched in 2001.

Over the last decades, dramatic changes in the *structure of the distribution and retail trade* have altered the market interface between the agro-food and trade sectors, with four chains representing some 95% of consumer markets. Agro-food producers now face a more active policy stance and purchasing power of a few large retail and distribution chains, with increased use of, for example, OEM-like own-branding systems.

Markets for agro-food are increasingly becoming Nordic, perhaps even Baltic, in part due to the "Nordicisation" of retail and distribution chains. Restructuring of agro-food production at the Nordic and European levels is beginning.

The food cluster is faced with both the need and opportunities for restructuring. The remaining part of this chapter attempts to outline how the sector is addressing these challenges through innovation activities. The table shown in annex outlines in somewhat more detail the institutional features of the Norwegian agro-food production system, corporate structures and associated dominant processes of change.

Innovation patterns in food industries

This section outlines some of the main characteristics of innovation patterns in the food industries. The presentation is based on the use of a wide range of data. A core data set for such analyses is the national innovation surveys, based on the Harmonised Questionnaire of the EUROSTAT Community Innovation Survey. The last national innovation survey was conducted in 1998, with respect to innovation performance in 1997. In addition to the specific data sources mentioned in the text, this presentation is based on detailed field work and related analysis of the agro-food industries, performed in STEP during 1998-2000.[4]

Innovation in any company is a function of the skills and competences available within the firm, through:

- Technological innovation.
- The skills and qualifications of the labour force.

- The abilities of the company and its organisation to master core competences in the industry in which it operates, and the company's capability to mobilise these resources.

This basis is a core requirement for competitive innovation performance, and hence for the wider industry impact of innovation activities. In a successful exploitation of technological and market opportunities, the mobilisation of these resources is integrated within the firm across functional and organisational boundaries, with systematic use of information and competences from external sources or repositories. Hence, innovation must be seen as both a multi-functional and a multi-organisational process.

Below, we will first outline some aspects of this *innovation basis* in food industries. The structure of interaction between this intramural basis and mostly extramural sources and repositories of competences will be discussed under the general heading of *innovation styles* that we deal with subsequently.

Innovation basis

Technological innovation in agro-food

Agro-food production systems in advanced economies afford an interesting contextual study of innovation dynamics. Agro-food production is generally a large, low-tech and mature industrial production system, dominated by the production of consumer goods and with a strong consumer orientation. Increases in income levels and social changes in consumer demand may have a direct impact on industry-wide behaviour and competition. It covers products with low, as well as high, income elasticities. High levels of exports in some product niches appear side by side with strongly domestic-oriented ones.

The notion that innovation in food industries is led by technical change embodied in capital machinery and equipment developed outside the food industry predominates. According to this view, the food industries are considered to be essentially technology-using rather than technology-developing. The agro-food cluster is described as being dominated by "implicit" modes of technical change, primarily through embodied use of capital goods and intermediate inputs.

However, a main conclusion of this chapter is that this conclusion is not warranted, either for innovation performance or for the structural impact of innovation activities. The structure of competition and innovation today does not reflect this simple picture. The food industries today place a strong emphasis on product innovations addressing new and differentiated demands, as well as health, safety and quality concerns. The food industries are innovative industries, even when mapped through technological product and process (TPP) innovations.

Today's innovation dynamics in food industries are not simply related to capital embodied technical change and process innovations. A dominant mode of innovation dynamics is directly related to market dynamics.

Employment and educational attainments

Employment in the food industries is dominated by non-tertiary qualifications. Any attempt to describe the "knowledge" or competence intensity of the food industries on the basis of employment of qualified scientists and engineers is doomed from the outset to characterise these industries as low-intensity industries. However, a major premise of our work is that, in analysing competence structures and intensities, the structure of the labour force will reflect the nature of the underlying competences. Formal qualifications are springboards for the formation of the industrial competencies and skills that are the foundation for innovation. The structure of employment in terms of qualifications will, at least in part, reflect the state-of-the-art requirements for building the necessary absorption and organisational capacities to become a competitive innovator, adjusted to the level of change in innovation frameworks and market and technological opportunities for innovation. The organisational mobilisation of these capacities cuts across qualification levels. If the required competences and skills for successful innovation performance are not dominantly science- or engineering-intensive – in a sense, if the industry is predominantly based on "old" science and engineering – we should not require companies to be intensive users of Qualified Scientists and Engineers (QSEs) in order to be qualified as successful innovators.

Other things being equal, a stable innovation framework should be reflected in a stable structure of employment. Rapidly changing employment structures are thus a potential signal of adaptation to the changing frameworks and requisites of innovation. There is a rapid process under way whereby the structure of employment in the food industries is changing. However, the main dimensions of this process are below the tertiary/non-tertiary threshold. In this process, large segments of formally unskilled labour are substituted for with increased use of skilled labour. From a situation where these industries were among the major employers of unskilled, blue-collar labour, the ticket for admission into employment has escalated in value.

It is noteworthy that, due to the substantial expansion of the education system at secondary and tertiary levels over the last few decades, the level of education attainment has risen. Our analysis to date does not allow us to disentangle the demand effects from the supply effects generated by an expansion motivated more by social welfare considerations than by industrial needs. Economy-wide employment data, through differential industry-level rates of change at industry levels, point to demand effects. However, even a supply-led process has consequences for potential innovation capabilities and capacities to restructure.

Table 2 describes the education levels of employees in the food industries. In the food industries, almost 75% of the labour stock in the mid-1980s were unskilled, including employees who had completed only the first year of secondary school. By 1999, this share had fallen to below 50%. The share of employees who had completed their secondary education, including vocational training, doubled between 1986 and 1999, rising to more than 37% in 1999. Although the share of labour with tertiary education remains low, it has been increasing rapidly. There has been a dramatic structural change in the employment structure over this period. However, employment in the food industries is still strongly biased towards lower educational attainments compared to employment in other manufacturing industries. The process is generally led by substitution through the recruitment of young people. The share of food-industry employees with an educational level in any year attained in the preceding three years, with 12-13 year secondary training, increased from 33% in 1986 to 64% in 1999, while the share of unskilled labour fell from 25% to 9% over the same period.[5]

Table 2. **Employment by highest level of educational attainment**

Food industries	1986	1990	1994	1999
Primary and low secondary (up to 10 years)	42.4%	36.0%	29.2%	21.6%
Upper secondary basic (11 years)	31.0%	31.6%	30.3%	27.8%
Upper secondary further (12-13 years)	19.2%	24.6%	31.1%	37.3%
Tertiary levels (14+ years)	4.3%	5.1%	6.6%	8.0%
Unspecified	3.1%	2.7%	2.8%	5.4%
Employment	55 028	51 378	52 297	52 698

Note: Primary and lower secondary education; Upper secondary education, including vocational training; Tertiary levels: college and university degrees, including post-grad education up to and including Master/Licentiate equivalents; PhD: doctoral degrees or equivalent.
Source: Statistics Norway Sysselsettingsfiler/Employment Registers 1986-99/STEP.

Complex knowledge bases and core competences

Various studies and indicators show that food industries have complex and changing knowledge bases. Increased "knowledge intensity" in employment is reflected in the complex, activity-specific knowledge bases underpinning agro-food production.[6] Our mapping of knowledge bases has been based on identifying relevant industrial activities and the competences underpinning them. We require these activities to be specific to the individual *industries*, *i.e.* excluding generic activities, such as management, finances, etc., as well as excluding firm-specific activities. Nine categories of industry-specific activities were identified in agro-food production. Four described different stages in the process of transforming the raw product, while two were product-related. The last three activities include activities in support of the actual products and production process.

Thus, a range of skills, scientific disciplines and competence areas are needed to carry out the functions and activities required in the food processing industries. Most of this knowledge can be categorised into two main knowledge areas. *Food science* is related to composition of food materials, and the behaviour of food stuffs under the conditions to which they are subjected. *Food technology* is the application of food science to the operational treatment of food stuffs to convert them into food products of the kind, quality and stability required, packaged and distributed to meet the need for safe, nutritious and attractive foods.

We have already noted the need for integrating analysis of competences and innovation across distinctions between technological, market and organisational capabilities. This approach is therefore a partial one. However, even with this technological, activity-specific approach, the mapping of core competences shows a complex picture of industry-wide competence bases. In conclusion, there is absolutely no doubt that food production is a highly knowledge-intensive production system.

Organisation and formalisation of innovation strategies

As noted above, the capability to mobilise internal resources is a basic requisite for successful innovation. Generally, we expect to see these mobilising capabilities reflected in the apprehension of innovation as competitive strategies of management and in the formalisation and organisation of innovation activities and planning in the firm.

We may broadly distinguish between two categories of innovators: innovators with formalised approaches to and strategies for innovation; and innovators that have an accidental or *ad hoc* approach to innovation. The level of formalisation may vary and the range of alternative models for formalisation is large. The actual model chosen by a formal, or systematic, innovator depends in principle on market and industry characteristics, and the nature of innovation opportunities. There appears, however, to be considerable variance in this; within an industry, several alternative models may be feasible and appropriate. A general result that may be drawn from the few studies which have attempted to map this aspect of company innovation activities (see, for example, Hauknes, 1998; Hauknes, Broch and Smith, 2000) is that the major difference in terms of innovation rates is between formal and informal innovators, rather than between different types of formalisation of innovation. The decisive aspect is not which type of innovation strategy a company chooses, but that it develops an innovation strategy at all.

In the food industries, unsurprisingly, the degree of formalisation of innovation activities within the organisation varies with company size. Large companies, both corporations and co-operatives, tend to formalise product development through active use of market departments, specific R&D departments, external and internal market analysts and external R&D and other competence suppliers. This is an indication not only of a formalisation of the innovation process; it is also a sign of a highly professionalised innovation process. At the other end of the scale, we find small companies with rather informal and more *ad hoc* innovation processes.

Wehave observed that innovation and the character of innovation styles must be seen as a function of companies' understanding of market characteristics and incentives. This understanding will, through the implied identification of "core competencies", be reflected in companies' internal organisation of capabilities and their use of external suppliers of knowledge, competences and information.

Innovation styles

R&D expenditures

Over the last decades, there has been a substantial increase in these industries' investments in R&D. For two decades up to the mid-1990s, R&D expenditures in Norwegian food industries increased almost ten-fold, albeit from a rather small base. Total R&D expenditures in the food industries in 1997 amounted to about 4.5% of total R&D expenditures in manufacturing industries.

The value-added-based R&D intensity, shown in Figure 1, was by the 1990s comparable with most other Nordic countries. The Netherlands, a major exporter of food stuffs, had an R&D intensity of over 2.5% relative to value added in these sectors, while another major agro-food exporter, Denmark, had an intensity below the Norwegian one. Compared to Canada, an economy that in terms of resource intensity has many similarities to Norway, the Norwegian R&D intensity in food production is almost three times as high.

Product-cycle-based arguments of innovation in food industries imply that R&D is dominantly development- and process-based, oriented towards incremental improvements in relevant process technologies. The national R&D surveys give a very different picture. Companies characterise nearly 75% of their R&D expenditures as having a product rather than a process orientation. According to these data, more resources are expended on basic research in food industries than in the chemicals and pharmaceuticals industries, while more than 30% of food-related R&D expenditures are characterised as (basic or applied) research, rather than as development costs. In terms of R&D activities, when classified according to technological areas, a mere 8% is oriented towards process technologies as ICT, materials, environmental and energy technology development. In particular, ICTs are marginal; the share of ICT-related R&D is less than 2%.

Figure 1. **Business-funded R&D expenditures as share of value added in the food industry (NACE 15+16), 1993**

Selected OECD countries, percentages

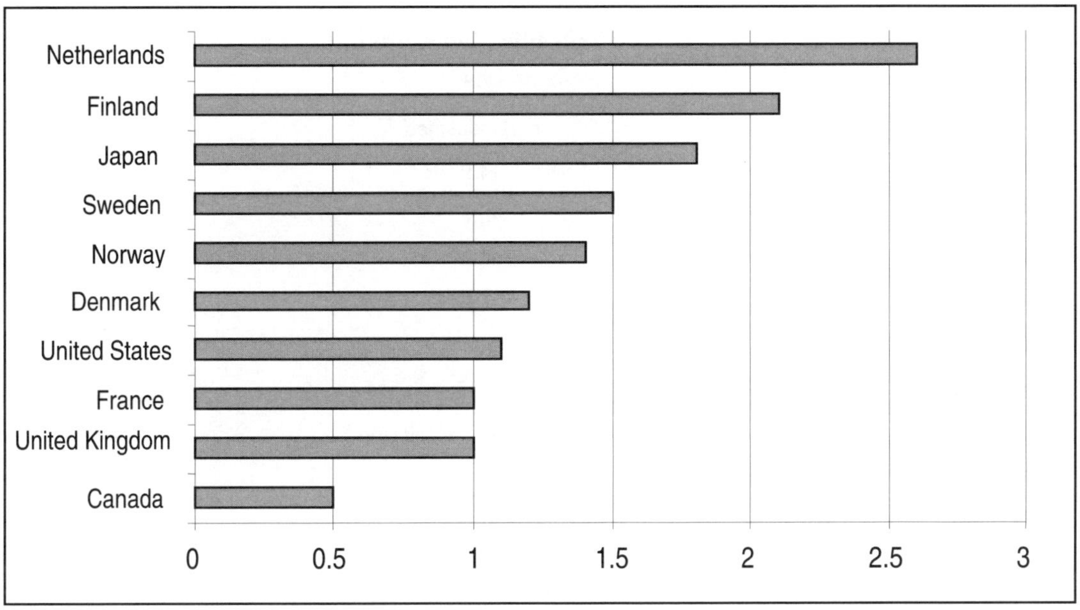

Source: Research Council of Norway (1999).

Thus, although R&D expenditures are not a dominant part of innovation expenditures in food industries, the increase in such expenditures may reflect more substantial shifts in the character of innovation activities and their focus. The structure of R&D performed in or acquired by these companies does not support a view of capital-embodied process-oriented innovation dynamics. However, the available statistics do not allow a mapping of the more detailed structure of these R&D activities and the underlying reasons for the increase in these expenditures.

With innovation rates high, R&D seems to be of lesser importance for food-related innovation than for other products; food companies have other sources of innovation. Still, it should be noted that R&D activities in food industries are a dominant feature of national industrial R&D performance.

Access to external repositories of technological knowledge

A complex knowledge system implies that access to and the capabilities to use internal and external repositories of knowledge are crucial. The national 1997 innovation survey, based on the CIS II Harmonised Questionnaire, distributes innovation costs in food industries and total manufacturing as shown in Table 3. Total innovation costs in food industries are estimated at some NOK 1.2 billion, about 11% of innovation costs in manufacturing industries. Costs related to access to external repositories of knowledge and other innovation inputs dominate innovation costs in the food industries, with more than 75% of innovation costs related to external R&D, acquisition of machinery and equipment and other external technology linked to the TPP innovation activities.

Acquisitions of machinery and equipment dominate TPP innovation costs for Norwegian food companies, suggesting traditional innovation styles based on adaptive innovation on the basis of capital-embodied technical change imported by food companies from technology-intensive machinery sectors.

Table 3. **Innovation costs, 1997**

Percentages

	Food industries	Manufacturing
Internal R&D	14.9	30.6
External R&D	4.3	14.3
Machinery and equipment	66.6	40.6
Other external technology	6.5	4.1
Industrial design	3.1	3.6
Competence development	2.1	3.5
Market introduction	2.7	3.5

Source: SSB/STEP 1998.

Orientation of innovation and innovation objectives

However, it is clear that these and other data do not characterise the nature of the links between food-producing companies and their machinery suppliers. The goals of innovation activities and innovation information sources indicated by the food companies may provide an indirect characterisation. If innovation activities were dominated by an adaptive style based on imports of more or less pre-designed production technology, we would expect to see the objectives of innovation in food companies dominated by process-oriented and cost-reducing innovation. The three major objectives listed by the food companies for their innovation activities are, in decreasing order:

- Product quality improvements.
- Reducing production costs.
- Opening new markets or increasing market shares.

These objectives are characterised as being very important by at least half of the innovators in the food industries. Unsurprisingly, when ranking information sources for innovation by importance, both product and supplier market nexi rank high on the list. The three most important information sources for innovation in food companies are, again in decreasing order:

- Other establishments or enterprises within the company or corporation.
- Customers and clients.
- Suppliers of equipment, material, components or computer systems.

Acquisition of information from sources within the same corporation dominates the picture. Most other sources of information score low on this measure. Less than one in ten innovators point to alternative sources as being important, the only exception being competitors. In particular, supporting organisations, such as universities and technology extension services, as well as consultants and public information, are rated weakly. Innovation is evidently strongly tied to the business environment of food companies.

Technological collaboration and networks

Food companies collaborate extensively. More than half of the innovating companies take part in some form of collaboration on innovation, with active participation by all the parties concerned. Collaborating companies are, on average, involved with more than three organisations and institutions, mainly on the domestic market. The distribution of these collaborations across types of partners in innovating food companies is compared to other manufacturing companies in Table 4.

Table 4. **Innovation collaboration in manufacturing industries**

Share of collaborating companies, percentages

Partner type	Food industries	Other manufacturing industries
Suppliers of equipment, computer systems, etc.	63.3	56.3
Other enterprises within corporate group	50.6	48.0
Clients and customers	40.5	59.5
Industrial RTD institutes	39.2	43.5
Universities and other HEIs	30.4	38.2
Consultancies	25.3	32.4
Competitors	16.5	17.3
Share of innovators with collaboration	**51.6**	**52.2**

The distribution reproduces features that should be well known by now. We note, however, that in terms of active collaboration, the roles of the national system of technological and industrial research institutes, higher education institutions (HEIs), and consultants have become more prominent. One in three innovating food companies collaborates with RTD institutes and HEIs, respectively. Note that this measure by construction excludes ordinary outsourcing of R&D services, as such links do not involve active participation by the collaborating company.

These data suggest that there are strong absorption capabilities for knowledge and competences from suppliers of scientific inputs and of machinery, as well as from other extramural sources of TPP-related competences in food companies.

Interaction with public and private knowledge infrastructures

Interactions with a range of knowledge and technology suppliers contribute to the shaping of innovation activities in food companies. Mapping the role of such suppliers, food industries' use of such suppliers and the structure of the interface between the suppliers and clients in the food industries, will aid our understanding of innovation styles in the food cluster. Here, we will briefly identify three categories of these suppliers' supply of:

- Machinery and equipment.
- R&D and other "knowledge products" (including RTOs and KIBS).

- Services and competences related to consumer relations and market research (including consulting companies supplying consumer information, such as suppliers of trend analysis and market surveillance).

Today, these knowledge suppliers are all frequently used by companies in the food industries to support company and product development, as well as innovation more generally. While suppliers of machinery remain the most frequent supplier, increased use of the two latter types of suppliers can be noted. We argue here that food industries and markets for food products are increasingly dependent on the use of both technological and market-related intangible assets, a shift that is expressed in the changing location of the "hot spots" alluded to above. Our analysis of innovation indicators has shown an increased use of R&D and R&D-based inputs in these industries, a process that is linked to the expansion of intramural capabilities for R&D and to frequent use of external R&D suppliers. The increasing dependence on market and consumer-interfacing competences is well illustrated by the recent developments at the national R&D institute for food research. The institute was established to serve industry needs, primarily on the basis of contract R&D with industry partners or of publicly funded long-term R&D aimed at building its capabilities to serve these needs. Until recently, the institute's R&D focused almost exclusively on food science and technology. On the basis of its close ties with the major companies and relevant industrial organisations, the institute built up a major R&D activity in market research and analysis in the late 1990s; "contributing to market-oriented product development" is now ranked first among the institute's four main research strategies.

Supply of machinery and equipment

Food industries are often argued to fall into the Pavitt class of industries where supplier-dominated innovation strategies predominate.[7] The relationship between suppliers and users is interactive; development of machinery is a collaborative effort between food companies and machinery suppliers. Supplier industries of machinery have grown out of food production as a dynamic process of specialisation and evolving divisions of labour.

Three major suppliers to agro-food industries illustrate this. These suppliers are organised with co-operative ownership within the industry. The Norwegian Agricultural Purchasing and Marketing Co-operative organises processing and trade in feed concentrates, fertilisers, machinery, and so on. L.O.G. – the horticultural seed and equipment association – sells related inputs and equipment to growers. Landteknikk supplies processing equipment, packing materials, machinery and other operating inputs to the food-processing industry.

The food industries are substantial users of advanced technologies. In the EIMS survey of innovation in food and beverages, Christensen, Rama and von Tunzelmann (1996) suggest that the scale of these industries implies that they are major markets for the specialised suppliers of capital equipment. The "size of this industry implies that many of its firms will be especially responsible for making use of innovations developed in other technologically more advanced industries....". Moreover, "(t)here is a high degree of dependence [in the food industries] on those developments in high-tech areas, like information technology, biotechnology and advanced materials".

R&D supply

Although the level of R&D expenditures in food companies was characterised as low, we noted a rapid growth in R&D expenditures in food industries, accompanied by an increase in the use of extra-mural supply of R&D and related services. This is reflected in a substantial web of R&D-performing supply institutions, partly organised as public institutions, partly as independent contract research organisations and partly owned by industry associations and co-operatives. Today, there is a substantial and far-reaching R&D sector oriented towards food industries (Table 5).

Table 5. **Norwegian R&D supply to agro-food industries**

Institutional sector	Institutions of higher education and research	Contract research institutes, etc.	Industrial R&D organisations
Major institutions	Norwegian Veterinary College (NVH) Agricultural University of Norway (NLH)	Matforsk (food research) Jordforsk (soil and environmental research) Planteforsk (crop research) Kontrollinstituttet for meieriprodukter (dairy) Norwegian Agricultural Economics Research Institute	TINE Norske Meierier (dairy products) Various national breeding associations Felleskjøpets fòrutvikling (grain fodder) Potetindustriens Laboratorium (potatoes) Norsk Kjøtt (meat and meat products)
Major funding pattern	Government Public R&D funds	Tax/fiscal Public R&D funds Industry	Industry
Ownership	Public	Mainly autonomous foundations	Industry

Market orientation and consumer relations

Innovation in the food industry is consumer-oriented, with consumer demand a core driver of innovation in the food industries today. A central characteristic is adaptation to increasingly differentiated consumer demands, and a strongly expressed need for reliable systems for informing consumers about the quality, safety and origin of products.

As in many consumer markets, branding, label building, etc., has become a major tool for these industries, with amplified use of market research and analysis, design, etc., aimed at mapping and influencing trends, needs and tastes and their development. We must understand these industries as strongly market-oriented sectors, the food industries today interact extensively with market patterns, tastes, changes in consumption, new sociological and economical settings, etc.

The main driving force in food industrial innovation is no longer the supply of machinery and equipment. Today, innovation styles are thoroughly integrated into these industries' market orientation. Customer needs and consumer trends are increasingly exerting a strong influence on the operations of the food industries and innovation. We agree with the conclusions of the EIMS survey of innovation in food and beverages:

"[The] "supplier-dominated" label is no longer adequate [for these industries]. Firms in this industry assess product innovation as being as important as process innovations in their goal of innovation, and see market developments as more important than either. Clients and customers – not suppliers – are regarded as the most important single source of information leading to innovation." (Christensen, Rama and von Tunzelmann, 1996)

However, this conclusion needs to be sharpened. The food industries are *demand-oriented*, and knowledge inputs about markets and trends are central elements in the innovation modes of these industries. The marketing director of the Norwegian subsidiary of a globally renowned multinational illustrated this focus, describing their role as follows; "we are the local eyes and ears for the corporation".

This market orientation applies to all producers. Multinationals face specific and differentiated needs and demand patterns in their host countries, and must integrate these specificities into their activities, locally in the host country and generally in the home country. While domestic producers are faced with the same challenges, they also face increased import-based competition and intensified branding and differentiation strategies put in place by competitors.

Conclusions

The Norwegian food production system is the largest economic cluster in the country, with its core in the food industries. The size of these activities makes the mode of organisation and economic performance of vital importance to the national economy as a whole, its future growth, as well as to regional employment and innovation. There are two main conclusion of this chapter. First, food industries form vibrant parts of the present innovation economy, with complex innovation dynamics and substantial scope for innovation performance. Second, cluster-specific innovation styles in food production have changed dramatically, the market link of innovation is strong and is seen as the fundamental driver for innovation patterns in the food industries. The dominant position of these industries in the national economies of the industrialised world alone suffices to highlight the importance of these conclusions for overall economic and innovation performance.

Undue neglect of low-tech industries in growth policies and innovation studies has led to a misguided view of their innovation styles and performance. The food industries, one of the oldest manufacturing industries, are clearly "traditional". However, there is no evidence to show that they are "mature" industries. However, to map innovation styles in food industries requires us reconsider our approach to innovation in these sectors and to develop ways of describing innovation that avoid the distinction between "technological" characteristics, on the one hand, and "subjective" market characteristics of products and processes, on the other.

Innovation styles in food industries are today dominantly oriented towards the market interface and customer-producer links. We argue that there has been a radical shift in the innovation approach of these industries over the last decades. This process reflects the fact that food producers are responding to and shaping customers' enhanced focus on food quality, safety and health, as well as general changes in the overall economic and innovation frameworks of these industries.

Three sets of policy conclusions follow from this. First, innovation policies must recognise the market and innovation dynamics of sectors like agro-food production. This concerns both the consumer orientation and the requisites for successful innovation that follow from this, and the importance of these large sectors of national economies as the generating locus of major parts of social utilisation and benefits from enabling technologies.

Second, major challenges in several regions arise from industrial restructuring and the related structural unemployment. A general reply to these challenges is to turn to sweeping substitution strategies based on new services or other technology-intensive activities. In many areas, we firmly believe that an in-depth understanding of innovation and the related potential for regional growth in food industries should be regarded as an option.

Lastly, the traditional organisation of agro-food sectors based on co-operative ownership and a strong regulatory role of co-operatives will increasingly come under scrutiny. Vertical integration and corporatisation of co-operatives, as well as internationalisation of food markets, will all put strains on regulatory frameworks and production systems where co-operatives dominate. The differentiation of food demand, the enhanced share of processing and related activities will also contribute to this. With co-operatives expanding into more competitive downstream activities, their position as market regulators, on one hand, and their supply obligations, on the other, may come into conflict, opening up the potential for anti-competitive behaviour. The multi-functionality rationale of national policies for agriculture and downstream agro-food production will no longer hold up, with multi-functionality becoming restricted to agriculture proper. Alternatively, it may force a divestiture of vertically integrated co-operatives.

Further opening up of international trade, and a "nordicisation" of food markets may have implications for organisation and ownership of production. With increased ownership and production at Nordic level, co-operatives will go far beyond their national role. If this occurs, it is difficult to see their national market regulating role being upheld, even if co-operatives pull out of actual production and revert to playing a role mainly as ownership-based associations. With the dominant position of co-operative production structures today, these are important challenges for industrial policy.

Annex

NORWEGIAN AGRO-FOOD PRODUCTION: BROAD STRUCTURE OF SELECTED NORWEGIAN AGRO-FOOD INDUSTRIES

	Corporate structure	Markets: consumer trade, institutional households, private brands, intermediate consumption	Processes of change
Meat and meat products (cattle and pigs)	Co-operative Gilde Norsk Kjøtt, of formally independent producers dominating Increasing competition from two private groups of meat producers	Domestic Import competition from differentiated products	Co-operative faces increased competition from domestic suppliers Retail chains negative towards co-operative Corporate integration of Gilde Norsk Kjøtt Branding important counter-force for co-operative Building long-term contracts to retail chains important for independent producers Labelling strategies Food safety
Dairy products	Co-operative TINE Norske Meierier dominates as quasi-monopoly A few rapidly growing private competitors, especially in processed dairy products	Mainly domestic Strong brand/product positions Export of few internationally branded products Competition from imported differentiated products	Decreasing domestic consumption with over-capacity Former regulated monopoly of TINE on consumer milk. De-monopolisation started in 1998 Corporate integration of TINE Substantial restructuring, increasing plant size Disentangling of price regulation TINE as market regulator under pressure New products and product differentiation – new demands and food quality Eco-food segment small but growing

	Corporate structure	Markets: consumer trade, institutional households, private brands, intermediate consumption	Processes of change
Eggs and poultry	Prior Norske Eggsentraler dominates. Prior owns four major subsidiary processing companies	Mainly domestic Price-sensitive markets, esp. catering Strongly regulated imports/exports	Corporatisation of Prior expansion into processing Food safety Vulnerability to chicken diseases (salmonella, Newcastle, avian influenza and fowl/human crossover) White meat is the fastest increasing meat segment New products and product differentiation – new demands and food quality Growing involvement of the Nordic countries in strategic alliances and markets
Fruit and vegetables	Co-operative Gartnerhallen/BAMA Distribution of fresh products mainly through three major wholesale distributors Dominant retail-based ownership of distribution	Mainly domestic Intermediate products to beverages and other food products Substantial imports, but substitutability low	Rapidly increasing role of retail-based distribution, due to chain formation Trade regulation from quotas to decreasing tariffs (GATT) Increasing import-based competition
Grains, cereals and milled products	Imports and distribution of grains formerly state monopoly. De-regulated 1995 (GATT). STATKORN (=CERMAQ) established as a plc (ASA) Two multi-plant producers, one Swedish owned, one partly owned by STATKORN	Mainly domestic; intermediate inputs and consumer Substantial import competition on cereals markets	Grain Act abolished 1995. Millers shifted from renting production capacity to grain owners to own production based on new grain markets Monopoly on grain imports abolished Major reorganisations of plant ownership and structure Increased efforts to capture cereals markets
Beverages	Two main brewing companies, both owned by Scandinavian MNCs Some small, mainly local breweries	Beer mostly national brands Domestic production, increasingly Nordic/Baltic market approach Marginal imports Soft drinks mostly international brands produced on licence	Rapid structural change, new forms of ownership, etc. Ban on domestic advertising for alcoholic beverages Ownership and markets are becoming increasingly Nordic
Other food products	Dominated by large corporations. Subsidiaries of multinationals.	Domestic, increasingly Nordic and international	Branding important for maintaining consumer interest and thereby maintaining relations with retail stores Nordic countries increasingly becoming a single market?

<table>
<tr><th></th><th>Corporate structure</th><th>Markets: consumer trade, institutional households, private brands, intermediate consumption</th><th>Processes of change</th></tr>
<tr><td>General</td><td>---</td><td>Mainly domestic, but slowly increasing international trade

Regulated upstream markets, competitive downstream markets

Co-operatives as market regulators

Co-operatives with supply obligation to downstream producers

Vertically integrated retail chains generally and their private brands, dominant development

“Internationalisation” of demand

Concerns for food safety and health</td><td>Multi-functionality of agriculture and agro-food under scrutiny?

Increased willingness to pay for food quality

Millennium Round of WTO and the Built-in Agenda

CAP and EU agro-food policies

High prices on Norwegian markets relative to its trading partners – increased policy and consumer focus</td></tr>
</table>

NOTES

1. A fuller discussion of these issues can be found in the paper on which this sections draws (Hauknes, 2000a).

2. On these and related issues, see Bilal and Pezaros (2000).

3. Godt Norsk (Good Norwegian) is a foundation initiated by the Minister of Agriculture in 1994 as part of the public programme for the restructuring of domestic agro-food production. The system covers the whole chain "from earth to table" ("fra jord til bord"), equivalent to the UK "farm to fork" approach; the slogan coming into use around 1990 to denote the enhanced value chain and consumer approach of national agro-food industrial policies. This reorientation of national policies provides a pertinent illustration of the shift in innovation dynamics and industry understanding presented in this chapter.

4. The main results of this field work are reported separately in Braadland (work in progress).

5. Case study information confirms this; the manager of a medium-sized food company in the dairy industry told us that they "require completed vocational training today. The days when young people could hope for a job here directly from lower secondary are gone".

6. The work summarised here is more fully reported in Knudsen, Isaksen and Smith (1999).

7. Note, however, that in Pavitt's original analysis (Pavitt, 1984), the food industries were not included among the supplier-dominated sectors. Pavitt's analysis classified them as scale-intensive industries, a classification that probably continues to describe the dynamics of technological change in many parts of this sector. The characterisation given here goes far beyond this.

REFERENCES

Bilal, S. and P. Pezaros (2000), *Negotiating the Future of Agricultural Policies: Agricultural Trade and the Millenium WTO Round*, Kluwer Law International, The Hague.

Braadland, T.E. (work in progress), *Innovation in the Norwegian Food System*, to be published in the STEP Working Papers series.

Christensen, J., R. Rama and N. von Tunzelmann (1996), *Innovation in the European Food Products and Beverages Industry,* European Innovation Monitoring System Publication 35, IKE Group, Aalborg University, Aalborg.

Hauknes, J. (1998), *Services in Innovation – Innovation in Services: SI4S Final Report*, SI4S Synthesis Report 1, STEP Group, Oslo.

Hauknes, J. (1999), *Norwegian Input-Output Clusters and Innovation Patterns* in OECD (1999).

Hauknes, J. (2000a), *A Service-based Approach to Innovation: A Need for New Concepts?*, paper presented at the Conference on the Economics and Socio-economics of Services: International Perspectives, organised by CNRS/IFRESI and the Université de Lille, STEP Group, Oslo.

Hauknes, J. (2000b), "Dynamical Innovation Systems: What is the Role of Services?", in M. Boden and I. Miles (eds.), *Services and the Knowledge-based Economy*, Continuum, London.

Hauknes, J., M. Broch and K. Smith (2000), "SND og bedriftsutvikling – rolle, virkemidler og effekter" [SND and Firm Development – Role, Instruments and Impact], STEP Report R-04/2000, STEP Group, Oslo.

Knudsen, T., A. Isaksen and K. Smith (1999), "Innovation and Knowledge Use in the Norwegian Food Industry", in O. Borch and E.P. Stræte (eds.), *Matvareindustrien – mellom økonomi og politikk* [The Food Industry: Between Business and Politics], Tano Aschehoug, Oslo.

Pavitt, K. (1984), "Sectoral Patterns of Technical Change: Towards a Taxonomy and a Theory", *Research Policy* 13, pp. 343-373.

OECD (1999), *Boosting Innovation: The Cluster Approach*, OECD, Paris.

OECD/EUROSTAT (1997), *Proposed Guidelines for Collecting and Interpreting Technological Innovation Data – Oslo Manual*, OECD/EUROSTAT, Paris/Luxembourg.

Research Council of Norway (1999), *Det norske forsknings- og innovasjonssystemet - statistikk og indikatorer 1999* [The Norwegian Research System – Statistics and Indicators 1999], Norges forskningsråd, Oslo.

Chapter 9

THE CONSTRUCTION CLUSTER IN DENMARK

by

Michael S. Dahl and Bent Dalum[*]
DRUID/IKE-Group, Department of Business Studies, Aalborg University

Introduction

During the 1990s, the concept of a construction mega-cluster (C-MC), was developed and implemented in the production of official statistical data by the Danish Agency for Trade and Industry (EFS) and Statistics Denmark. The C-MC is defined in the following section, and its relative size *vis-à-vis* a series of mega-clusters in the Danish economy is shown in Figure 2. The present analysis is based on the mega-cluster concepts and borderlines developed by the EFS and Statistics Denmark in various analyses since 1993. In accordance with the EFS definition, the C-MC is focused on construction of *buildings* and the entire set of auxiliary industries (building materials, engineering consultancy, etc.). The definition does not include construction of major physical infrastructure facilities, such as roads, bridges, tunnels and airports.

The point of departure of the analysis is that the C-MC thus defined plays an important role in Danish business activities, accounting for 25% of employment but only 12% of exports. The core building sector is labour-intensive and is basically oriented towards the domestic market. For more than 30 years, the productivity performance of the building sector has been considered rather poor. Although in terms of such dimensions as design, interior climate and (low) energy consumption, the Danish building sector is generally considered to be first-rate, overall productivity development indicates a mega-cluster suffering from weak long-term economic performance. This has been related to problems with the finish and quality of new buildings. The mediocre productivity development has, however, been somehow "hidden" by the economic upswing of the 1990s. The building sector flourished, as it did in most of the OECD countries, during this unusually long upswing. The aim of the present analysis is to go beyond this short-term development to focus on some of the more basic structural problems which will become more visible when the general business cycle returns to a more moderate level.

Among the OECD countries, the relative size of construction, as conventionally defined in the national accounts, varies substantially. Denmark is ranked ninth, with a construction share of 6.9% of total value added in the business sector (it ranked seventh in 1995).[1] The Swedish and Finnish shares

* The authors would like to thank Mikkel Andreas Thomasson of the Copenhagen Business School for production of Table 7, and Anker Lund Vinding of the DRUID/IKE-Group for the data presented in Table 5.

are at the same level, while Norway's share is rather lower, at 4.7%. However, the presence of a large oil and gas sector substantially upsets the Norwegian picture.[2] Countries with higher shares than Denmark in construction are Canada, Japan, Korea, Spain and Greece. No simple pattern appears from these observations. The northern OECD countries – Canada and the Nordic countries, as well as Austria – all countries with hard winters, have fairly large construction sectors, which is also the case for Spain and Greece with lower per capita incomes and a milder climate. In 1985, the construction ranking was topped by Finland at 9.5%, followed by Sweden, Korea and Japan with around 8.5%. Three out of these four countries experienced very rapid industrial growth in the 1980s. In 1995, Korea ranked first at 15.5%, followed by Japan at 11.0%, and by Spain and Austria with around 9%-9.5%. Countries still in the catching-up phase of industrial growth thus typically have a high construction share, as is the case for the rich countries in the (cold) North.

The construction sector, therefore, appears to be rather nation-specific in terms of size as well as characteristics. In general, the sub-industries do not experience serious international competition. They are oriented towards domestic, if not local, market conditions. The sub-industries are often characterised by a dual structure with a certain amount of concentration, on the one hand, especially in building materials and integrative construction engineering activities, and a large amount of small craft-oriented firms, on the other. Public regulation plays a key role at all levels, ranging from the level of construction activity seen from a macroeconomic point of view to detailed specifications and control of building standards in private homes.

These features naturally call for a broader analytical approach than the standard IO structure-conduct-performance analysis from the discipline of industrial organisation.[3] The mere fact that the determinants of construction sector output consist of a complicated web of economic as well as institutional and/or political factors calls for a broader approach. The cluster approach, as developed in previous chapters, is an appropriate framework which can serve to widen the scope and integrate a broader set of entities – such as related and supporting industries, public sector bodies, knowledge institutions – to enable a better understanding of the salient features relating to the development of construction.

The chapter is divided into six sections. The following section provides a practical definition of the cluster, while the third gives a brief historical overview. The fourth section contains the bulk of the empirical analysis of the Danish Construction mega-cluster, while the fifth contains a description of innovation and R&D. The final section brings the conclusions and policy perspectives.

Identifying and defining the C-MC

The definition of the Construction mega-cluster, C-MC, includes all activities where the end product is a *building*. The C-MC is thus defined as *activities which together contribute to the construction, maintenance, management and demolition of buildings – no matter what the use of the buildings*. As mentioned above, construction activities related to the broad physical infrastructure are not included in this definition,[4] although the infrastructure activities are conventionally considered to be part of the construction sector in, for example, national account definitions. The present analysis has followed the (Danish) convention developed by the EFS with a focus on buildings.[5]

The tradition adopted in Denmark during the 1990s has been to analyse business sectors more broadly, using a cluster-based approach.[6] The basic idea is to include important interactions which may not be captured by traditional industrial classification schemes. The explicit goal is to promote discussions of policy implications at the cluster level. Recently, the EFS launched a second round of cluster studies. A so-called "benchmarking" report has been published which provides an overview of the basic statistical trends of the clusters.[7] Both rounds of cluster studies have included a report on the

C-MC. In this process, the entire Danish economy has been divided into mega-clusters. The C-MC has thus been defined as group of statistical categories (four-digit NACE codes). The "mega" level indicates that it may be relevant also to focus upon "small-scale clusters".[8] Obviously, such a method of defining clusters has some disadvantages. Industries such as furniture and household appliances have been important for the development of the C-MC, as has bridge building. However, in the EFS classification, furniture is part of a mega-cluster of its own, the furniture/textiles mega-cluster; while household appliances has been left in a residual group of "other businesses" in the EFS classification.

The present study has followed the EFS borderlines for the C-MC, which have become standard in, for example, publications of various annual statistical indicators. The EFS analysis (2000b) presents three levels. The cluster statistics are divided into four main economic sectors. These are related to five analytical areas[9] and a series of sub-areas, characterised by different groups of actors (Table 1).

Table 1. **Main categories of the construction mega-cluster**

Main sector	Analytical areas	Sub-areas
Primary sector	Industrial area	Raw materials
Manufacturing sector		Manufacture of building materials
Service sector		Wholesale and retail of building materials
	Building area	Contractors
		Craftsmen
		Consultants
	Building owner/awarding authority	Building owner/awarding authority
		Building managers
	Supporting activities	Other services
Supporting sectors		Supporting activities
	End-users	

Note: The main sectors are defined by the NACE codes presented in Annex.
Source: EFS (2000b).

The *industrial* area does not basically differ from other manufacturing-related activities. The *building* area includes consultants (architects, engineers and project consultants) as well as contractors and craftsmen. It differs from the industrial area in that the activities do not have a permanent building or production site. Distances to temporary building sites are usually an important factor, as a significant share of the building materials is hard to transport.

The third area includes the *building owner/awarding authority* and *building managers*. The building owner/awarding authority is the person or firm who pays and bears the risk of the project. These actors are not the direct users of the building; they sell or rent square metres. The *supporting activities* include knowledge institutions and a variety of supporting functions in the building process, such as rental and leasing of machinery and equipment. Other kinds of supporting activities are services, which deliver products directly to the building owner/awarding authority and the end-users, such as real estate agencies and mortgage institutions. The final area consists of *end-users* in terms of the owners or tenants of the buildings.

Main features of the history of the C-MC

The economic activity of the C-MC is characterised as sensitive to changes in the business cycle as well as rather labour-intensive with low export and import ratios. Construction sector activities have often been used as a fiscal policy instrument to stimulate the entire economy through such measures as public investments in major building projects. Among other policy measures widely used in Denmark since 1973 are tax incentives to stimulate the renovation of old buildings or to encourage energy saving in houses (insulation and energy-saving technologies in domestic electrical appliances, etc.).

Up until the 1950s, skilled craftsmen dominated the sector. There were only a few large contracting firms, primarily in infrastructure construction. In most OECD countries until the late 1950s, the building industry was mainly oriented towards the home market. However, the widely used international calls for tender allowing foreign architects to submit projects for major buildings were an important diffusion mechanism for new fashions in design as well as new building technologies. Since the 1950s, construction has been influenced by the general trend towards internationalisation which has swept through all industries in the OECD countries; however, at the beginning of the new Millennium, it remains oriented towards domestic markets. In the Danish context, however, during the 1930s, a group of contractors in bridge construction (which as noted previously is not part of the C-MC), emerged as internationally oriented firms with significant competitive strength. The strength of the group was based on the internationally recognised reputation of the construction departments at the Technical University in Copenhagen.

In the 1950s and 1960s, there was an extensive demand for housing in Denmark due to the very rapid increase in per capita incomes resulting from the deliberate efforts to create a Nordic welfare state. In this context, good-quality housing for ordinary wage earners was given very high priority. Meeting this demand called for the industrialisation of the construction sector. This led the government to set up new regulations promoting this process. Another goal was to increase the amount of unskilled labour in the building process. At first sight, this appears to be self-contradictory. The skilled craftsmen in construction, such as carpenters, bricklayers and blacksmiths (manufacturing water supply and heating systems) enjoyed – and continue to enjoy – a good reputation. However, the rapid "industrialisation" of the building process called for a certain amount of "Taylorisation". Given the general rapid economic growth and lack of labour, it was considered a necessity to employ far greater numbers of unskilled labourers.

By subsidising both private and publicly funded housing projects and specifying that this should include no more than a 15% share of skilled labour, production moved from the building sites to manufacturing plants or factories. This movement of the physical activities was necessary in order to receive the subsidy. In this way, the Danish Government succeeded in promoting the industrialisation of the sector. Ever since, the cluster has been highly influenced by regulative legislation.

This industrialisation process has been a key factor in the internationalisation of the cluster in recent decades. Current exports from the industrial part of the C-MC amount to around DKK 40 billion (see below). Until the beginning of the 1970s, the C-MC was characterised by radical innovations based on industrialisation. Among the results was the emergence of industries producing building elements (such as concrete panels) and advanced building materials, of which some later found their way to the international markets. However, in the 1970s, maintenance problems in the large-scale concrete apartment houses emerged in most countries. There were serious problems with the quality of the buildings, and in Denmark the strategy was to move towards one-storey houses and away from the unattractive – often dull and grey – concrete apartment houses of the past.

During this period, activity in the C-MC slowed down as a result of the two oil crises. These recessions had a direct influence on the direction of technological development. New policy measures promoted innovation of new isolation and energy-saving methods. This was the beginning of a series of policy regulations on energy consumption in buildings which led to considerable growth in business activity. In particular, the production of insulation materials benefited from this development; this is the main reason why this industry has such a strong international position today.[10] The declining trend in demand from 1973 is illustrated in Figure 1.

Figure 1. **Investments in construction in Denmark, 1954-95**
DKK billions, in 1980 prices

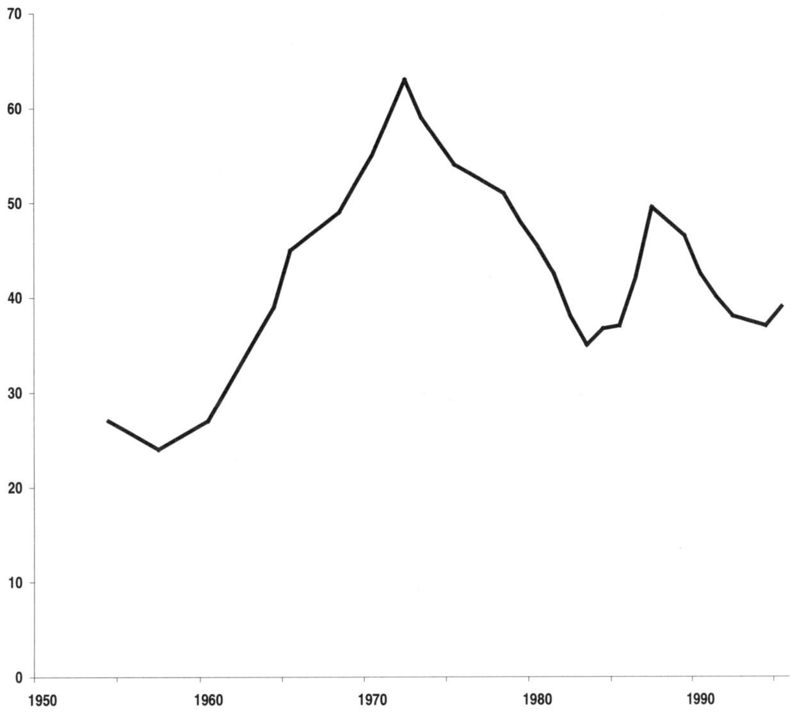

Source: EFS (2000b).

As of 1980, combating inflation, mainly through drastic increases in interest rates, was given high priority in economic policy across the OECD – with clearly negative effects for the interest-sensitive building sector. The Danish upswing in the mid-1980s was cut off by the government as early as 1986 to prevent overheating of the economy. Construction activity, as well as general economic growth, remained subdued until the next upswing in 1993-94. The foot on the brake during 1987-93 was the outcome of severe balance payments problems in 1985-86, and illustrates how demand conditions have to a significant extent been determined outside the cluster itself.

Two other factors also played an important role. The first is the *specific structure of finance* of houses and flats through dedicated mortgage institutions. Typically, the debt of a house stays with it, even after it has been traded. There is very little evaluation of the individual person's credibility when borrowing money for a house. Only during the 1990s have mortgage institutions introduced the practice of evaluating the creditworthiness of house buyers. It remains the exception that a loan does not stay on the house after it has been traded; the new buyer simply takes over the existing loan on the house. Therefore, the house may be just as important as the characteristics of its owner.

Secondly, the fact that a large share of the population are house-owners has been closely related to the development of the furniture and domestic appliance industries. In Denmark, a large market for *Scandinavian Design* in such areas as furniture, domestic appliances, lighting and fittings, etc., has existed for decades. Design, taste and fashion are important factors for competitiveness in a wide variety of "equipment" for buildings. The majority of consumers are willing to pay high prices for the "right" bathroom, kitchen (including domestic electric appliances), furniture, etc., for their homes. A certain "design" culture pervades these fields, of which some are not conventionally classified as belonging to construction.

Mapping the Danish C-MC

The C-MC is mapped according to five dimensions: *i)* relative size compared to other mega-clusters; *ii)* the internal segments; *iii)* export performance and regional distribution of employment; *iv)* labour productivity; and *v)* the system of public subsidies for housing.

Relative size

The relative size of the Danish mega-clusters, as defined by EFS, is shown in Figure 2. The C-MC is the largest employer, with almost 25% of private sector employment and around 12% of total exports of goods and services. The biggest exporter is agro-food, with more than 25%, although this cluster has an employment share considerably lower than that for the C-MC. Compared with these two big clusters, representing mature and rather slow-growing economic activities, the ICT mega-cluster amounts to around 13% of employment and 12% of exports. In terms of employment, some 40% of the entire business sector is concentrated in two mega-clusters, construction and food, generally characterised by low-income elasticities and, thus, a rather low growth potential. This structure represents a major challenge for the Danish business sector as a whole.

Segments

In 1996, there were approximately 50 000 firms in the C-MC, employing 225 000 full-time workers. The average firm size is some 4.5 employees. The different segments are shown in Table 2.

More than 3 000 firms employ a mere 1 320 persons in the production of *raw materials*; this segment has come under pressure in recent years due to the increased attention being paid to recycling and environmental issues. An increasing share of building materials is recycled in the construction of new buildings. Manufacture of *building materials* is strongly internationalised. The original foundation for this internationalisation was the requirements for contractors and craftsmen. However, since the industrialisation of the C-MC in the late 1950s, exports from this segment have developed independently of the other parts of the cluster, and this is the only segment in which exports have increased rapidly. Foreign companies have acquired large shares in the firms.

Nearly 4 000 firms employing 27 000 workers are involved in *wholesale and retail* of building materials. These firms hold key competencies related to the logistics and planning of building projects. Finally, the *building* segment consists of 33 840 firms, employing some 110 000 full-time personnel. This segment is primarily composed of more than 28 000 small firms, with a total of almost 80 000 employees, most of which are craftsmen.

Figure 2. **Employment and export shares for the Danish mega-clusters in total private employment, 1997**

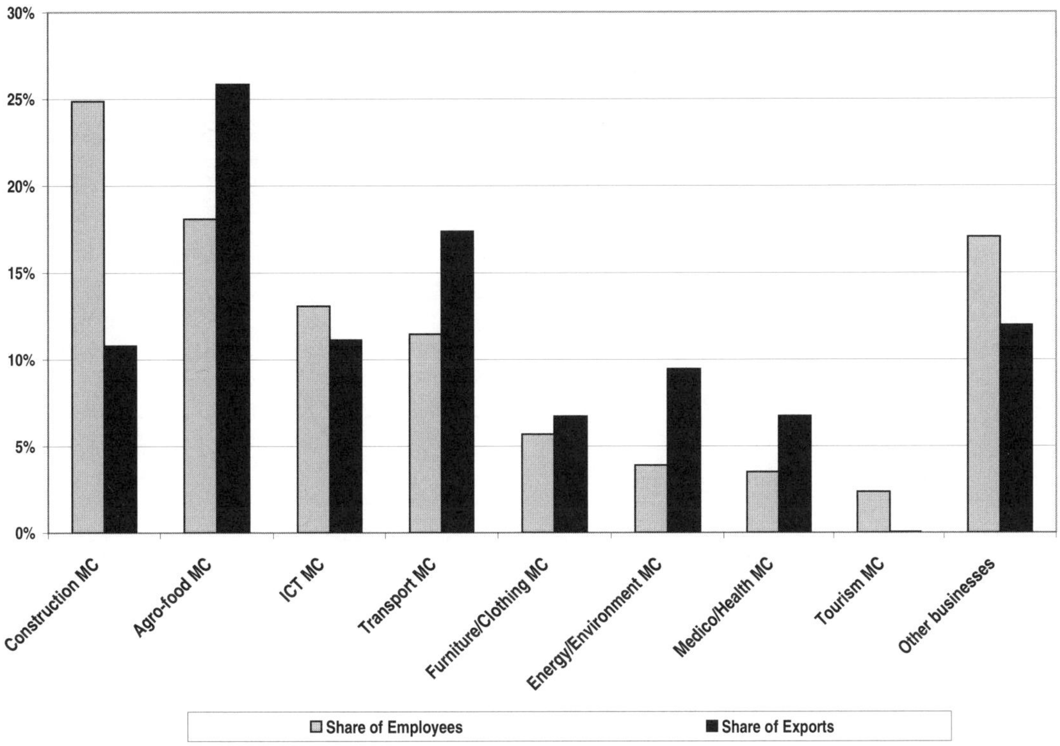

Note: Data on employment covers full-time employees. Exports include trade in goods and services.
Source: Statistics Denmark (The Danish Mega Cluster Statistics).

Table 2. **Main segments of the Danish C-MC, 1996**

	Number of firms	Total turnover (DKK millions)	Exports (DKK millions)	Number of full-time employees
Raw materials	3 156	1 766	364	1 321
Manufacturing of building materials	3 744	(68 789)	(25 697)	74 898
Wholesale and retail of building materials	3 874	(69 133)	(6 724)	27 084
Total industrial segment	10 774	(139 688)	(32 785)	103 303
Contractors	2 336	23 207	1 229	1 9279
Craftsmen	28 128	44 078	1 485	7 9154
Consultants	3 376	11 702	2 983	12 417
Total building segment	33 840	78 987	5 697	11 0850
Building owner/awarding authority	2 130	3 117	24	4 059
Supporting sectors	582	4 159	1 970	3 261
Total construction MC	**47 326**	**(225 951)**	**(40 476)**	**221 473**

Note: It has not been possible to completely consolidate the values in turnover and exports. Thus, double counting cannot be excluded in these categories.
Source: EFS (2000b).

Export performance

The development of Danish C-MC exports compared with OECD average imports is shown in Figure 3. Danish C-MC exports grew at close to the OECD average until the beginning of the 1990s. Since 1993, the cluster has been losing OECD market shares.

Figure 3. **Danish C-MC exports and OECD imports, 1988-96**
Index: 1988 = 100

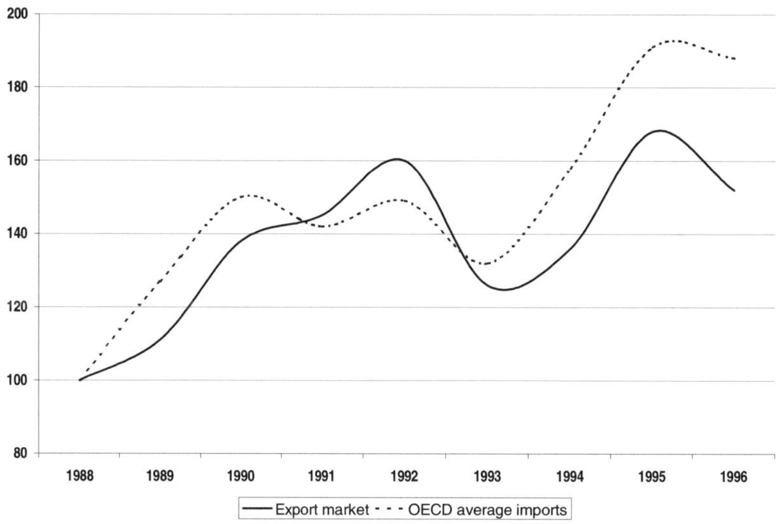

Source: EFS (2000a).

One of the often-discussed causes underlying this pattern is the rapid increase in domestic demand which occurred during the 1990s. With a thriving home market, the incentive to conquer market shares abroad decreased. The increase in market share in the early 1990s can probably be attributed to the reunification of Germany, which compensated for the subdued domestic construction demand at the turn of the decade. According to the benchmarking report (EFS, 2000a), the Danish market has experienced a boom since 1993, increasing by more than 30%. Furthermore, export market shares have decreased by 18% since 1988, while the market has grown by 17%.

In its most recent version, the C-MC statistics from Statistics Denmark contain cluster-by-cluster export data at the four-digit NACE level. Table 3 ranks the six largest four-digit NACE codes of the construction cluster (all of which exceed DKK 2 billion).

Table 3. **Ranking of C-MC exports, 1997**

NACE code		Exports (DKK billions)
7420	Architectural and engineering activities and related consulting	7 974
2913	Taps and valves	7 566
5153	Wholesale trade: wood, building materials & paintings	4 737
2923	Cooling and air-condition equipment for business	2 931
4521	General construction of buildings and civil engineering works	2 089
2430	Paint, varnishes and similar coatings & mastics	2 085
Total C-MC exports		**46 193**

Source: Statistics Denmark (The Danish Mega Cluster Statistics).

These six fields account for almost 60% of construction exports. The two largest exporters are very different in nature. The architectural, engineering and consulting activities appear to be historically related to the early internationalisation of infrastructure construction (bridges).

Labour productivity

A basic problem contributing to the weak economic performance of the C-MC has been *stagnation of labour productivity for more than 30 years*, as shown in Figure 5. C-MC productivity remains at the level of the mid-1960s, although total manufacturing productivity in Denmark more than doubled in the period 1966-96.[11] Figure 5 also shows the development of productivity for the metal industry and for food and food processing, which respectively doubled and trebled during the period. However, Figure 6 indicates that the productivity problem is concentrated in specific parts of the cluster. Supply to construction – *i.e.* the industrial segment of the C-MC – has shown productivity development in line with total manufacturing and the metal industry. This major productivity problem appears, thus, to be concentrated in the *building part* of the C-MC.

Figure 5. **C-MC productivity, 1966-94**
Index: 1966 = 100

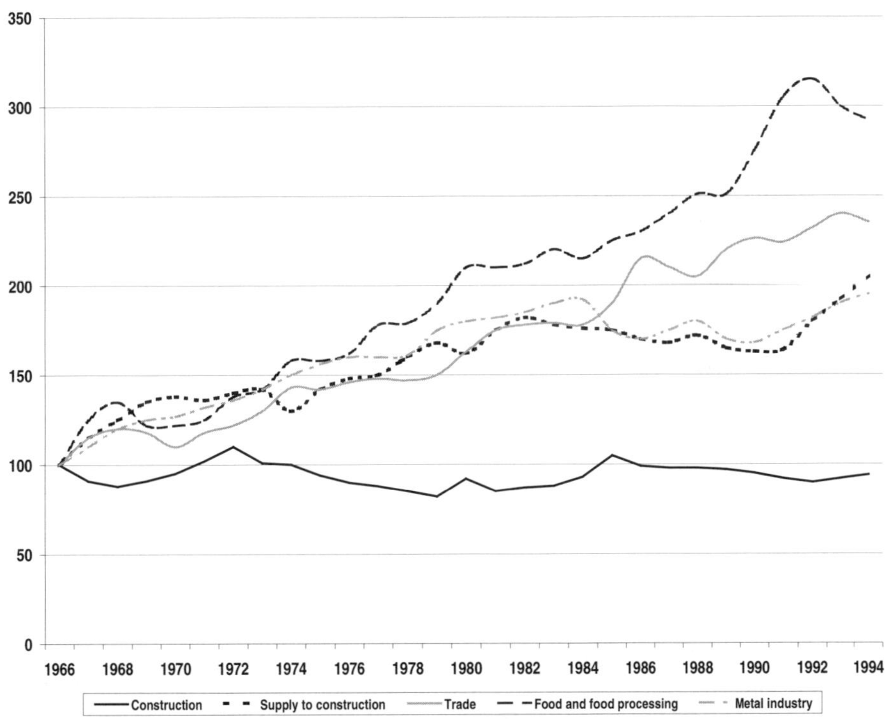

Source: EFS (2000b) p. 82.

In spite of high standards of living and, seen from the outside, a very high standard of housing, the Danish construction sector has shown stagnant productivity growth for more than three decades and has suffered from serious complaints of bad quality from housing consumers. These features have recently been extensively analysed in a series of government-initiated studies, in particular because the industry experienced substantial growth in production and earnings during the boom of the 1990s. However, productivity was increasing at a fairly high rate in the period 1948-57,[12] a time of rapid industrialisation in the building of new apartment houses in Denmark as an integral part of the philosophy of the emerging welfare society.[13] The momentum of this process was stopped in the late 1960s.

A joint ministerial report published in 2000 shows that in 1970-95, labour productivity grew by approximately 100% in Sweden, 60% in Germany and around 25% in the Netherlands, but hardly grew at all in Denmark, except for a short period in the late 1980s.[14] This poor Danish productivity performance has been attributed to the lack of an "innovation culture" in the construction industry, combined with a "lock-in" to a more traditional division of labour among a rather large number of insufficiently co-ordinated partners. The latter include architects, independent engineering consultants, building companies ("construction entrepreneurs" in the Danish vocabulary) and a series of subcontractors.[15] Among the decisive features are the lack of R&D departments in the construction industry (see below) and the lack of a co-operative culture among the firms possessing complementary knowledge. New developments are usually organised on an ad hoc project-to-project basis and the necessary knowledge accumulation is not sufficiently embedded in the construction firms themselves.

Public subsidies

Finally, a factor of major importance for the C-MC is that a large portion of demand is dependent on public expenditure and consumption subsidies. The total amount of public subsidies to private building projects was DKK 15 billion in 1997 (Table 5).

Table 5. **Direct subsidies for housing, 1997**

	DKK millions	Percentage
Total individual rent subsidy	***8 165***	***55.1***
Subsidy for senior citizens	6 191	41.7
Rent security subsidy	1 975	13.3
Supported building of houses	***5 304***	***35.8***
Interest rate support	4 342	29.6
Foundation capital	961	6.5
Renewal subsidy	***1 324***	***8.9***
Other housing subsidy	***35***	***0.2***
House-saving contracts	35	0.2
Total direct subsidy	**14 828**	**100%**

Source: EFS (2000b).

One-third of the direct subsidies comprise financial support to the building of new houses, which means that the subsidy increases with the cost of the project. Indirect subsidies consist of the value of tax deductions on the interest paid.[16] Total direct and indirect subsidies are very high compared to, for example, Germany. The value of tax deductions per capita is more than eight times larger in Denmark. The direct subsidies per capita are also higher, but only twice as high, as shown in Table 6.

Table 6. **Direct and indirect subsidies in Denmark and Germany, 1990**

	Denmark (DKK per capita)	Germany (DKK per capita)
Value of tax deduction of interest costs	9 060	1 085
Direct subsidies	4 570	2 400

Source: EFS (2000b).

Direct subsidies are also high in the Nordic context, with only Sweden having a higher level (Annex Table A2). Subsidies have increased rapidly in Denmark and Sweden, while they have remained fairly stable in Finland and Norway. The basic motives for the subsidies appear to have been oriented towards support to housing consumption as part of the general welfare policy. Until recently, an explicit innovation dimension had not played a major role in the debate. The Danish building sector has thus been caught in a peculiar trap of stagnant productivity, high prices, a high level of subsidies and complaints about the poor quality of new buildings. Breaking this "vicious" circle has been given top priority on the policy agenda (see concluding section).

Research and development, education and innovation

R&D and education

Denmark experienced high growth in aggregate R&D expenditure during the first half of the 1990s – with an annual growth rate (in current prices) of around 11%, compared with an OECD average of 5.5%. This pattern cannot, however, be found for the C-MC, which saw annual growth in R&D expenditure of -11%. The OECD average has been 1%, as shown in Figure 6.

Figure 6. **Annual average growth in R&D expenditure, R&D employment and patents, Danish C-MC,1989-95**

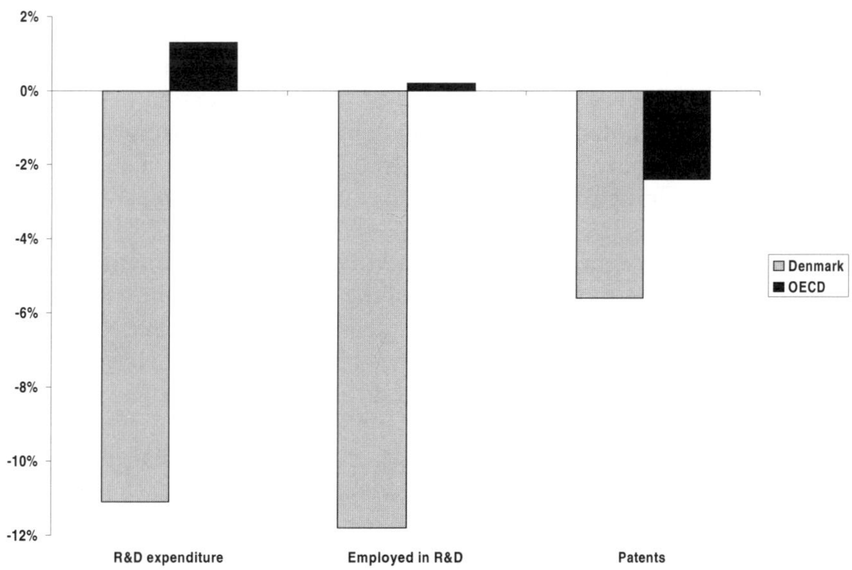

Source: EFS (2000a).

More recent data further strengthen this analysis. While R&D expenditure in the whole Danish business sector (as well as in manufacturing) approximately doubled in current prices during 1989-98, the level of construction R&D was halved. Taking into account large variations across industries, construction (and the related stone and glass industry, not shown) is the only Danish industry with such a negative R&D development.

Table 7. **R&D expenditure for selected business sectors, 1989-98**
Current prices. Index: 1988 = 100

Industry	1989	1991	1993	1995	1997	1998
Total manufacturing	100	120	124	149	173	205
Meat and dairy	100	190	143	286	167	243
Food and beverages	100	82	87	83	109	112
Pharmaceuticals	100	137	143	199	247	335
Telecom equipment	100	147	154	180	240	279
Computer services	100	155	281	369	562	821
Engineering consultancy	100	151	159	141	126	215
Construction	***100***	***111***	***82***	***39***	***44***	***48***
Total business sector	100	126	140	163	203	235

Source: Analyseinstitut for Forskning (2000), pp. 86-87.

Nevertheless, Denmark has traditionally focused on educating highly qualified engineers in construction. There are two technical universities: the Technical University in Copenhagen (DTU) and Aalborg University (AAU), which produce bachelors (diploma engineers), masters and PhDs. DTU (founded in the early 19th century) has a very long tradition and well-established reputation in the field. AAU was founded in 1974 and is already a reputed institution in terms of research as well as teaching. Furthermore, there is a tradition of a decentralised and regionally diversified system of smaller engineering schools producing bachelors in construction. The latter institutions were created as a higher education opportunity for skilled workers who followed tailor-made admission programmes to enter these schools. In 1994, there were six of these engineering schools. An evaluation made by the Danish Centre for Quality Assurance and Evaluation of Higher Education (1994) found some of them to be rather weak. This led to the number of independent institutions being reduced to four – one has been closed, while another has been merged into AAU. There is, thus, in the international context, a fairly large capacity for the training of construction engineers in Denmark.

The engineering school system is supplemented by a series of publicly financed knowledge institutions in construction. The most important is the Danish Building Research Institute (SBI), financed by the Ministry of Housing and Urban Affairs. SBI is broken down into six divisions, collaborates extensively with international research establishments and has a multi-skilled research staff of architects, engineers, psychologists, sociologists and economists. The total staff is 135 people.

Denmark is also characterised by a well-established system of formal training of skilled workers and a set of institutions for training technicians between the level of skilled workers and engineers. The fairly high level of training in construction is illustrated in Table 8. Furthermore, the eight EFS mega-clusters seem to fall into three groups according to the share of highly educated employees. Construction is slightly above the Danish average. This also applies for the Energy and environment MC, which employs many construction engineers, as well as the "Other businesses" residual. The C-MC is also above average for the medium-educational group.

It has not been possible using the available data to show changes in the educational profile of the labour force. However, the present level does not at first sight point towards an explanation for a "technological downturn" in terms of a lack of capacity of the education system. In terms of the quality of the engineering schools, the most recent evaluation report (Danish Centre for Quality Assurance and Evaluation of Higher Education, 1994) does not contain any systematic international comparisons. A number of institutional changes were recommended and have now been implemented. According to our knowledge, no evaluation of Danish construction research in the public sector is available.

Table 8. **Education level of the employees in the mega-clusters, 1997**
Percentages

	Low	Medium	High	Total
ICT MC	25	57	19	100
Medico/health MC	26	55	19	100
Construction MC	33	57	10	100
Energy/environment MC	34	56	10	100
Transport MC	37	59	4	100
Furniture/textiles MC	39	57	4	100
Tourism MC	48	49	4	100
Agro-food MC	49	47	4	100
Other businesses/Rest of the economy	31	58	12	100
Total	36	55	9	100

Note: High is defined as BSc and MSc.
Source: Statistics Denmark (The Danish Mega Cluster Statistics).

Features of innovation

The Danish Community Survey Data (CIS II data) on innovation activity have not been used due to the low response rates of that survey. However, the DISKO project contains a data set from which Table 9 is constructed. As a rough illustration of the variation in innovativeness across clusters, firms reporting product innovations in the DISKO data have been aggregated to five EFS mega-clusters.

Table 9. **Relative number of innovative firms in five EFS mega-clusters, 1993-95**

N = 1 910	Product innovation	No product innovation	Total	N
ICT MC	75.1	24.9	100	249
Medico/health MC	57.8	42.2	100	93
Agro-food MC	52.4	47.6	100	248
Construction MC	38.3	61.7	100	517
Total economy	50.4%	49.6%	100%	1 910

Note: The DISKO data set contains questionnaire data from a sample of 1 910 firms, representative of the private sector of the Danish economy. Further information on the DISKO project is available at http://www.business.auc.dk/disko.
Source: DISKO panel data 1993-95.

According to this source, the C-MC is well below the Danish average, with only 38% of firms reporting product innovations during the study period. This pattern supports the negative picture drawn in the EFS (2000a) benchmarking report.

Patenting is one of the most frequently used indicators for innovative output. It is well known that the propensity to patent – and innovative activity in general – varies from one industry to another. However, the absolute numbers of patents on a country-by-country basis may be of interest in analysing the characteristics of innovations. Table 10 shows a ranking of the top 20 patenting firms cumulated over the period 1982-93.

Table 10. **The top 20 patenting firms in Denmark, cumulated 1982-93**

Rank	Name	Number of patents
1.	Novo Nordisk A/S	917
2.	Danfoss A/S	520
3.	F.L. Smidth & Co. A/S	179
4.	Hans H. Haraldsted	155
5.	Velux A/S (V. Kann Rasmussen Industri A/S)	125
6.	Rockwool International A/S	122
7.	Haldor Topsøe A/S	118
8.	Slagteriernes Forskningsinstitut	109
9.	MAN-B&W Diesel	108
10.	Coloplast A/S	86
11.	Lego A/S	84
12.	Niro Holding A/S	77
13.	Dansk Industri Syndikat A/S	76
14.	Radiometer A/S	62
15.	A/S Ferrosan	61
16.	Eskofot A/S	55
17.	NKT A/S	50
18.	Brdr. Hartmann A/S	49
19.	Maskinfabrikken Taarup A/S	46
20.	Grundfos A/S	44

Note: Firms belonging to the C-MC are shaded in gray.
Source: IKE-Danish Patent Database.

It is striking that C-MC firms make up four of the six top-ranked firms. Two firms clearly patent the most: Novo Nordisk in pharmaceuticals, and the mechanical engineering firm Danfoss. The latter is attributed to the C-MC in the EFS mega-cluster statistics, probably because some of its main products are applied in heating systems in buildings. Ranked at number three is F.L. Smidth, a Copenhagen-based engineering firm which is one of the world's leaders in delivery of turnkey cement manufacturing plants. F.L. Smidth also dominates in concrete building materials as well as in cement and roof plates. Among the leaders is a group of fairly large building material firms, as well as a leading engineering firm from North Jutland, specialised in machinery for concrete foundry. The historical background for this group of firms, dominated by F.L. Smidth, is the presence of large quantities of raw materials in the region. There are indications of a cluster in building materials in North Jutland, where the second technical university, AAU, with a fairly large research capacity in the construction disciplines, is located. Ranked at numbers five and six are Velux in specialised double-glass roof windows, and Rockwool in isolation material for houses.

It is noteworthy that demand for the main product from Danfoss (thermostats for radiators), as well as for Velux and Rockwool products, is highly related to energy-saving measures in housing. Energy saving has been given high priority in a significant number of government-initiated policy measures, be it economic incentives for consumers or more technology policy oriented programmes.[17]

The DISKO panel data make it possible to go a step further into the degree of innovation of different parts of the C-MC, as shown in Table 11.

Table 11. **The degree of innovation in various C-MC segments 1993-95**

	No innovation	Innovation already known on the national market	New innovation on the national market	New global innovation	Total
Consultants	41	54	0	5	100
Contractors	78	20	2	1	100
Material producers	44	28	17	11	100
Total (DISKO)	49%	39%	7%	5%	100%

Note: The DISKO data set contains questionnaire data from a sample of 1 910 firms, representative of the private sector of the Danish economy. Further information on the DISKO project is available at http://www.business.auc.dk/disko.

Source: Based on calculations on the DISKO Database made by Mikkel Andreas Thomasson, IVS at the Copenhagen Business School.

Consultants have above-average rates of innovation. An above-average number of consulting firms in construction implement new products or services already known on the Danish market. Conversely, a high percentage (78%) of *contractors* have not introduced any new innovations. Of the innovating firms, only 2-3% have introduced entirely new innovations. This picture differs for producers of *building materials*, of which an above-average share have innovated. A high number of these firms have introduced new innovations at the national as well as the global level.

The general pattern of the R&D and innovative performance of the C-MC is, thus, rather poor. However, a more detailed analysis requires some differentiation. The problems are basically concentrated in the building (or "contractor") segment. Table 10 indicates that, at least until 1993, there was some visible patenting activity in firms closely related to the C-MC (Danfoss, F.L. Smidth, Velux and Rockwool). However, these firms do not belong to the building core of the C-MC.

A recent evaluation of a major EFS programme involving four building consortia on Products and Processes in the Building Sector (PPB), pointed to the lack of an "innovation culture" (EFS, 2001a; EFS, 2001b). The problem is so basic that it ought to become a major policy issue. Serious doubts have been expressed about whether the fragmented Danish building sector is capable of organising the required inter-firm co-ordination, let alone of developing the necessary R&D departments in the building companies which are currently lacking. In contrast to the manufacturing sector, there has been no tradition among building companies of organising R&D as a separate activity. *In a sense, innovation has always been organised on an ad hoc basis.* Skilled craftsmen have innoved through learning-by-doing, *but organised R&D activities have been fundamentally absent. Lack of persistency in the innovation process and insufficient systematic accumulation of knowledge over time at the company level are considered to be the two fundamental problems.*

Large multinational construction companies have, until now, appeared to be necessary mediators in such a process. This is one of the important factors underlying the recent foreign take-overs of some of the largest construction firms (such as Skanska Denmark and NCC Denmark). These multinational construction firms, of which Skanska has emerged as one of the big global players, will apparently enable a solution to be found to the problems of internalising R&D activities.[18] These firms are apparently capable of creating continuous knowledge accumulation – the usual way of organising innovative activities in manufacturing for most of the last century. At the broad economic level, the doubling of the Swedish productivity level between 1979 and 1995 described above is seen as an outcome of the considerable consolidation of the Swedish construction industry into large companies which occurred during this period.[19] In this context, more mergers and acquisitions by foreign companies are expected in the Danish construction industry.

Strategic challenges and policy perspectives[20]

The C-MC faces considerable challenges. Prices are viewed as being too high and the quality of the buildings too low. The C-MC has lost market shares on OECD markets as well as on the home market; and labour productivity in the building part of the cluster has stagnated for the last three decades. R&D expenditure in construction decreased significantly during the boom period of the 1990s. These rather serious problems are correlated, calling for comprehensive solutions rather than simply focusing on each of the factors in turn. This forms a major barrier to performance improvements.

The cluster analysis emphasises that different segments of the cluster stagnate in productivity and innovation. For small firms, product development and innovation may be a hard task to perform. Most improvements in quality and process of the building segments depend on innovative activity and product development in the industrial part of the C-MC. The poor performance of the building segments in R&D and innovation may, however, show the entire cluster in a worse situation than is actually the case. The weak performance may be connected with structural differences in terms of the relative size of the small craftsmen-based building firms in the international context.

Three major, and interconnected, *weaknesses* of the cluster are summarised in EFS (2000b), Ministry of Trade and Business and Ministry of Housing and Urban Affairs (2000), and EFS (2001a; 2001b). First, demand from end-users is not being properly met. In a market, utility and price are compared in order to meet demand. However, in construction the true price of a project is not known until long after the demands from the end-users are formulated. Furthermore, specification of demand is often made by third persons, such as consultants or architects. The ultimate end-users are often not involved in the formulation. Secondly, prices are too high. The splitting of project formulation and construction makes it very difficult for end-users to gain knowledge of the market, because no exact knowledge of the prices is available at the time the project is specified in more detail. Thirdly, the new buildings contain too many errors. This is the type of problem that arises when a construction project involves 15-25 different professions. The risk of construction errors increases as the process moves from one profession to another. Building errors also result when firms are not able to meet the agreed price. Once they realise this, they have to cut costs, which may give rise to more errors.

These three problems call for a larger degree of co-ordination and a strengthening of the management capabilities of the building firms. To solve these problems, it is necessary that all partners are aware of the problems and take part in the process of improving C-MC performance. Policy initiatives should involve all the main actors of the cluster. Such complex problems cannot be solved through traditional one-way industrial policies aimed at isolated parts of the cluster: policies should aim especially to improve the ways in which the different parts of the cluster interact. There appears to be a fundamental need for new ways of organising, operating and, not least, co-operating in building projects which require the involvement of many different partners and professions. In a knowledge or learning economy context, there is a need for "un-learning" some of the institutions and routines of the C-MC before a new learning process can be instigated (Lundvall and Johnson, 1994). As it is now, the Danish C-MC is locked into an unhealthy trajectory which may be very hard to change.

On the other hand, the C-MC has some visible *strengths*. The industrial part contains a series of strong international competitive positions related to energy saving and building materials and components, such as insulation materials, machinery for the production of building materials, windows, doors, components for heating systems, etc. The R&D carried out by these parts of the C-MC is considerable.

Another strong Danish competence is the more technical part of the building process. The competencies of the skilled professions (ranging from skilled craftsmen to engineers) have always played an important role in the strong position of Danish firms. However, in recent years the industry has experienced some problems in maintaining this position, mainly due to the operation of the decentralised education system, which may have weakened the direct focus on key competencies of the labour force.

Danish architecture is also a strong part of the C-MC. Danish architects have proven their competitive strength by winning several international architectural competitions during more than a century. In addition, the general level of architectural quality of buildings in Denmark is considered to be high. This position is strongly connected to a general perception of *Scandinavian Design*, which is an important feature in the competitive strength of the architectural qualities of the buildings and, not least, of their interiors. This strength in design is a very significant competitive factor behind such related industries as kitchen furniture, living rooms and offices; lighting accessories (*e.g.* PH lamps) and a series of consumer electronics goods (Bang & Olufson) and domestic electrical appliances. Concerning the latter, the design of an oven is considered to be an important part of a kitchen, which again is an key meeting point in family houses. The strength of architects is not only in their competence, but also in the capability to co-operate closely with the other professions involved in a building project. The architectural understanding among the other professions in construction is considered to be high in Denmark compared with other countries. According to EFS (2000b), town planning and design of building sites are considered to be strengths of the C-MC. Compared with, for example, Germany, the planning process of Danish projects is more advanced, leading to a shorter building periods. Finally, there has been a strong tradition among Danish construction companies of participation in large international bridge building process. The 1990s saw substantial domestic demand in this field with the construction of two large bridges linking Sealand and Funen, and Copenhagen and Sweden, respectively. The participating Danish contractors basically belong to the C-MC because they are also active in the building sector.

Historically, the C-MC has been a highly regulated field with considerable economic success. Important private sector actors have been able to influence the formulation of policy in the past. As a result, regulations have played an important role in the development of key competencies. The major improvements in building materials in the 1960s were the outcome of policy regulations implemented in the 1950s and early 1960s, in which the Danish Association of Engineers played a significant role. Although public regulation has been successful in the past, in recent decades the construction sector has not managed to improve labour productivity and exploit its great potential. The complexity of the problems indicates a "lock-in" situation,[21] with the C-MC locked into a specific pattern of co-operation and a specific technological paradigm. The latter has emerged over a long period of continuous social and technical change, and the elements of the paradigm are closely connected. Major changes to such a paradigm involve a complex and integrated process of change in science, technology, work practices, social organisation, etc. Changes in paradigms cannot be accomplished by public policies alone. Nor can the actions of individual agents overcome "lock-in". There is a need for co-ordinated incentives, development of technological alternatives and nurturing of emerging technological systems in order to shift the paradigm. If the C-MC is to move out of this "lock-in" situation, there is a need for fundamental change in the way the construction industry is organised today (Smith, 1996; EFS, 2000b).

The *policy initiatives* which followed from the mega-cluster analysis of construction are still in the development phase. The Ministry of Trade and Business is co-operating with the Ministry of Housing and Urban Affairs to formulate new policy initiatives to cope with the complex problems facing the C-MC. This clearly reflects a cluster-based approach, with two ministries having different responsibilities working together to improve the C-MC.

Initiatives for the future are now in the stage of being formulated (see especially EFS, 2000b, 2001a and 2001b, and the joint ministerial *Task Force Report* from the Ministry of Trade and Business and the Ministry Housing and Urban Affairs, 2000). Policy makers are recommended to focus on six primary areas, which are the main targets for solving the problems of the C-MC:

- Increasing the degree of co-operation between the different professions in the C-MC.
- Collecting and publishing information on firm performance in order to improve transparency. It is expected that the Competition Authorities will increase their efforts to strengthen competition.
- Making the Danish Building Defects Fund more active. The fund holds a great deal of information on market performance of the firms which may be used to improve transparency.
- Increasing the degree of industrialisation of the building process with the aim of lowering prices.
- Increasing the demands for quality in future public building projects could lead to an increase in the quality of the entire C-MC.
- Improving the structure of the knowledge institutions to increase the exploitation of the existing knowledge base and the degree of co-operation between firms and knowledge institutions.

NOTES

1. OECD (1998) Annex Table 2.3; 1995 data. To adjust for differences in the size of the public sectors of the OECD countries, the analysis is based on shares of value added in the business sector.

2. Mining and quarrying (including oil and gas) amounts to 15.8% in Norway. Australia is ranked second, with 4.1%, and Canada third, with 3.8%. Adjusted to a more "average" OECD share (1%-2%), Norway would be among the upper third in terms of the share of construction.

3. Represented, for example, by Scherer and Ross (1990) and Martin (1994).

4. See EFS (2000b), p. 17. The statistical definition of the C-MC used in EFS (2000a) and EFS (2000b) is shown in the Annex.

5. It is not easy to separate construction of infrastructure from the C-MC definition. In practical terms, some of the activities appear to be included in the C-MC data.

6. See Chapter 4 of this volume, on the ICT Cluster in Denmark, for a brief theoretical background of the cluster studies. For more detailed information on the mega-clusters and the Danish cluster studies, see Drejer *et al.* (1999) and Holm Dalsgaard (Chapter 18 of this volume).

7. Danish Agency for Trade and Industry (2000), referred to as EFS (2000).

8. See also Holm Dalsgaard (Chapter 18 of this volume).

9. The analytical areas are defined by grouping related actors of the C-MC.

10. This area has been one of the primary technological areas of research in the C-MC in the period 1987-96. According to a bibliometric study referred to in EFS (2000b), 30% of scientific publications focused on the environment, energy and internal climate.

11. Manufacturing productivity increased in 1966-96 from an index of 100 to 226, while construction productivity (according to the conventional definition in the national accounts) decreased from index 100 to 97 (Dalum, 1999). Measured in absolute terms, the level of manufacturing productivity for construction was 70 in 1966 and 134 in 1996. The level of manufacturing reached the construction level in the late 1970s. Since the early 1980s, growth of Danish manufacturing productivity has slowed down, but it has grown faster than for construction in the 1980s and 1990s.

12. Council of Economic Advisors (1997), Table V.7, p. 242.

13. Bertelsen (1997) contains a historical analysis of 25 years of industrialisation of the Danish building industry in the Copenhagen area from the late 1940s to the early 1970s (the era of high growth in productivity).

14. Ministry of Trade and Business and Ministry of Housing and Urban Affairs (2000).

15. These issues have been analysed by EFS (2000b; 2001a; 2001b).

16. Interest deductions have decreased through a series of policy measures since 1987.

17. Disentangling the different causes of this patenting pattern would require a detailed qualitative study of the patent accounts. However, the "macro-pattern" appears to be sufficiently clear, and the very nature of the cluster approach is geared towards dealing with such interactions.

18. Biannual surveys of the 500 largest Danish companies are published currently as *Børsens Magasiner Guld* – the most recent edition was published in November 2000. The data are ordered according to turn-over and include gross and net profits as well as employment. The companies are subdivided into a series of industries, among which data for "construction and engineering consultants".

19. Productivity increases are also reported for the Netherlands and Germany. They are not as large as the Swedish increase, but they clearly differ from the Danish pattern.

20. Some of the judgements presented here draw on EFS (2000b), written by Tage Draebye.

21. For detailed descriptions of the concept of "lock-in", see David (1985) and Arthur (1989).

Table A1. **Definition of the Danish construction mega-cluster**

Primary	
201	Saw-milling and planing of wood, impregnation of wood
1411 - 1413	Quarrying of stone, limestone, gypsum, chalk and slate for construction
1421 - 1422	Operation of gravel and sand pits and mining of clays and kaolin
1450	Other mining and quarrying n.e.c.
Manufacturing	
1753	Manufacture of non-wovens and articles made from non-wovens, except apparel
2010	Saw-milling and planing of wood, impregnation of wood
2030 + 2051	Manufacture of builders' carpentry and joinery and manufacture of other products of wood
2124	Manufacture of wallpaper
2416	Manufacture of plastics in primary forms
2430	Manufacture of paints, varnishes and similar coatings, printing ink and mastics
2462	Manufacture of glues and gelatines
2521	Manufacture of plastic plates, sheets, tubes and profiles
2523	Manufacture of builders' ware of plastic
2611 - 2612	Manufacture, shaping and processing of flat glass
2614	Manufacture of glass fibres
2615	Manufacture and processing of other glass including technical glassware
2622 - 2624	Manufacture of ceramic sanitary fixtures, ceramic insulators, insulating fittings and other technical ceramic products
2626	Manufacture of refractory ceramic products
2630	Manufacture of ceramic tiles and flags
2640	Manufacture of bricks, tiles and construction products, in baked clay
2651 - 2653	Manufacture of cement, lime and plaster
2661 - 2665	Manufacture of articles of concrete, plaster and cement
2670	Cutting, shaping and finishing of stone
2681 + 2682	Manufacture of other non-metallic mineral products
2721	Manufacture of cast iron tubes
2744	Copper production
2811 + 2812	Manufacture of structural metal products
2822	Manufacture of central heating radiators and boilers
2862 - 2863	Manufacture of cutlery and tools
2873 - 2874	Manufacture of wire products, fasteners, screw machine products, chain and springs
2913	Manufacture of taps and valves
2921	Manufacture of furnaces and furnace burners
2923	Manufacture of non-domestic cooling and ventilation equipment
3120	Manufacture of electricity distribution and control apparatus
3613	Manufacture of other kitchen furniture
Supporting sectors	
2952	Manufacture of machinery for mining, quarrying and construction
4550 + 7132	Renting of construction or demolition equipment with and without operator
Service sectors	
202	Forestry and logging related service activities
4511 + 4512	Site preparation
4521	General construction of buildings and civil engineering works
4522 + 4525	Erection of roof covering and frames and other construction work involving special trades
4531	Installation of electrical wiring and fittings
4532	Insulation work activities
4533	Plumbing
4541 - 4545	Building completion
5113	Agents involved in the sale of timber and building materials
5153 - 5154 + 5162	Wholesale
5246	Retail sale of hardware, paints and glass
7011 + 7012	Real estate activities with own property
7020	Letting of own property
7031 - 7032	Real estate activities on a fee or contract basis
7420	Architectural and engineering activities and related technical consultancy
7470	Industrial cleaning

Source: EFS (2000b).

Table A.2. **Direct subsidies for housing in the Nordic countries as a percentage of GDP**

	1980	1985	1990	1991	1992	1993
Denmark	0.45	0.52	-	0.57	0.72	0.90
Finland	0.57	0.49	0.56	0.63	0.73	0.57
Norway	0.32	0.22	0.42	0.37	0.36	0.47
Sweden	0.95	1.42	1.45	1.87	2.62	2.66

Source: EFS (2000b).

REFERENCES

Analyseinstitut for Forskning (2000), *R&D Statistics*, Analyseinstitut for Forskning, Aarhus.

Arthur, W.B. (1989), "Competing Technologies Increasing Returns and Lock-in by Historical Events", *The Economic Journal* 99, pp. 116-131.

Balassa, B. (1965), "Trade Liberalization and 'Revealed' Comparative Advantage", *The Manchester School of Economics and Social Studies* 32(2), pp. 99-123.

Bertelsen, S. (1997), *Bellahoej, Ballerup, Broendby Strand*, Danish Building and Urban Research, Copenhagen.

Council of Economic Advisors (1997), *Dansk Oekonomi – Foraaret 1997* [The Danish Economy – Spring 1997], Council of Economic Advisors, Copenhagen.

Dahl, M. and B. Dalum (2001), "The ICT Cluster in Denmark", Chapter 4 of this volume.

Dalum, B., K. Moeller, U. Joergensen and F. Valentin (1991), *Internationalisering Og Erhvervsudvikling*, Ministry of Trade and Industry, Copenhagen.

Dalum, B. (1999), "Erhvervsstruktur og Erhvervspolitik" [Industrial Structure and Industrial Policy], in T. Andersen *et.al.*, *Beskrivende Oekonomi*, DJOEFS Forlag, Copenhagen.

Danish Centre for Quality Assurance and Evaluation of Higher Education (1994), *Evaluering Af Bygningsingeniøruddannelser* [Evaluation of the Education System for Construction Engineers], Danish Centre for Quality Assurance and Evaluation of Higher Education.

David, P.A. (1985), "Clio and the Economics of Qwerty", *American Economic Review. Papers and Proceedings* 75(2), pp. 332-337.

Drejer, I., F.S. Kristensen and K. Laursen (1999), "Studies of Clusters as the Basis for Industrial and Technology Policy in the Danish Economy", in OECD (1999), *Boosting Innovation: The Cluster Approach*, OECD, Paris.

EFS (1993), *Ressourceområdet Bygge/Bolig - En Erhvervsanalyse* [The Construction Mega Cluster – A Business Analysis], Danish Agency for Trade and Industry (EFS), Copenhagen.

EFS (2000a), *Benchmarking Af Dansk Erhvervsliv* [Benchmarking Danish Businesses], Danish Agency for Trade and Industry (EFS), Copenhagen.

EFS (2000b), *Bygge/Bolig – En Erhvervsanalyse* [Construction – A Business Analysis], Danish Agency for Trade and Industry (EFS), Copenhagen.

EFS (2001a), *Proces- og Produktudvikling i Byggeriet – Analyserapporten* [Process and Product Development in Construction – Analysis]", Danish Agency for Trade and Industry (EFS), Copenhagen, forthcoming.

EFS (2001b), *Proces- og Produktudvikling i Byggeriet – Perspektivrapporten* [Process and Product Development in Construction – Perspectives], Danish Agency for Trade and Industry (EFS), Copenhagen, forthcoming.

Holm Dalsgaard, M. (2001), "Danish Cluster Policy: Improving Specific Framework Conditions", Chapter 18 of this volume.

Lundvall B.-Å. (1999), *Det Danske Innovationssystem* [The Danish Innovation System], Erhvervsfremmestyrelsen, Aalborg.

Lundvall, B.-Å. and B. Johnson (1994), "The Learning Economy", *Journal of Industry Studies* 1(2), pp. 23-42.

Martin, S. (1994), *Industrial Economics: Economic Anakysis and Public Policy*, Prentice-Hall Inc., New Jersey.

Ministry of Trade and Business and Ministry of Housing and Urban Affairs (2000), "Byggeriets Fremtid – Fra tradition til innovation" [The Future of Construction – From Tradition to Innovation], Ministry of Trade and Business and Ministry of Housing and Urban Affairs, Copenhagen.

OECD (1998), *Science, Technology and Industry Outlook*, OECD, Paris.

Scherer, F.M. and D. Ross (1990), *Industrial Market Structure and Economic Performance*, 3rd edition, Houghton Mifflin.

Smith, K. (1996) "Systems Approaches to Innovation: Some Policy Issues", note for the European Commission as a part of the TSER project, STEP Group, Oslo.

Vernon, R. (1966), "International Investment and International Trade in the Product Cycle", *Quarterly Journal of Economics* 80(2), pp. 190-207.

Chapter 10

INNOVATION IN THE DUTCH CONSTRUCTION CLUSTER

by

Pim den Hertog and Erik Brouwer[*]
Dialogic Innovation & Interaction and Department of Innovation Studies, Utrecht University

Introduction

As is the case in most other countries, the Dutch construction cluster is a fairly well-established and mature cluster.[1] It is unique in terms of the actors involved, its institutional organisation, its culture and innovation style. Compared with manufacturing, there was much earlier development of specialisation, outsourcing and strategic co-operation, allowing innovation to take place outside the construction firms themselves (Hillen, 2000, p. 10). Technological innovation is therefore mainly imported and absorbed through supplies of manufactured goods and services. Well-developed interfaces between the various elements of the value chain are a prerequisite for innovation. This requires not only an effective technology transfer mechanism, but also a well-developed knowledge base at the absorption end, a working culture conducive to innovation and well-functioning links between producers and end users. In other words, innovation in construction is a combination of technical, organisational and market innovation factors. Studies of the economic and innovation performance of the Dutch construction cluster conducted over the years have indicated the need for further improvement in the interplay between the various actors, despite the sector's apparent resistance to change (see, for example, AWT, 2000). The policy measures taken to date to overcome this problem have been increasingly initiated from a cluster perspective.

The analysis presented here is restricted to the mega-cluster level. We are aware that there is considerable variety in the characteristics, innovativeness and performance of the individual sub-clusters within this mega-cluster. For example, the internationally competitive dredging and related construction activities, and constructions in water and soft soil function as sub-clusters within the wider construction mega-cluster. On a similar note, the construction of residential and non-residential buildings and the construction of civil engineering works can be labelled as sub-clusters operating at a lower level of aggregation. Although the larger multi-product firms in particular operate in a number

---

* The authors can be contacted at Dialogic, Wilhelminapark 20, 3581 ND, Utrecht, the Netherlands (e-mail: denhertog@dialogic.nl). The research on the construction cluster reported here is partly based on the ClusterMonitor Construction conducted for the Dutch Ministry of Economics in co-operation with Berenschot (Dr. Maurits Verwey), Technopolis (Dr. Patries Boekholt) and CBS (Dr. Robert Goedegebuure). The empirical statistical work was carried out at the Centre for Research of Economic Microdata (CEREM) of Statistics Netherlands. The views expressed in this chapter are those of the authors and do not necessarily reflect the policies of Statistics Netherlands. The map was kindly provided by Peter Louter (TNO-Inro).

of these sub-clusters at the same time, policies aimed at removing barriers to innovation in the construction sector are most likely to be implemented at this lower level of aggregation.

This chapter first introduces the Dutch construction cluster, identifying relevant segments. Some basic statistical information is provided on the main characteristics and trends of the cluster. We follow up with a description of the innovation dynamics and innovation style of the construction cluster, using so-called ClusterMonitor data, before moving on to a discussion of the various policy initiatives aimed at improving the functioning of the construction cluster. Finally, a number of conclusions are drawn.

An introduction to the Dutch construction cluster

Main characteristics

The construction process can be described rather crudely as an activity in which different inputs are brought together, generally on building sites, in order to construct final products or objects (houses, offices, roads, bridges, tunnels and other types of civil engineering works). In addition to these individual objects, combinations of objects or the redevelopment of urban or industrial areas or dredging works could also be labelled as a final output of the construction cluster. The traditional major market segments are civil and commercial and industrial building, civil engineering and installation. However, other activities outside the construction industry play a fairly important role in the production of the final output mentioned above. Apart from mining and quarrying, the manufacture of all kinds of materials and pre-fabricated elements, as well as general and specialised machinery, is equally important. These supplies are partly sold through wholesalers. Furthermore, companies providing services as diverse as engineering and architectural services, construction management services, real estate development and brokerage and specialised financial services, also enter into the construction process (Figure 1). The role of service companies has increased significantly since the early 1970s, when the government began to withdraw from direct intervention in the construction market, leaving more room for private initiatives, *e.g.* in project development. As we will see, manufacturing and service companies, in particular, play a relatively large role in the innovation process in the construction mega-cluster.

Figure 1 does not include the extensive intermediary structures present in the construction industry. It goes without saying that such an established cluster is surrounded by a well-developed intermediary structure of schools, training institutes and knowledge institutes, industrial organisations and various organised groups of clients. The three technical universities of the Netherlands have reputed faculties of architecture and civil engineering (although these are not necessarily well integrated). Further intermediary structures such as research colleges and institutes such as TNO Construction and various long-ranging strategic research programmes are also present. However, this does not mean that the interface between education, research and industry does not offer possibilities for improvement (see, for example, AWT, 2000). The myriad of organisations within the industry only adds to the complexity. In a well-known Porter study of the Dutch construction cluster dating from 1992, it was observed that there were no less than 220 industrial organisations in the construction cluster (Jacobs *et al.*, 1992). Although some co-ordination and consolidation has since taken place, a considerable number of interest groups are still present. The same applies to the demand side. Private home owners, for example, have joined together in what has developed into a fairly powerful organisation (*Eigen Huis*).

Figure 1. **Main players in the construction cluster**

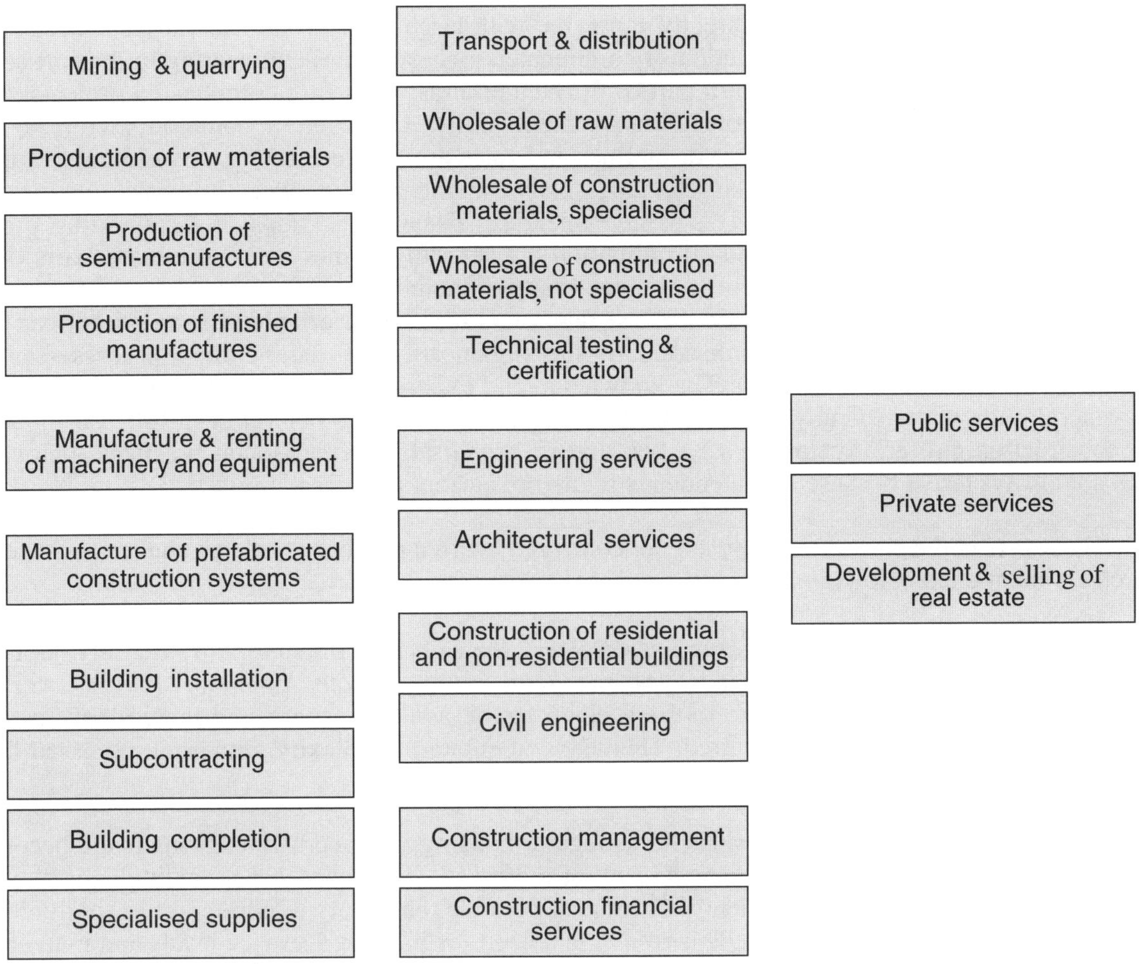

Source: Verweij *et al.*, 2000, p. 9.

The Dutch Government has its own agency responsible for the quality of public buildings (*Rijksgebouwendienst*) and also has considerable capacity in terms of house engineering (*Rijkswaterstaat*, part of the Ministry of Traffic, Transport and Water Management), responsible for procuring and co-ordinating many of the major infrastructural and civil engineering works. These two bodies are important players in the construction cluster. That being said, it should be noted that, more generally speaking, the nature of government involvement is manifold but has evolved over the years. In an analysis of the economic dynamics of the Dutch construction sector, Romijn (2000) indicated that there were several reasons for the highly interventionist government involvement in the construction sector during the 1945-70 period. These include rebuilding the country after the war, population growth and the expansion of production capacity to fuel economic growth. After 1970, and especially after 1985, government involvement was wound down. There was stimulation of corporate investment in residential and non-residential buildings and many of the large-scale investments in new projects concerned with large civil and hydraulic engineering works in particular (including many major investments in water management, the "Delta works") came to an end and were replaced by extra attention for maintenance and extension of the capacity of existing infrastructures. It was not until the end of the 1990s that major new infrastructural works were put back on the agenda. Decision making on such matters is notoriously slow, and government increasingly encourages co-investment by the private sector, resulting in attempts to realise public/private partnerships in this field.

Construction is almost by definition a project-based activity, although projects obviously differ in their complexity, in both technical and organisational terms. The construction of a standard house or civil engineering project cannot be directly compared with large-scale urban renovation projects, large infrastructure projects or the construction of complicated pieces of civil engineering in the soft soil so typical of the Delta area of the western part of the Netherlands. These are complicated projects which require advanced planning and project management skills. However, a common factor in most construction projects is that distinctive disciplines and roles are divided among a considerable number of parties, and that co-ordination largely takes place on the building site. Informal relations and networks are therefore as important as vested interests. The various steps of the building process involve separate responsibilities, and co-ordination between disciplines and the various parts of the value chain is not always easy to achieve. In fact, many of the proposals for improving the functioning of the Dutch construction cluster over the last few years have been oriented towards improving the value chain management and optimisation of the organisation of the building process (see, for example, AWT, 2000; Hillen, 2000). The knowledge and experience amassed during one-off projects are generally not systematically stored as knowledge management has not yet been fully developed in the construction cluster. Against this background of organisational and social practices, innovation does not always thrive because of its tendency to disrupt current practices.[2]

The actors in the construction cluster are confronted with a number of related challenges, a few of which are mentioned below:[3]

- *Scarcity of space requires creative solutions.*[4] Space is in short supply in a densely populated country like the Netherlands. At the same time, various functions (living, working, transporting, recreation, etc.) all lay claim to the available space and increasingly creative solutions are therefore required. This leads to greater complexity, requiring increased levels of co-operation (Verweij *et al.*, 2000, pp. 9-10).

- *More demanding end users.* As far as regular housing is concerned, consumers are becoming more and more demanding. As a result of increased economic welfare, individualisation and the desire to live independently as long as possible, end users require comfort, individuality and high-quality buildings. This results in a need for increased service levels and market-pull innovations, as well as client-oriented construction concepts (Verweij, 2000, p. 14). The market for offices is characterised by a similar need for adaptability (*e.g.* plug-and-play buildings). Improved fine-tuning among the players involved in the building process would enable the industry to gain a better understanding of the needs of the end user (more effective communication) while reducing "failure costs" (Hillen, 2000, p. 14).

- *Changing role of government.* Government has to wear various hats in construction projects; it has a prime responsibility for spatial planing, the initiation of large-scale infrastructure projects, regulation of the basic requirements for constructions and buildings, while also being a major client for buildings and offices, and acting as a financier and regulator for at least part of the housing stock, etc. New European legislation is on the cards and, probably most importantly, experiments are underway to find new types of procurement, such as innovative procurement policies and public private partnerships, in the building and operation of infrastructure works, for example.

- *Increased competition on construction markets.* With the opening up of European markets, the international market for mega-infrastructure projects, and hence international competition, is on the increase, requiring that construction firms achieve a certain scale. The main Dutch players are relatively small in the international context.[5] Competition from non-construction companies, such as innovative suppliers of construction materials and service companies, is also on the rise. The former offer complete components and inputs which are manufactured in a semi-standardised manner and therefore move up in the value chain. The latter are mainly professional service firms, such as engineering firms, real estate developers and construction consultants (with the exception of banks and pension funds, for example, which have their own real estate development departments), which are taking the lead in the management of complex construction and infrastructure projects and the operation of complexes, for example. Most of the construction companies still mainly focus on managing their capacity.

From the above, it is clear that these are challenging times for players in the construction cluster. However, it will be the capacity to co-operate along the value chain, to jointly innovate and manage the building process more effectively, that will count in the long run. New forms of co-operation are required to ensure that the above challenges are met and innovation realised without damaging the essential elements of competition.

Before looking at the innovation and innovation style of the construction cluster in greater detail, some basic economic parameters are given below.

Basic economic parameters

The economic importance of the construction cluster as a whole is undisputed. In the Netherlands, as in most other countries, the construction process plays an important role in terms of employment and added value. Furthermore, buildings, roads and other products and services provided by the construction cluster are an important input factor (and hence influence costs) for firms, households and government in its capacity as producer of public services. It is important that the construction cluster be both efficient and innovative, if only because of its sheer size.[6] Tables 1 and 2 provide some insight into how the construction industry developed over the 1995-99 period. Residential and non-residential building and building installation (the majority of "Construction other") are the two largest segments, jointly employing over 400 000 persons in 1999. Also remarkable is the large number of firms – more than 60 000 in 1999. This is mainly due to the fact that the share of firms without personnel increased steadily to 57% in 1998 as experienced workers increasingly began to start up their own firms. This trend is particularly noteworthy in the residential and non-residential building segment. Approximately 5% of GNP is realised in the construction industry. This is a fairly stable percentage, although the growth of the construction industry lagged somewhat behind the overall growth of the Dutch economy in 1995-99. In 1999, there were clear signs of labour scarcity in some segments. In mid-1999, for example, the number of unfilled vacancies amounted to 18 000.

Table 1. **Number of firms and persons employed in the construction industry, 1995-99**

	Number of firms[1]					Persons employed ('000)				
	1995	1996	1997	1998	1999	1995	1996	1997	1998	1999
Total construction	43 945	48 400	51 405	58 340	62 335	437	445	464	474	485
- Residential & non-residential building	20 545	22 730	24 060	27 445	29 445	176	175	183	185	188
- Civil engineering						72	74	74	76	78
- Construction other	23 400	25 670	27 345	30 895	32 890	189	195	206	213	219

1. Including firms without personnel.
Source: CBS Statline.

Table 2. **Production value and value added of the construction industry, 1995-99**
In NLG millions, current prices

	Production value					Value added				
	1995	1996	1997	1998	1999	1995	1996	1997	1998	1999
Total construction	93 532	99 513	104 080	110 524	120 121	33 531	34 331	36 299	38 695	43 091
- Residential & non-residential building	48 136	49 633	51 700	53 687	57 772	14 541	14 394	15 072	15 526	17 208
- Civil engineering	17 212	19 406	20 470	21 575	21 973	6 492	6 983	7 371	7 767	8 712
- Construction other	28 184	30 474	31 910	35 262	38 376	12 498	12 954	13 856	15 402	17 171

Source: CBS Statline.

This being said, a study over a somewhat longer time period shows that the construction industry generally under-performs compared with other sectors in the economy. Romijn (2000, p. 36) observed that, over the 1969-97 period, the construction industry had not grown in real terms (*i.e.* after correction for inflation). This contrasts markedly with the decades following World War II.

If our analysis is extended to include the manufacturing, wholesale and service firms which play a role in construction, thus adopting a cluster perspective, the economic importance of construction increases even further. Using the ClusterMonitor[7] data set, we see that manufacturers of building materials and machinery are a particularly important factor in the construction cluster (Table 3). The size distribution of the various segments of the construction cluster indicates that, relatively speaking, pure construction firms are more often found in the smaller scale classification.

Table 3. **Some basic characteristics of the Dutch construction cluster in 1997[1,2]**

Cluster segments	Total number of persons employed in 1997 (1 000*FTE)	Total net turnover in 1997 (NLG millions)	Total gross value added 1997 (NLG million)
Total cluster	637	186 864	60 581
- Manufacturing	80	28 216	9 502
- Construction	300	80 948	25 771
- Wholesale	36	25 215	3 488
- Services	56	9 420	5 399
Total all firms	**2 931**	**1 217 207**	**293 727**

1. For the tables on retailing derived from the ClusterMonitor Database, we used the standard weighting factors applied by CBS. These weighting factors vary according to the industries (two-digit) and firm size. We did not change the weighting factors when analysing the results at a more disaggregated level.
2. The data mainly refer to 1997 for production and 1996 for innovation. Data for 1998 for these two factors are now available and will be linked to the ClusterMonitor data set in due time.
Source: ClusterMonitor Database CBS/Dialogic, 2000.

Although international competition is certainly on the increase for large infrastructure projects, competition is felt most strongly at the national, and even more likely, at the regional and local levels, by the majority of firms. In 1999, there were only six building concerns with a turnover of more than EUR 1 billion.[8] It is noteworthy that, even in times of economic upswing, such large construction firms still have fairly low profit margins although they take considerable financial risks. All the more so since construction is a capacity-driven economic activity which is sensitive to trends in the business cycle. Low margins also imply that limited resources are spent on R&D and innovation, as will be observed below.

Table 4. **Percentage of businesses by size according to sub-cluster in the Netherlands, 1996**
Percentages

Cluster segments	10-19 employees	20-49 employees	50-99 employees	100-199 employees	200-499 employees	500 and more employees
Total cluster	49	34	11	4	2	1
- Manufacturing	39	38	12	7	3	1
- Construction	52	34	9	3	1	0
- Wholesale	59	28	8	3	2	0
- Services	56	29	8	3	3	0
Total all firms	**50**	**32**	**10**	**4**	**2**	**1**

Source: ClusterMonitor Database CBS/Dialogic, 2000.

As construction is a fairly ambulant activity, mapping out the construction industry in geographical terms is of limited value. Major construction projects (both residential and infrastructural) seem to be concentrated somewhat more in the western provinces of the Netherlands. This may simply be because this is the most densely populated part of the country, where both the main ports (the Port of Rotterdam and the area around Schiphol) are located. A closer look at the location of the construction industry (see Annex 1) reveals that construction firms are often located in neighbouring provinces, such as the provinces of Noord-Brabant and Gelderland. It is well known that many of the workers on sites in the western part of the country are based in these areas.

Innovation dynamics and innovation style

The construction cluster is a mature cluster, based on a combination of traditional skills, industrial technologies and a well-established informal culture of the "way things are done". Although it includes highly innovative actors, the Dutch construction industry as a whole is a fairly traditional world with an in-built preference for proven technology and organisational concepts. Construction firms are mostly technology followers which absorb innovations originating from manufacturing industries and increasingly from wholesale and service industries. New building technologies and concepts[9] are experimented with, but it takes time before new approaches are accepted as common practice. This can be partly explained by a tradition of risk aversion among the clients and the detailed procurement and planning undertaken by these clients. Detailed technical requirements and designs instead of more functional designs leave little room for innovative solutions, and the price factor has therefore become the dominant weapon in competition. In some markets such as road construction, urban redevelopment and major infrastructure projects, government is the major client. Only recently have attempts been made to use this position to challenge the construction cluster, partly as a result of cluster policies. This has resulted in the development of innovative procurement policies and experiments with public/private partnerships (a form of risk sharing allowing the creation of extra leverage for investments in infrastructure).

Another explanation could be that the talent for absorbing out-of-industry innovation needs to be better developed, or should at least not be automatically perceived as a threat to one's position in the value chain. A number of new technological and organisational concepts are difficult to realise in practice owing to this factor. Contractors play a key role in the construction process,[10] for example. On the one hand, they must become more sensitive to demanding customers. On the other hand, they run the risk of being squeezed by innovative manufacturers of building and installation materials, while increasingly capable subcontractors and (more) assertive service firms monopolise the relationship with the client. Since various players can integrate forwards or backwards in the value chain and roles are in a state of flux, innovations can be seen as threatening.

It is worth noting that analyses of, for example, technology transfer in the UK construction industry show considerable resemblance to the Dutch situation. Barrett *et al.* (2000) observe that "present construction industry technology transfer endeavours are being severely hampered by a lack of proper understanding of technology transfer processes. Current approaches tend to view technology transfer as a mechanistic 'pick-and-mix' exercise – identifying new technologies from other industries, and trying to insert them in their existing form into an (unsurprisingly) unreceptive construction industry." They stress that the ability to absorb out-of-industry technologies is affected not only by the level of prior related knowledge and expertise, but also by "each industry having its own unique 'recipe' of assumptions, knowledge bases, technologies and practices. These 'recipes' considerably erode the ease with which technologies can be transferred from one industry to another by creating 'incompatibility barriers'."

In conclusion, in order for innovation in the core construction industries to occur, a good understanding is required of client needs (including the required communicative skills), on the one hand, and collaboration with industries further upstream in the value chain, on the other hand.[11] Innovation in the construction cluster should, therefore, as in most other clusters, be seen as a multifaceted phenomenon which includes clear technological, organisational and market aspects. In the following section, we shall look at some, admittedly crude, innovation indicators of the Dutch construction cluster using ClusterMonitor data (see Annex 2).

R&D and innovation indicators

Established ways of "doing things" are relatively important in the construction cluster. Risk aversion and a tendency towards optimisation on a project basis are more or less the norm. While innovation may not be second nature to construction firms, the manufacturing and service firms within the construction cluster outperform all other firms in terms of innovativeness and R&D activity (Table 5). The wholesale segment also scores better than the pure construction firms on average. If we focus on R&D intensity (measured in terms of R&D man-years as a percentage of all employees within the construction cluster), the construction segment scores a meagre 0.1% in 1996 (equivalent to some NLG 32 million), while the manufacturing and services segments score 2.5% and 2.1%, respectively, and the construction cluster as a whole scores 0.6%. This is almost half the score of all firms in the Dutch economy. A similar picture emerges in terms of the number of firms with R&D activities. Whereas in the construction cluster, an average of 10% of firms perform R&D on a permanent basis and 8% on an incidental basis, the figures for the pure construction firms are 2% and 4%, respectively. This again indicates that innovations are not brought about by the construction firms themselves, but are more likely to originate at the manufacturing or service firms in the cluster.

Table 5. **Total number of businesses, innovative businesses and businesses engaged in R&D activity, according to sub-cluster, 1996**

Cluster segments	Total number	% of innovative firms	% of firms with R&D activities
Total construction cluster	13 728	32%	18%
- Manufacturing	1 534	54%	36%
- Construction	7 091	19%	6%
- Wholesale	920	31%	14%
- Services	1 279	54%	45%
All firms	**46 933**	**37%**	**20%**

Source: ClusterMonitor Database CBS/Dialogic, 2000.

Table 6 provides more detailed insights into the wider category of innovation expenditure. Aside from the fact that this expenditure is relatively low as a percentage of turnover in the construction segment, it is striking that purchase of equipment is a dominant factor within innovation expenditure in the construction segment. Training is also relatively important compared with the other three segments within the construction cluster.

Table 6. **Total and average innovation expenditure for the Dutch construction cluster, 1996**
In NLG millions

Cluster segments	Manufacturing	Construction	Wholesale	Services	Total construction	Total
Total innovation exp.	2 113	485	65	423	3 898	22 783
- Purchase of equipment	1 412	365	33	69	2 175	8 959
- Own R&D	514	54	12	202	1 028	8 031
- Outsourced R&D	33	11	7	72	168	1 548
- Industrial design	44	11	3	30	153	994
- Licences, advice	30	4	2	14	120	829
- Marketing	59	8	3	11	144	1 062
- Training	21	33	4	24	112	1 361
Total innovation expenditure as a % of turnover	**6.0%**	**0.6%**	**0.2%**	**2.5%**	**1.5%**	**1.6%**

Source: ClusterMonitor Database CBS/Dialogic, 2000.

If we extend our analysis to include non-technological innovation and its various forms (Table 7), the picture does not change dramatically. The construction cluster as a whole scores below average for all firms and the construction firms once again record the lowest scores of the construction cluster. The scores for strategy and (re-)organisation in particular are far below average. These are further indications that aspects such as client orientation and organisational capabilities are not considered very important within the construction core.

Table 7. **Firms with non-technological innovation for the Dutch construction cluster, 1996**
Percentages

Cluster segments	Non-technological innovation	Innovative strategy	Innovative marketing	Innovative reorganisation	Innovative management
Total construction	49	37	18	26	11
- Manufacturing	51	38	24	30	12
- Construction	41	30	9	21	10
- Wholesale	57	44	24	30	6
- Services	78	66	49	41	19
Total all firms	**55**	**42**	**26**	**30**	**12**

Source: ClusterMonitor Database CBS/Dialogic, 2000.

Networking and outsourcing

Since the level of investment in R&D and innovation is not particularly high within the construction cluster, and especially not in the construction segment, it is interesting to consider which information sources for innovation are used that might be perceived as a compensation mechanism for this low level. Table A.1 in Annex 3 provides an overview of the degree to which a selection of information sources for innovation are used. Most striking is the fact that construction firms score considerably below the average for the cluster as a whole for three sources. Pure construction firms are less likely to derive innovative ideas from their own firm/concern. This might be an indication of the fact that, due to building projects being the major organisational form, good ideas are not always retained and transferred to other parts of the organisation. Other studies indicate that knowledge management is one of the challenges confronting the construction cluster, in combination with more advanced use of ICT (AWT, 2000; Hillen, 2000). The score on innovative ideas stemming from clients is considerably lower in the construction segment of the wider cluster. This may be the result of the behaviour of clients who are not very innovation oriented,[12] but may also be an indication of firms' less developed capacity to adopt ideas from clients. The third source for which pure construction firms score considerably lower than service firms are ideas stemming from universities and allied institutes (while the score for ideas stemming from semi-governmental research is more in line with the overall average). Although well-developed construction-related knowledge infrastructure does exist, this raises doubts as to whether this interface is working effectively. More often than not, universities complain that construction firms do not show much interest in strategic or long-term research, but rather seek solutions for more acute problems.

Other networking indicators (Table 8) indicate average levels of R&D co-operation for the construction cluster as a whole, but show wide variety within the cluster. Construction and wholesale score below average, since the share of firms with any form of R&D is very low. If, for example, a pure construction firm has some form of R&D activity, it is more likely to deploy R&D outsourcing. This seems to be further proof of the marginality of the autonomous R&D base in the construction cluster and particularly in the construction segment. As observed at the beginning of this section, while construction firms may be able to depend on other segments for their innovations, they must invest in their own knowledge base if they are to be in a position to absorb and adapt such innovations.

Table 8. **Share of firms with some form of R&D co-operation or R&D outsourcing in the Dutch construction cluster, 1996**

Percentages

Cluster segments	Share of all firms with national and/or international R&D co-operation	Share of all firms with private and/or public R&D co-operation	Share of all firms with R&D activities that outsource R&D
Total construction	8	10	44
- Manufacturing	15	18	37
- Construction	3	4	64
- Whole sale	4	4	36
- Services	22	30	36
Total all firms	**9**	**11**	**51**

Source: ClusterMonitor Database CBS/Dialogic, 2000.

This becomes even more apparent when we consider the relative importance of foreign firms in the Dutch construction cluster (Table 9). Such firms are limited in number, but are jointly responsible for 25% of the R&D in the construction cluster as a whole. Foreign firms are relatively important in the wholesale segment, but do not perform any R&D. In the construction and services segment of the construction cluster in particular, foreign firms do not play a very important role, although they do perform six to ten times as much R&D as would be expected purely on the basis of the number of employees. Foreign firms possibly recognise opportunities to gain know-how and perform R&D in an attractive environment.

Table 9. **Relative importance of foreign firms in the Dutch construction cluster, 1996**

Percentages

Cluster segments	Share of foreign firms (number of firms with headquarters abroad)	R&D of foreign firms as a % of all R&D man-years in the Netherlands	Share of employees of foreign firms as a % of employees in the Netherlands
Total construction	4.9	24.8	8.2
- Manufacturing	8.2	17.4	19.7
- Construction	0.7	25.5	2.5
- Wholesale	18.6	0.0	23.9
- Services	4.5	30.7	5.3
Total all firms	**7.7**	**23.6**	**12.3**

Source: ClusterMonitor Database CBS/Dialogic, 2000.

Performance

Innovation is not performed for its own sake; its goal is better corporate performance. Although the relationship between innovation and corporate performance is a complex one, descriptive tables may help in assessing the importance of innovation, in this case in the construction cluster. The percentage of firms in a given sector or cluster which report improved sales through innovation is an indication of the importance of innovation in corporate performance (or rather how it is perceived). The construction cluster as a whole already scores 5% lower than the score for all firms (22% and 27% respectively view innovation as being of considerable importance for improving sales), which can be largely attributed to the low score of the construction segment within the cluster. This raises questions regarding the causes of this observation. Is innovation less important (than price, for example) in the construction segment for firms seeking to improve their position on the sales market *vis-à-vis* their competitors? Or is innovation not perceived as important and therefore less often reported as being a key factor for corporate performance?

Table 10. **Number of businesses in the Dutch construction cluster for which sales have improved due to innovation, 1996-94**[1]

Percentages

Cluster segments	Manufacturing	Construction	Wholesale	Services	Total cluster	Total all firms
Little importance	38	55	41	19	39	34
Slight importance	43	34	37	43	40	39
Considerable importance	19	10	22	38	22	27
Total	100	100	100	100	100	100

1. This measure was chosen because the usual indicators, "percentage of sales achieved from products new to the company" and "percentage of sales achieved from products new to the branch", were not available for all service sectors. The latter are included in the innovation survey for 1998, which will be linked to the ClusterMonitor data set.

Source: ClusterMonitor Database CBS/Dialogic, 2000.

We indicated above the various challenges with which the construction cluster is faced and which will require innovative solutions to be found. The indicators presented in this section show that the construction cluster scores below average as far as (mostly technological) innovation is concerned.

Some caution should be exercised here, however, as it is mainly the pure construction firms which score poorly in terms of innovation. Within the wider construction cluster, manufacturing firms and service firms, in particular, appear to be considerably more innovative. Another reason for avoiding premature conclusions is the fact that the available data mainly measure technological innovation whereas organisational and market innovations were observed to be of considerable importance in the construction cluster. However, the Porter analysis of the construction cluster in the early 1990s (Jacobs *et al.*, 1992, p. 121) observed that the Dutch construction cluster is "not so much in need of technological innovation as in need of social-organisational innovation: the way competition and co-operation is balanced needs revising, other project networks, different visions on innovation as well as less of a wait-and-see approach towards innovation and major client groups." More recent studies point in similar directions, especially in terms of "barriers to innovation in construction" (AWT-advice, p. 4) and "reasons why innovations do not materialise in construction" (Verweij, 2000, p. 39; NRK, 2000, pp. 48-49). The barriers which are most often cited are:

- Difficulties in co-ordination between the various disciplines involved in the construction process as well as the various phases of this process (and lack of trust) leading to segmentation, and fragmentation of markets. This implies difficulty with more integrated construction processes and true client orientation. The culture of projects implies, in practice, that innovative solutions to problems are not always well retained and that learning is limited.

- A risk averse culture and a focus on cost rather than on value added and innovation opportunities. Most innovations are still of the technology-push type (not least from innovative suppliers) and less of the market-pull type.
- Detailed procurement does not leave much freedom for innovative solutions.
- Well-developed order portfolios provide no incentive to move forwards with innovative constructions and experiment, while low margins imply that little finance is available for investing in R&D and innovation.
- As there is no real market pressure to innovate, there is plenty of room for improvement in knowledge transfer towards firms and articulated demand by firms towards the knowledge infrastructure. A number of intermediary organisations are active in this field.

These qualitative statements, which indicate a considerable need for organisational and market innovations, cannot, however, be substantiated quantitatively using the standard innovation data[13] given in the ClusterMonitor database.

Cluster-relevant government policies

Traditionally, various strands of policy making affect the structure and dynamics of the construction cluster. We will begin by mentioning a few of these "contextual policies" before moving on to focus on the use of innovation policy instruments by firms in the construction cluster. We shall then provide a few examples of specific actions directed towards facilitating the functioning of the construction cluster. It should be kept in mind, however, that some of these policies clearly focus on or affect specific segments of the wider construction cluster. In the area of residential housing, for example, there is a long tradition of detailed regulation, established for reasons other than supporting innovation, *e.g.* minimum quality levels, keeping housing affordable, ensuring that enough houses are built, systematically investing in renovating older city areas, detailed spatial planning to allow for a variety of spatial functions, and fighting urban sprawl while using the limited space available as efficiently as possible, etc. These goals may be important in themselves but may also lead to a certain construction paradigm or culture which can hinder innovation. However, in a field such as civil and hydraulic engineering, performance *vs.* compliance with standards leaves much more room for innovation. Construction activities in a typical soft-soil Delta area have led to the construction of innovative buildings and the creation of a unique knowledge base. Here, the demanding customer role played by the government is vital if innovation is to occur. These two examples illustrate how regulatory controls have different effects on each component of the wider construction (mega-)cluster.

"Contextual" policies

Construction as an activity is greatly affected by government policies in the fields of spatial planning, traffic and transport, housing and the environment. These are policy fields which have been in a state of flux over the last few years, not least because economic growth has increased the need to invest in infrastructure, accessibility, housing and the quality of the environment in the widest possible sense. Most recently, the white paper on spatial planning has once again focused the discussion on the role of government in spatial planning. This is an important document since many spatial functions compete for limited space. Which regions will be allowed to invest massively in new housing or industrial estates? To what extent can the Green Heart of the urban regions of the Netherlands be protected from the pressing need for space? To what extent can government steer spatial and infrastructure developments? How much will government invest in keeping the western part of the Netherlands, in particular, accessible? Similarly, white papers on traffic and transport or the *Action Plan on Spatial Economic Policy* serve a comparable function in discussing ways in which the numerous spatial and accessibility problems, the functioning of the main ports (Schiphol area, Port of Rotterdam), the planning of industrial estates and the vitality of urban areas can be improved. The recent white paper on Housing (Ministry of Housing, Spatial Planning and the Environment, 2000) will affect housing policies in a similar way. Most noteworthy for the construction cluster are the attempts to simplify regulations for housing construction and at the same time to introduce new regulations regarding safety and energy efficiency of housing, for example. Furthermore, there is a call for more client-oriented building concepts since it has become apparent that the construction industry cannot always comply with changing consumer preferences. Also included are initiatives to support sustainable buildings, a subject to which greater policy attention has been given in recent years and which, together with the various environmental policy initiatives, has led to a series of experiments (sustainable buildings, energy-saving programmes, etc.) which serve as a source of innovation for firms in the building cluster.

Current use of innovation policy instruments

Participation in the kind of experimental programmes mentioned above, which are related to sustainable housing or new housing concepts for specific target groups (*e.g.* the elderly), is a source of innovation which should be welcomed, particularly as participation in the regular innovation programmes, as shown in Table 11, is rather bleak for the pure construction firms in particular. In line with what has been said above concerning the dynamics of the construction cluster as a whole, it will come as no surprise that, even when only innovative firms are considered, manufacturing and service firms seem to benefit far more from the existing innovation support instruments. This could be due to the design of the instruments, but it is more likely that the lack of priority afforded to innovation by construction firms plays a role as well.

Table 11. **Businesses in the Dutch construction cluster which received government grants related to innovation, 1996**
Percentages

Cluster segments	As a % of all firms	As a % of all innovative firms
Total cluster	9	27
- Manufacturing	24	45
- Construction	2	10
- Wholesale	4	13
- Services	17	31
Total all firms	**9**	**25**

Source: ClusterMonitor Database CBS/Dialogic, 2000.

Cluster-inspired policy initiatives

In addition to the regular innovation programmes, various policy initiatives have been undertaken to spur the construction cluster to become more innovative and competitive.[14] These initiatives were partly inspired by the cluster policies introduced in the second half of the 1990s, which were then structured as framework policies, demanding end user or innovative procurement policies, and brokerage policies (Roelandt *et al.*, 1999, p. 329). Examples of each are provided below.

First, serious attempts are underway to simplify the regulations concerning spatial planning, housing and construction. Such regulations have evolved into a complex system which increases costs and could, for example, deter foreign firms from entering the Dutch market (CPB, 1999), thus limiting market competition. It has been suggested, for example (see CPB, 1999), that the way in which the larger real estate developers and building companies in particular are allowed to anticipate the changing use of space, may well affect competition and thus innovativeness. These firms buy the right to build as it were, but it should not be taken for granted that these are also the most efficient and innovative firms.

Second, government has started to experiment with innovative procurement. The public sector spends more than NLG 50 billion annually on purchases of goods, services and construction work. Industry is challenged to greater creativity and improved price-quality ratios through the strategic co-ordination of government procurement and tendering policies, imposition of functional requirements, consistent application of European tendering rules and the encouragement of innovative alliances (Ministry of Economic Affairs, 1999, p. 35). A task force on innovative procurement has initiated detailed analysis of various markets, including construction markets, to take stock of the possibilities for innovative procurement. In a similar way, experiments are underway with public/private partnerships to ensure that certain infrastructure works are realised on time. Quite often, this entails a shift away from a one-off investment decision based on price towards forms of procurement in which there is more room for innovative solutions (even more so when operational costs are part of the equation). Both innovative procurement and public/private partnerships are examples of policy experiments which substantially increase the room for innovation in the construction cluster since government is a major acquirer of buildings, urban reconstruction and infrastructure works. It is, however, too early to judge whether these policies will pay off in terms of innovation.

Government also plays a brokerage role in the construction cluster in various ways. For example, through steering the relevant knowledge infrastructure into place for the construction sector. The large research organisation, TNO, includes an important institute for the construction discipline, for example. This institute is partly financed by government, and some degree of co-financing is now mandatory for such government-financed programmes in order to ensure that the results are relevant for industry. In the early 1990s, research programmes were established to stimulate industry, university and research institute co-operation in research[15] on themes such as subsoil building and IT in constructions, and civil engineering.[16] Notwithstanding these and other efforts, the Advisory Council on Science and Technology recently concluded, on the basis of a foresight study into the construction sector, that the interaction between the practice of construction and the knowledge infrastructure could be further improved (AWT, 2000).[17] The exchange of knowledge remains overly focused on practical problem solving. Among other issues, the Council recommended heavier investment in knowledge and skills related to the construction process, to ensure that TNO construction not only invests in the direct technical problem solving in which industry is most interested, but also in strategic research for the somewhat longer term (AWT-advice, 2000, pp. 5-6; AWT-foresight, 2000, p. 41). Another example of typical brokerage policies aimed at facilitating the functioning of the cluster are a series of three strategic conferences organised together with an industry association and a consultancy firm (NRK, 2000). Players from all over the construction cluster participated in these conferences, which were aimed at encouraging reflection on innovation, increasing insight into major trends in construction, facilitating network building and establishing contacts between possible collaborators working on the development of innovation trajectories. Other examples of brokerage policies include performing a ClusterMonitor study on construction (Verweij *et al.*, 2000). This study not only informs policy makers, but also functions as a document for analysing the strengths and weaknesses of various segments in construction and, more generally, as a device for stimulating discussion on innovation and upgrading in the various segments of the construction cluster.

Summary and conclusions

As a mature cluster, the construction cluster is based on a combination of traditional skills, industrial technologies and an informal culture of the "'way things are done". While sometimes including highly innovative parties, the Dutch construction cluster is a fairly traditional cluster, a project-based cluster in which informal networks and relations are important and knowledge management practices need further development. Proven technology and organisational concepts prevail as well as risk-averse clients, detailed procurement and a clear focus on prices and costs. Current challenges include scarcity of space, requiring creative solutions; more demanding end users; a changing role for government; and increased competition on construction markets. The latter comes particularly from non-construction firms in the cluster that have started to integrate further backwards and forwards in the value chain. It is against this rather complex organisational background that innovation in the construction cluster has to be examined. The functioning and innovation style, or more likely innovation styles, in the various sub-clusters of the construction mega-cluster, are specific to the cluster. In well-established, mature clusters in particular, "recipes" for innovation cannot be copied or changed overnight.

As in most other clusters, innovation in the construction cluster is a multifaceted phenomenon involving clear technological, organisational and market aspects. The quantitative data presented in this contribution mainly focus on technological innovation in the Dutch construction (mega-) cluster. Innovation in the overall construction cluster was found to be higher than for the pure construction industries at the core of the cluster. The latter score considerably lower in terms of (mainly technological) innovation, innovation networking and the degree to which innovation is thought to have contributed to added success on sales markets. The same data show that the manufacturing and service firms active in construction markets perform much better. For their (technological) innovation, construction firms are dependent on innovation in manufacturing firms both upstream and downstream in the value chain. This need not be a problem if the ability to absorb out-of-industry technology and the links with end users are well developed. It is precisely these factors that qualitative studies highlight as being important barriers for innovation. Such barriers include the need to co-ordinate disciplines and the various phases in the building process, a limited sensitivity to the needs of end users, detailed procurement and the current lack of market pressure to move forward with innovative products. However, these qualitative assertions, which mainly point to a considerable need for organisational and market innovations, could not be substantiated quantitatively using standard innovation data as such data mainly measure technological innovation.

Policies relevant to the cluster are manifold. Spatial planning, traffic and transport, housing and environmental policies provide the context in which firms in the construction cluster have to manoeuvre. However, these policies can have different effects on each component of the wider construction (mega-)cluster. Furthermore, it can be concluded that (room for) innovation, in this case in the construction sector, is shaped as much by general policies as by specific cluster or innovation policies. Regular innovation schemes were found to be better used by the non-pure construction firms within the cluster. Partly as a result of the introduction of cluster policies, new initiatives have been taken, such as simplification of regulations (combined with new regulations), experiments with innovative procurement and public/private partnerships and various brokerage policies. The latter include initiatives to improve the industry-science interface, performing ClusterMonitor studies and the organisation of strategic conferences, one of the targets being to facilitate the ability of firms to work together to develop ideas for innovative products and concepts.

On the one hand, cluster policies have been designed as a result of the particular characteristics of the cluster in question, which become most visible when procurement policies are concerned. On the other hand, the introduction of cluster policies has functioned as a relatively new way of policy determination aimed at revitalising mature clusters like the construction cluster. As a number of the sub-clusters within the wider mega-cluster varies considerably, cluster policies aimed at removing barriers to innovation in construction are most likely to be implemented at a lower level of aggregation.

Annex 1

RELATIVE REPRESENTATION OF THE DUTCH CONSTRUCTION INDUSTRY

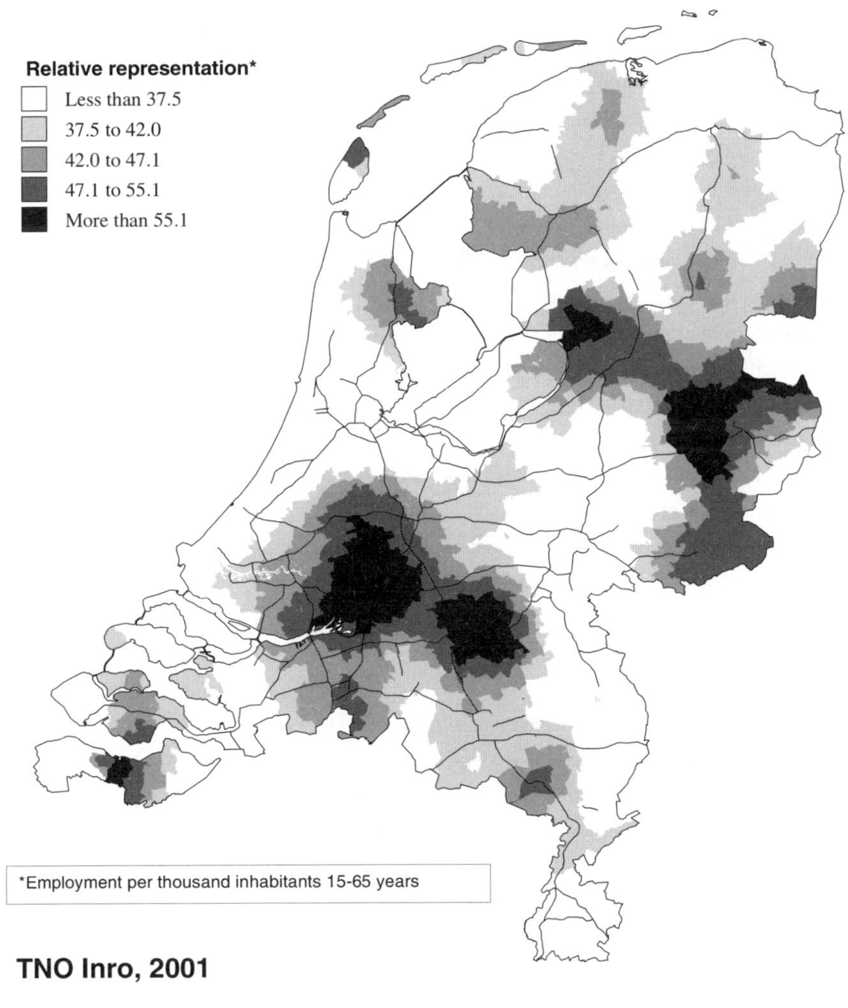

Annex 2

THE CLUSTERMONITOR DATABASE: SELECTION OF INDUSTRIES AND ANALYSES PERFORMED

Manufacturing	
14210	Operation of gravel and sand pits
14220	Mining of clays and kaolin
20101/2	Saw-milling and planing of wood; impregnation of wood
20200	Manufacture of veneer sheets; manufacture of plywood, laminboard, particle board, fibre board etc.
20301/2	Manufacture of builders' carpentry and joinery
24300	Manufacture of paints, varnishes and similar coatings, printing ink and mastics
25230	Manufacture of builders' ware of plastic
26110	Manufacture of flat glass
26120	Shaping and processing of flat glass
26140	Manufacture of glass fibres
26150	Manufacture and processing of other glass, including technical glassware
26220	Manufacture of ceramic sanitary fixtures
26230	Manufacture of ceramic insulators and insulating fittings
26300	Manufacture of ceramic tiles and flags
26400	Manufacture of bricks, tiles and construction products, in baked clay
26510	Manufacture of cement
26520	Manufacture of lime
26530	Manufacture of plaster
26611/2	Manufacture of concrete products for construction purposes
26620	Manufacture of plaster products for construction purposes
26630	Manufacture of ready-mixed concrete
26640	Manufacture of mortars
26650	Manufacture of fibre cement
26820	Manufacture of other non-metallic mineral products n.e.c.
28110	Manufacture of other non-metallic mineral products n.e.c.
28120	Manufacture of builders' carpentry and joinery of metal
28220	Manufacture of central heating radiators and boilers

Construction	
45111/2	Demolition and wrecking of buildings; earth moving
45120	Test drilling and boring
45211/2/3	General construction of buildings and civil engineering works
45220	Erection of roof covering and frames
45231/2	Construction of highways, roads, airfields and sport facilities
45240	Construction of water projects
45251/2/3/4	Other construction work involving special trades
45310	Installation of electrical wiring and fittings
45320	Insulation work activities
45331/2	Plumbing
45340	Other building installation
45410	Plastering
45420	Joinery installation
45430	Floor and wall covering
45440	Painting and glazing
45450	Other building completion
45500	Renting of construction or demolition equipment with operator
71320	Renting of construction and civil engineering machinery and equipment
Wholesale	
51531-8	Renting of construction and civil engineering machinery and equipment
51541/2	Wholesale of hardware, plumbing and heating equipment and supplies
51620	Wholesale of construction machinery
Services	
65221	Other credit granting
70110	Development and selling of real estate
74141	Renting of personal and household goods n.e.c.
74201-8	Architectural and engineering activities and related technical consultancy

Use of the ClusterMonitor database for analysing innovation in the construction (mega-)cluster

At the end of 1999, the Dutch Ministry of Economic Affairs had an analytical instrument developed, the so-called ClusterMonitor, which included a large database formed in association with Statistics Netherlands (CBS). In order to produce this data set, various production and innovation statistics were linked at the micro-level (for firms with more than ten employees). The database contains 46 900 firms and allows for a comparison of firms belonging to a certain cluster versus all firms in the economy. This data set allows us to truly adopt a cluster approach and, aside from the construction firms, also includes data on manufacturing, wholesale and service firms which are part of the wider construction cluster, to an important degree. Mining and quarrying firms were not included in the innovation statistics and are therefore not included in the ClusterMonitor database.

The four sub-groups are significantly different in terms of their innovativeness, and the sub-clusters have therefore been carefully chosen. However, we are aware of the fact that the somewhat aggregated representation of data cannot provide the full detail of cluster behaviour and need not reveal important differences in innovation behaviour within categories used. In order to test this, we have observed the differences within the separate sub-clusters, and found marked differences in innovative behaviour, as we had indeed expected. Within the sub-cluster of the core construction firms, for example, "building completion" is less innovative than "installation firms". Similarly within manufacturing, some of the suppliers of specialist building materials and machinery outperform the manufacturers of more standard materials. However, we are not allowed to publish more detailed micro-data and therefore cannot present more disaggregated tables.

Further, larger firms are more likely to innovate, but there may be differences between the four sub-clusters. We therefore additionally estimated the chance that a firm is innovative or not for all firms in the Netherlands with at least ten employees. A probit model was used to assess the effect of a firm's size on the probability that a firm is innovative, when compared with the average probability of a randomly selected firm from this sub-cluster. The dependent variable is the dummy variable "innovative or not". The independent variables are the five sub-clusters of the overall (mega-)cluster construction. The referent group is the whole economy except the firms included in the (mega-)cluster construction. The highest elasticities are found in manufacturing, indicating that the probability of innovation increases most rapidly for large-scale manufacturing firms. Elasticities are lowest in the construction sub-cluster. The sub-clusters were also crossed with "number of employees". If we look at the elasticity of number of employees on innovation, the elasticity is highest for "manufacturing".

Variable	Coefficient (+ = significant at 95% level)
Constant	-0.2267+
Firm size dummy sub-cluster manufacturing	0.044
Firm size dummy sub-cluster construction	-0.997+
Firm size dummy sub-cluster wholesale	-0.429+
Firm size dummy sub-cluster services	0.578+
Firm size dummy "alternative selection"	-0.244+
Number of employees "non-construction"	0.001+
Number of employees dummy sub-cluster manufacturing	0.008+
Number of employees dummy sub-cluster construction	0.001+
Number of employees dummy sub-cluster wholesale	0.004+
Number of employees dummy sub-cluster services	0.003+
Number of employees dummy "alternative selection"	0.006+

Annex 3

USAGE OF INFORMATION SOURCES BY FIRMS IN THE DUTCH CONSTRUCTION CLUSTER, 1996, AS A PERCENTAGE OF ALL INNOVATIVE BUSINESSES

	Manufacturing	Construction	Wholesale	Services	Total construction	All firms
Innovation ideas within own firm or within same concern						
Source not used	52%	71%	56%	47%	58%	54%
Somewhat important	11%	11%	4%	7%	10%	10%
Important/very important	37%	19%	40%	46%	32%	36%
Innovative ideas through clients						
Source not used	30%	47%	42%	26%	35%	32%
Somewhat important	32%	24%	15%	33%	27%	25%
Important/very important	38%	29%	43%	41%	39%	43%
Consultancy and research firms in the market sector						
Source not used	74%	68%	82%	62%	68%	64%
Somewhat important	15%	17%	5%	25%	19%	21%
Important/very important	11%	16%	13%	13%	13%	15%
(Semi)-governmental research institutes						
Source not used	68%	67%	77%	59%	68%	73%
Somewhat important	18%	21%	9%	24%	20%	16%
Important/very important	14%	11%	14%	17%	12%	11%
Universities, including allied institutes						
Source not used	87%	87%	94%	54%	81%	81%
Somewhat important	10%	10%	5%	20%	12%	12%
Important/very important	4%	3%	1%	26%	8%	7%
Innovation ideas through trade journals and trade exhibitions						
Source not used	35%	29%	37%	16%	29%	28%
Somewhat important	56%	56%	52%	43%	54%	51%
Important/very important	9%	14%	11%	40%	17%	21%

Source: ClusterMonitor Database CBS/Dialogic, 2000.

NOTES

1. See Dahl and Dalum (Chapter 9 of this volume) and Vock (Chapter 11 of this volume) for an analysis of the Danish and Swiss construction clusters, respectively.

2. For an informative study on how recent organisational and technological change might affect the already complex organisational and social practices in the British construction industry, see Miozzo *et al.* (2000).

3. In a recent publication by NRK (2000, p. 20), quality of living, sustainability, energy management, safety, maintenance, costs of construction, flexibility, environmental and (European) construction regulations were listed as drivers for change in construction. In an advisory paper on the interaction between the construction industry and the knowledge infrastructure in construction, from the Advisory Council on Science and Technology (AWT), published in mid-2000, three major challenges were identified: the changing role of government; the need to switch from managing projects to processes; and the changing content or functionality of the demand for buildings and constructions (AWT, 2000).

4. This challenge is particularly visible where the mobility puzzle in the urban areas is concerned or in huge infrastructure projects such as the construction of express train links and extension to harbours and airports, redevelopment of inner cities and establishment of large new building sites close to the major agglomerations.

5. The first attempts at internationalisation, made by two large Dutch builders HBG and NBM Amstelland, both of which have been struggling with their German subsidiaries, have not been promising.

6. As noted by Hauknes for the agro-food cluster (Chapter 8 of this volume), this is one of the main reasons for not underestimating the potential effects of innovation in so-called mature clusters.

7. The Annex to Chapter 7 describes the ClusterMonitor instrument. The data set only includes firms with ten or more employees. Annex 2 shows the industries included in the construction cluster and some additional information on how the ClusterMonitor database was used.

8. In 1999, these were respectively HBG (almost EUR 5 billion), NBM-Amstelland (EUR 2.6 billion), VWS (EUR 2.4 billion), Ballast Nedam (EUR 2 billion), Heijmans (EUR 1.2 billion) and BAM Group (EUR 1.1 billion) (*Financieele Dagblad*, 19 February 2001).

9. These include "construction team" approaches; turnkey completion of projects; building and operation of constructions; pre-fabricated and modular construction systems; plug-and-play buildings; faultless construction trajectories; assembly at the construction site, etc.

10. An interesting comparative study among contractors in six European countries has been performed by Miozzo *et al.* (2001), focusing on the mechanisms of innovation at the firm level (especially the link between corporate governance and innovation). This study indicates, for example, that the ability of contractors to undertake R&D in production technologies varies widely across countries. In the Dutch case, corporate governance was found to include elements of both the Anglo-Saxon and Germanic models. It is further concluded that "although some contractors are active in product and process developments, it is inter-firm collaborations across the supply chain and with research institutions, in

many cases supported by the government, that has been the most important determinant for the development of cost-cutting innovations" (p. 33).

11. It is increasingly acknowledged that innovation in construction requires a value-chain approach, see for example the forthcoming First International Conference on Innovation in Architecture, Engineering and Construction, Loughborough University, United Kingdom, and the various activities of the International Council for Research and Innovation in Building and Construction (http://www.iabse.ethz.ch/lc/cib.html).

12. In the market for residential buildings, many home owners prefer pre-war style houses. In some newly developed projects, "retro" housing, combining the most characteristic elements of these houses with modern comfort have proven very popular.

13. Future Community Innovation Surveys should address this clear deficiency in the current innovation statistics.

14. We do not include the overall strengthening of competition policy since the early 1990s, which most likely affected the functioning of the construction cluster, in which attempts to limit rivalry were not uncommon (see, for example, Jacobs *et al.*, 1992).

15. Similar attempts are known to exist in other countries as well. In the United States, for example, similar public-private partnerships focusing on innovation in construction and building emerged during the 1990s. The National Science and Technology (NSTC) Sub-committee on Construction and Building has played a substantial role in co-ordinating federal R&D in collaboration with industries of construction. Removing barriers to innovation is the leading principle and has materialised in streamlining of regulations concerning building, supporting various innovation centres. Further, a Partnership for Advancing Technologies in Housing (PATH) and Partnership for Advancing Infrastructure and its Renewal (PAIR) were launched, along with other co-operative programmes (see NSTC, 1999).

16. These four research programmes of considerable size are financed through the so-called ICES/KIS means. This budget is financed through the profits made by the government from the sale of natural gas. The funds are used to invest in strengthening the physical economic infrastructure.

17. In some areas, new initiatives have been taken, such as the establishment of the Delft Cluster. Here, five knowledge institutes in the area of civil engineering in Delft have joined forces to create an open network for knowledge creation and diffusion through increasing the efficiency of research programmes as well as the quality of education, research and advice. The main goal is for the network to become a leading knowledge institute in the area of sustainable development of densely populated deltas or coastal areas.

REFERENCES

AWT (2000), "Bouwen op Kennis" [Building on Knowledge], report of the Foresight Committee for Construction, Background Study 17, Advisory Council for Science and Technology Policy, The Hague.

AWT (2000), "Advice Based on the Foresight Report [available in Dutch only], Advisory Council for Science and Technology Policy, The Hague, http://www.awt.nl/nl/briefbouw.html.

Barret, P. and M. Sexton (2000), "The Transformation of 'Out-of-industry' Knowledge into Construction Industry Wisdom", mimeo, Research Centre for the Built and Human Environment, University of Salford.

CBS (2000), *Bouwcijfers 1999* [Construction Figures 1999], Heerlen/Voorburg.

CPB (1999), "Woningbouw. Tussen markt en overheid" [Civil Construction. Between Market and Government], The Hague.

Dahl, M.S. and B. Dalum (2001), "The Construction Cluster in Denmark", Chapter 9 of this volume.

Hauknes, J. (2001), "Innovation Styles in Agro-food Production in Norway", Chapter 8 of this volume.

Hillen, M.Ph. (2000), "Bouwen op kansen. Een verkenning in elf signalementen" [Building on Opportunities. A Foresight in Eleven Sketches], SMO, 2000-2, The Hague.

Jacobs, D., J. Kuijper and B. Roes (1992), "De economische kracht van de bouw. Noodzaak van een culturele trendbreuk" [The Economic Power of Construction. The Need for a Cultural Breakthrough], SMO, 91-4, The Hague.

Ministry of Economic Affairs (1999), *Scope for Industrial Innovation. Industrial Policy Agenda*, The Hague.

Ministry of Housing, Spatial Planning and the Environment (2000), *Nota Wonen. Mensen, wensen, wonen. Wonen in de 21e Eeuw* [White Paper on Housing. People, Desires and Housing. Housing in the 21st Century], The Hague.

Miozzo, M. and P. Dewick (2001), "Innovation and Corporate Governance in European Construction: Mechanisms of Innovation at the Firm Level", mimeo, Manchester School of Management, Umist, Manchester.

Miozzo, M. and C. Ivory (2000), "Restructuring in the British Construction Industry: Implications of Recent Changes in Project Management and Technology", *Technology Analysis and Strategic Management*, Vol. 12, No. 4, pp. 513-531.

NRK-Bouw (2000), "De Bouw innoveert. Verkenningen van innovatieve kunststof- en andere materiaaltoepassingen in de bouw. Resultaten van 3 strategische conferenties" [The Construction Sector Innovates. Foresight on the Innovative Use of Plastics and Materials in Construction. Results of 3 Strategic Conferences], NRK, Leidschendam.

NSTC Sub-committee on Construction and Building (1999), *Construction and Building. Interagency Program for Technical Advancement in Construction and Building*, Washington, DC.

Roelandt, Theo, Pim den Hertog, Jarig van Sinderen and Norbert van den Hoven (1999), "Cluster Analysis and Cluster Policy in the Netherlands", in OECD (1999), *Boosting Innovation: The Cluster Approach*, OECD, Paris, pp. 315-338.

Verweij, Maurits, Hans Praat, Hubert Sturm, Ruurd Jansen, Robert Goedegebuure (2000), "Clustermonitor Bouw" [ClusterMonitor Construction], Ministry of Economic Affairs, The Hague.

Vock, P. (2001), "An Anatomy of the Swiss Construction Cluster", Chapter 11 of this volume.

Chapter 11

AN ANATOMY OF THE SWISS CONSTRUCTION CLUSTER

by

Patrick Vock*
Centre for Science and Technology Studies (CEST), Swiss Science and Technology Council

Introduction

The construction cluster is primarily a national – in this case a Swiss – phenomenon, with only the suppliers and supporting industries having international components. Although it is currently experiencing a recovery, it still has to deal with the effects of the severe recession of the 1990s, and the general base for a take-off and further restructuring and innovation is small. The construction cluster is perceived as mature and established, although there is considerable room for applying new technologies and products as well as for introducing organisational innovations. Due to the ongoing structural changes in the cluster itself as well as in its environment, firms are forced to orient themselves more towards the needs of their clients, to adapt to growing competition and to consider the integration of activities such as the bundling of previously separate activities (hence the increasing importance of the "one-stop shop" for planning, building and managing). To overcome the severe fragmentation of the value chain in the construction cluster (especially between planning and production), innovations in project handling and knowledge management for the development of more integrated solutions are needed.

The aim of this paper is three-fold: the first two sections are dedicated to the identification and description of the Swiss construction cluster, based on a combination of a (quantitative) input-output analysis and a more qualitative investigation. The main segments – building construction and civil engineering; installation and completion; suppliers; related services – are characterised by their economic profile, including employment, firm structure and market orientation. The chapter then goes on to discuss dynamics, innovation and co-operation. The core of the cluster can be characterised as being moderately innovative, with suppliers being an important source of innovations. A central issue concerns the fragmentation of the activities in the life cycle of a building, mirrored in the very many organisations involved, the distributed responsibilities and the traditional project handling, which all lead to sub-optimal – and usually not very innovative – solutions. Finally, cluster-related policies are discussed, emphasising the crucial role played by state agencies as procurers and brokers.

* The author gratefully acknowledges useful comments by the editors and by A. Berwert.

Identification of the Swiss construction cluster

In this chapter, construction and its related businesses are analysed using the cluster or value-chain approach,[1] with a view to identifying knowledge flows and innovation. This approach combines a more quantitative analysis of user-supplier linkages exploiting input-output data (transaction matrix) with a qualitative investigation of the value chains leading to the demand for buildings.

The analysis of the Swiss input-output data carried out by means of an algorithm[2] used in other OECD countries (Peeters *et al.*, Chapter 12 of this volume) shows that the cluster has two main strands: "building construction and civil engineering activities" (20) and "installation and completion activities" (21), with their corresponding preceding activities (see Figure 1). Both strands (20 and 21) are forward-linked to the "real estate business" (31). Upstream activities are, on the one hand, the "non-ferrous metals" (16) and "electricity, gas and water" (2) sectors and, on the other, the "woods and furniture" (8) and "wood industry" (9) sectors. Due to the level of aggregation of the input-output data (37 sectors), the analysis does not yield more detailed results.

Nevertheless, the synthesis of various elements (input-output analysis, value-chain approach, statistics, expert opinions, knowledge flows) confirms this basic cluster structure (depicted in Figure 2). Private and public customers demand a wide variety of buildings for living or industrial purposes. These core products of the cluster constitute the base for construction and all other complementary activities; namely, inputs from suppliers as well as supplementary installations and completion work. Various services accompany the different value chains. This economic tissue of the cluster is influenced by other organisations; namely, professional associations, educational institutions and state agencies. In reality, the construction cluster – even when perceived as a mature cluster – is not as clear-cut as the sketchy graph suggests. Some activities are closely tied to the core activities while others have much looser links.

Figure 1. **The Swiss construction cluster as the result of input-output analysis**

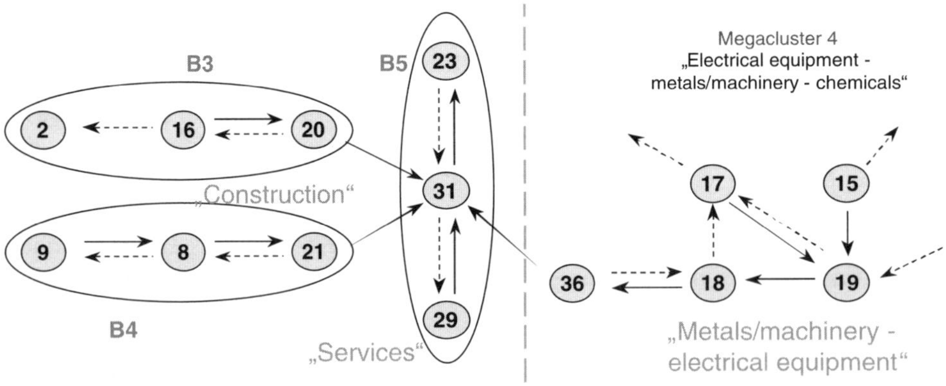

Key: [2] Electricity, gas and water; [8] Wood and furniture; [9] Wood industry; [15] Rubber and plastics; [15] Non-ferrous metals; [17] Metals; [18] Machine building; [19] Electrical equipment; [20] Construction (main activities); [21] Building (finishing and completion); [23] Retail trade; [29] Banking; [31] Real estate; [36] Government.
Source: Peeters *et al.*, Chapter 12 of this volume.

Figure 2. **The Swiss construction cluster**

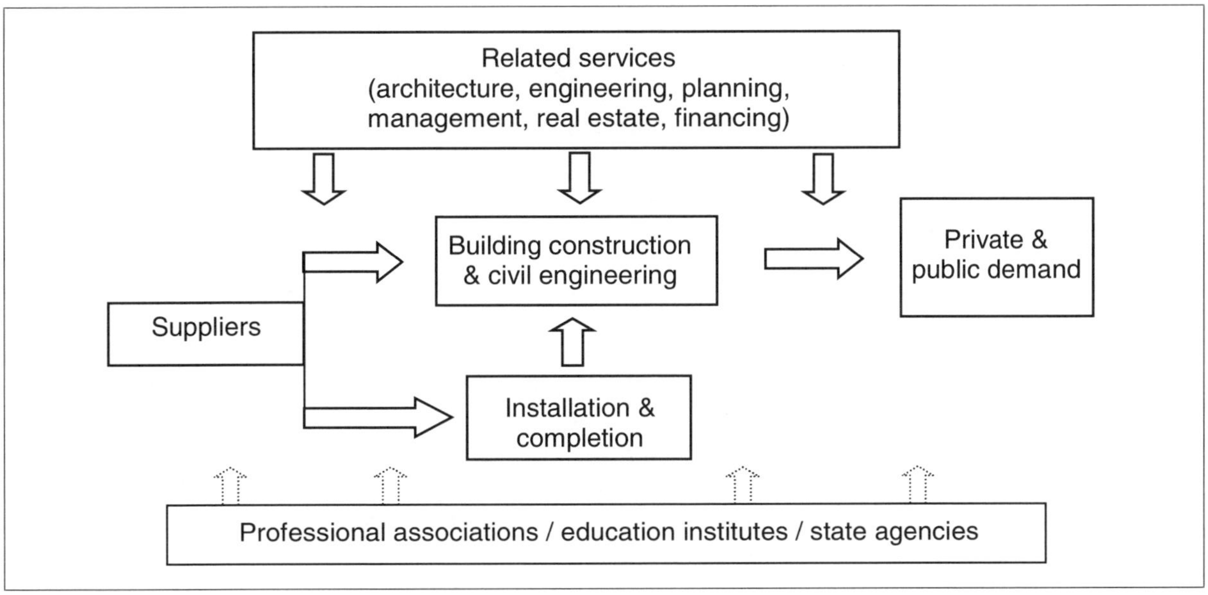

Profile of the Swiss construction cluster

The construction cluster constitutes a considerable part of the Swiss economy, contributing 6%-10% to Swiss GDP (depending on the indicator) and employing around 400 000 people (some 10% of the Swiss workforce). Table 1 summarises the economic tissue of the construction cluster based on employment data, firm structure and market orientation.

Table 1. **Synopsis of the economic tissue of the construction cluster**

	Employment 1998	Structure	Market orientation
Building construction and civil engineering	156 000	SMEs. 3 big firms (loose holdings), followed by 10 smaller firms.	Domestic.
Installation and completion	81 000 + 52 000	SMEs.	Sanitary install.: European. Rest: Domestic.
Suppliers	50 000 (estimate)	Some big (international) firms. Rest: SMEs.	Cement, construction chemicals, polymeric materials: International (specialised) players. Rest: Domestic.
Related services	32 000 + 17 000 + 18 000	SMEs.	Domestic. Some firms of international repute.

Note: Some figures are estimates.

The cluster is based on heterogeneous demand: the demand for houses as a final good; and the demand for buildings for industry as an intermediate demand, implying different owners, requirements, etc. The fragmentation of the construction market is based on technologies, know-how, education or machinery and is not much determined by demand criteria such as functions, quality, standards, etc. Private construction spending is twice as high as public spending, with house building being the major private construction activity. Unfortunately, the majority of clients, especially unprofessional house builders as well as conservative public procurers, do not usually play the role of demanding customers. Furthermore, the severe ten-year recession which started in 1990 limited already scarce resources (due to very low profit margins) for longer-term development and did not nurture a climate of long-term change and innovation. Business-cycle-corrected employment data (1985-98) show – against the background of a general shift away from the secondary to the tertiary sector – a reduction in the core construction business as well as in the segment of potential suppliers. A clear upward trend for "other business activities" (NACE 74) is not fully mirrored in the sub-segment of construction-related services (for data, see Annex Table 1). Currently, there are signs of a smooth recovery in the construction cluster as a whole.

The core segment of "building construction and civil engineering" (NACE 45.2) was exposed to the severe recession, competition and restructuring. It is dominated by SMEs, although there are three big building contractors (with a turnover of more than CHF 500 million), followed by ten smaller ones. The big firms (Zschokke, Batigroup, Marti) all result from the mergers of the 1990s and incorporate dozens of firms with several thousand employees. Their activities usually comprise construction as the core [all segments (building and civil engineering) with varying specialisations], but also include services such as planning and general contractor activities and, above all, engage in real estate. The segment is oriented towards the domestic market. Previous engagement in neighbouring countries has now practically ceased.

Where are the available opportunities for Swiss construction firms? Due to the project-oriented work and hence the fragmentation of the construction process, a better integration in terms of activities, knowledge and other resources is needed. SMEs will continue to operate in the national market, but a better profiling based on quality and performance will need to gradually replace pure price competition. Large construction firms have the opportunity to bundle their competencies to allow them to operate on the world market, especially in regions with booming demand. Strengths lie in tunnelling activities, but also in segments such as hydro-electric power plants, subway construction, and highway and railway systems in mountainous regions (Girmscheid, 1997).

Construction business is supported by building installation (NACE 45.3) and building completion (NACE 45.4)[3]. This segment which is fairly substantial in terms of employment is dominated by domestically oriented SMEs, although some companies have grown to international scale (see example in Box 1).

Box 1. **Geberit: sanitary technology for the European market**

The Geberit group with a turnover of CHF 1.1 billion employs 4 300 persons and is active in more than 70 countries, although its core markets are situated in Europe. The company invests nearly 3% of sales in innovative products and, as a result, more than 30% of sales are generated by new or improved products which have been brought to market within the last three years. Electronic components are becoming increasingly important in sanitary engineering. R&D in the group – which has its headquarters in Switzerland – is performed by cross-functional teams applying information technology.

Construction and supporting businesses rely on a variety of inputs from the following sectors: Quarrying (NACE 14.1, NACE 14.2); Renting of construction/demolition equipment with operator (NACE 45.5); Manufacture of non-metallic mineral products (NACE 26), especially glass, ceramic tiles and flags, bricks, cement, articles made of concrete, plaster and cement; Manufacture of metal products (excl. machinery) (NACE 28); Wood-working and processing (NACE 20); Chemicals; Other products and services (wholesale trade, transport, energy). Since these branches are defined by products and not by the use of their products, it is difficult to determine the employment share of the construction cluster, but estimates attribute around 50 000 employees. This segment is dominated by SMEs (sometimes of regional importance) which are oriented towards the domestic market. However, there are some – in their fields – big, international players (Box 2). Undoubtedly, these internationally competitive and successful companies contribute to the dynamism and innovation of the construction cluster.

Box 2. **Suppliers: some international players of Swiss origin**

Holderbank: Holderbank is one of the world's leading suppliers of cement, aggregates and concrete. With a turnover of around CHF 11 billion, it employs over 40 000 persons in more than 70 countries. Founded in 1912, investments outside Switzerland (and increasingly to growing markets) began in the 1920s, establishing a strategy which assumes that demographic expansion is the driving force behind cement consumption. For its cement business, two companies, Management and Consulting Ltd in Switzerland and Engineering Canada Ltd, constitute a turntable for group-wide transfer of know-how, experience and best practices. These firms support the group's activities with services, advice and management tools in the fields of environmental performance, development of human resources, IT, process technology, engineering, geology, research and logistics.

Sika: Founded in 1910 in Switzerland, Sika is a globally active company in the field of construction chemicals and industrial materials, which employs around 7 000 persons and yields a turnover of CHF 1.6 billion. As a producer of speciality chemicals, both for construction and other industries, the company builds on strong and independent R&D organised in multidisciplinary teams.

Sarna: The company is specialised in the processing and application of polymeric materials, employs 2 500 persons and realises a turnover of CHF 666 million. The competence in polymers is applied in different fields, with only one of the three firms belonging to Sarna being relevant for the construction cluster; namely, Sarnafil, which delivers polymer membrane waterproofing systems for use in high-rise construction and civil engineering. The other two firms deliver products such as IT peripherals, electronics, medical technology and components for the automotive industry.

ZZ Holding: The recent development of this industrial conglomerate illustrates the importance of the international scale in the supplier business. Starting in 1912 with the merger of local brick plants, it has developed into an industrial group which in 1998 employed 2 900 persons and realised a turnover of around CHF 1 billion in the fields of machinery and plant construction, building materials and building systems, consumer and industrial products and others. In 1999 it sold its brick, roof tile and insulation materials activities to the Belgian/Austrian Koramic/Wienerberger Group, an international/European ceramics and brick producer. In view of the increasing concentration in Europe, with fewer but larger suppliers operating throughout the region, the construction business of ZZ Holding had become too small to be competitive on its own.

The related services segment plays an integrating role in the cluster and shows considerable innovation potential. Activities are quite heterogeneous, either preceding, following or paralleling the core activities of building construction. In these fields, process innovations are important and constitute a major potential for increasing the efficiency of the construction cluster. Although the delineation is not very strict, parts of the following activities are deemed to belong to the construction cluster: Architecture and engineering firms (NACE 74.2); Real estate (NACE 70); Financing. Architecture and engineering firms are mostly small enterprises, some of international repute. The business climate is shaped by competitions and tenders. Architectural competitions have a long-standing tradition, with the first rules implemented in 1877 by the Swiss Society of Engineers and Architects. In addition to firms which concentrate their activities in a specific part of the value chain, others – through their services – span the whole cluster. Steiner, for example, styles itself as a total services contractor, offering conception, design and planning, realisation, and property management.

Box 3. **OLMeRO: linking via the Internet**

This is a newly launched business-to-business (B2B) Internet-platform which aims to become the most important virtual marketplace for the Swiss construction business. OLMeRO is a start-up which grew out of the business plan competition, "Venture2000", organised jointly by the Federal Institute of Technology Zurich (ETHZ) and McKinsey. Placing of orders and trade in all kinds of material via the Internet have the potential to increase market transparency and the efficiency of transactions.

Dynamics, innovation and co-operation

Changes and challenges

Over the coming decades, the construction cluster could benefit from positive growth rates in the economy and population as well as from speculative price trends in real estate markets. Since the 1990s, the situation has changed fundamentally, and the construction cluster is experiencing not only cyclical problems but also structural changes. One indication is the detachment of construction investment from GDP development, leading to fierce price competition which is pushing down the average number of employees per firm while leaving the number of firms constant.

The demand for residential construction is strongly influenced by changes in age structure, income distribution, family structure and modes of living together. Next to costly renovations, there is increased demand for cheaper, standardised houses. According to different cost studies, around one-quarter of total construction costs could be influenced by introducing "industrialised construction" (Girmscheid and Hofmann, 2000). The construction of industrial buildings shows a trend towards shorter planning and production times, lower costs, reduced space requirements for workplaces as well as increased variability of use. These tendencies could lead to a fragmentation between purpose-built constructions with high requirements and cheaper multi-purpose buildings with modular structures. Construction activities related to infrastructure are shifting from new construction to renovation, from building construction to civil engineering, from road to railroad construction.

These general trends towards more advanced and demanding construction activities are impeded by several factors: the traditional and regional orientation of markets hinders further specialisation and the establishment of bigger units; reduction of excess capacity is difficult since entry barriers for newcomers are low; above-average performance is not adequately rewarded since price competition and personal networks determine which companies win the bids for construction projects; lack of transparency regarding performance, quality and costs; ignorance of long-term costs; tender strategies of firms, which tend to be oriented towards short-term full employment of resources; collusive behaviour (in 2000, the national competition commission had to sanction some distribution, tender and price agreements).

Innovation and performance

The construction cluster is perceived as not being very innovative. However, a closer look shows that some segments – especially those with suppliers and services but even some businesses in the more core activities – are more dynamic and innovative. Looking at R&D, construction-related activities are small, with some private applied R&D performed by suppliers and companies carrying out completion work, and some basic and applied R&D performed by the Federal Institutes of Technology (ETH), the specialised colleges (*Fachhochschulen*) and the professional associations.

Data from the two relevant innovation surveys for 1996 (including industry and construction and services) confirm this picture. The share of innovative firms (for example, 42% for construction) shown in Table 2 does not seem to be particularly low, but when these figures are compared with the industry average, construction and real estate activities score well below average and suppliers show only mediate innovativeness. Nevertheless, some specialty fields such as construction chemicals are very innovative; this is in line with the overall innovativeness of the chemicals sector, where more than 90% of firms conduct some innovative activities.

Table 2. **Percentage of innovative firms, 1996**

Core business	
Construction (45)	42%
Potential suppliers	
Manufacture of wood and wood products (20)	52%
Manufacture of other non-metallic mineral products (26)	54%
Manufacture of fabricated metal products (28)	67%
Potential related services	
Other business activities (74)	63%
Real estate activities (70)	40%
Industry	78%
Construction / services	58.7%

Source: Arvanitis *et al.* (1998); NACE codes in brackets.

Patents are often used as an indicator of the success of innovative activities. An analysis of Swiss patenting at the European Patent Office (Schmoch *et al.*, 2000) shows that the construction sector accounts for a considerable 6% of all Swiss patents. Since the mid-1980s, the number of patents has increased in accordance with the international trend. Key patenting fields relate to building construction and completion. The specialisation index[4] rose from 26 in 1984-86 to 42 in 1989-91 and up to 45 in 1994-96, indicating a very solid and even increasing strength in Swiss construction patenting. This patenting is broadly based on a wide number of firms, especially medium-sized firms.

Another indicator of the importance of innovations is the percentage of turnover attributable to new or improved products. Table 3 shows a lower importance for the core construction sector in comparison with suppliers or total industry.

Table 3. **Percentage of turnover attributed to new or substantially improved products, 1996**

Core business	
Construction (45)	9%
Potential suppliers	
Manufacture of wood and wood products (20)	13.5%
Manufacture of other non-metallic mineral products (26)	14.9%
Manufacture of fabricated metal products (28)	13%
Potential related services	
Other business activities (74)	9%
Real estate activities (70)	1%
Industry	22.7%
Construction / services	9.7%

Source: Arvanitis *et al.* (1998); NACE codes in brackets.

This is in line with the perception of actors (Table 4). Only 29% of construction firms consider their product innovations as important for their economic success, which is less than the average of potential suppliers and of total industry. It is interesting to see that, despite the equal importance of product and process innovations in the total economy, construction firms attribute a higher economic significance to process than to product innovation. This, and the fact that innovations which are new to the sector and not only to the firm, are quite rare, are clear signs of a mature sector.

Table 4. **Percentage of firms which rank their product/process innovations as "economically important", 1996**

	Product innovation	Process innovation
Core business		
Construction (45)	29%	35.8%
Potential suppliers		
Manufacture of wood and wood products (20)	35.8%	46.1%
Manufacture of other non-metallic mineral products (26)	49.2%	59.2%
Manufacture of fabricated metal products (28)	76.5%	47.4%
Potential related services		
Other business activities (74)	38.6%	46.2%
Real estate activities (70)	0%	25.1%
Industry	59.4%	57.8%
Construction / services	44.4%	44.3%

Source: Arvanitis *et al.* (1998); NACE codes in brackets.

How is innovation linked to performance? Unfortunately, there are no econometric studies available for the construction cluster. Nevertheless, data suggest stagnating performance with overcapacity. In spite, or perhaps because, of the recession of the 1990s, the construction volume per employee in the construction business rose by 6% between 1991 and 1998, going hand in hand with increased capital intensity. However, labour productivity is low and barely improving (Table 5). Of course, there is a considerable variance between the different sub-sectors of the cluster and between individual firms. Sectors peripheral to the construction business, as well as firms with a pronounced innovation strategy, seem to perform better than others.

Table 5. **Gross value added per FTE**

In CHF thousands

	1997	1998	Percentage change
Core business			
Construction (45)	64	63	-1.4%
Potential suppliers			
Manufacture of wood and wood products (20)	78	74	-5.3%
Manufacture of other non-metallic mineral products (26)	108	114	+5.8%
Manufacture of fabricated metal products (28)	80	79	-0.7%
Potential related services			
Other business activities (74+71)	77	77	+0.2%
Real estate activities (70)	193	195	+0.8%
Other sectors (for comparison):			
- Hotels and restaurants (55)	44	44	+0.8
- Financial intermediation (banks) (65)	246	258	+4.9

Note: FTE = full-time equivalent; 1998: provisional data.
Source: Statistics Switzerland (BfS).

Innovation and co-operation

The requirements of modern innovation processes push firms to co-operate with each other (OECD, 1999). Data show that, in performing R&D or innovation activities, firms in the construction cluster collaborate slightly less than the industry average, with the exception of wood manufacturing and real estate, where collaboration intensity is well below average (Table 6).

More revealing is the pattern of co-operation partners (Table 7). In general, co-operative activities related to innovation or R&D are average or below-average, with supplier-producer relationships being the most important.

Table 6. **Percentage of firms with co-operations in R&D or innovation, 1996**

Core business	
Construction (45)	41.3 %
Potential suppliers	
Manufacture of wood and wood products (20)	28.3 %
Manufacture of other non-metallic mineral products (26)	40.2 %
Manufacture of fabricated metal products (28)	48.7 %
Potential related services	
Other business activities (74)	41.7 %
Real estate activities (70)	28.6 %
Industry	50.8 %
Construction / services	44.4 %

Source: Arvanitis *et al.* (1998); NACE codes in brackets.

Table 7. **Partners in R&D or innovation co-operation, 1996**

	Clients	Suppliers	Competitors	Other firms	Universities
Core business					
Construction (45)	35%	67.5%	50%	22.5%	7.5%
Potential suppliers					
Manufacture of wood and wood products (20)	53.2%	37.6%	46.8%	31.2%	46.8%
Manufacture of other non-metallic mineral products (26)	53.8%	23.6%	26.4%	0%	36.6%
Manufacture of fabricated metal products (28)	92.6%	22.2%	7.4%	22.2%	66.2%
Potential related services					
Other business activities (74)	47.7%	36.4%	47.7%	27.3%	27.3%
Real estate activities (70)	(50%)	(50%)	(100%)	(0%)	(0%)
Industry	51.7%	59%	28.5%	27.8%	52.6%
Construction / services	38.5%	50.9%	50.4%	19%	13.7%

Source: Arvanitis *et al.* (1998); NACE codes in brackets.

Firms in the core business show an above-average co-operation intensity with suppliers, and a below-average co-operation intensity with clients. This indicates that the innovation potential incorporated in forward and backward linkages is better used on the supplier than on the customer side. This, and the fact that the peripheral sectors of the cluster are more customer-oriented in their innovation co-operation activities, supports the view that construction firms draw on innovations stemming from suppliers. On the other hand, project-based organisation of work, fierce price competition on the end-product market and the usually not very demanding customers, can explain the low engagement with customers.

Another interesting observation is the very low co-operation rate with universities. Is this a sign of weakness and a potential area for improvement? What do universities have to offer the construction cluster? Is there a real need for co-operation? These questions bring up the question of the link between construction activities (or economic activities in general) and science at universities. Trying to provide an answer with data calls for patent analysis which provides an indicator for the linkage to science (Schmoch *et al.*, 2000). References to scientific journals in patent documents show that, on average, 0.85 scientific articles are cited per patent. For construction patents, only 0.2 scientific articles are cited, meaning that there is only a weak linkage of construction to science. Of course, this goes well with the low co-operation rate. Nevertheless, experts refer to important co-operative efforts with construction firms and the Federal Institute of Technology, especially in the field of construction engineering and management. It is likely that such co-operation activities are not well reflected in scientific journals and hence not in patent documents.

The peripheral sectors of the cluster are more customer-oriented in their innovation co-operations. The more innovative and dynamic the supplier firms, the better they link to the university sector (hence the higher percentage in construction chemicals or concrete technology). In addition, firms have strong ties to EMPA, a federal institute for research and testing which is affiliated to the Federal Institutes of Technology (ETH).

The need for co-operation and organisational innovation

So far we have looked at innovation and patenting and the situation for the cluster does not seem to be as bad as one would have expected. So, is there a problem? Yes, and it relates to the organisation of the construction process, its fragmentation and project orientation. A major inefficiency stems from the division of labour and poor co-ordination in the sequence of the activities during the life cycle of a construction project; namely, strategic planning, project planning, realisation (construction), use (including real estate management and renewal) and destruction (Girmscheid, 2000; Wettstein, 2000; Girmscheid, 1997).

As illustrated in Figure 3, the responsibilities for the different processes are most often organisationally separated since there are virtually no systems providers and only few total or general contractors. Furthermore, as in the case of the construction phase, responsibility is frequently split between several performers. This (over)fragmentation and poor co-ordination leads to a key deficiency: optimisation takes place only on activities under the same responsibility and is usually detached from the whole value chain. This also reflects the (somewhat inherent) project-oriented business approach (bringing together owner, planer, manager and construction firms) and the traditional way it is handled (mental attitudes; detailed specifications; almost no project-spreading knowledge use). The most severe divide exists between the planning process and the construction process. Often the useful experience and know-how of those involved in construction is not used sufficiently in the planning process. And even when producers are involved, it is often too late to implement any innovative, efficient and beneficial solutions. Nevertheless, there are promising attempts to improve this crucial interface, especially the joint initiative (SMART) of the associations for engineers, architects and construction firms for a concept of inter-sectoral collaborations which aim at the delivery of a single, integrated product in the place of several separate products.

Figure 3. **The construction process: organisational fragmentation and optimisation**

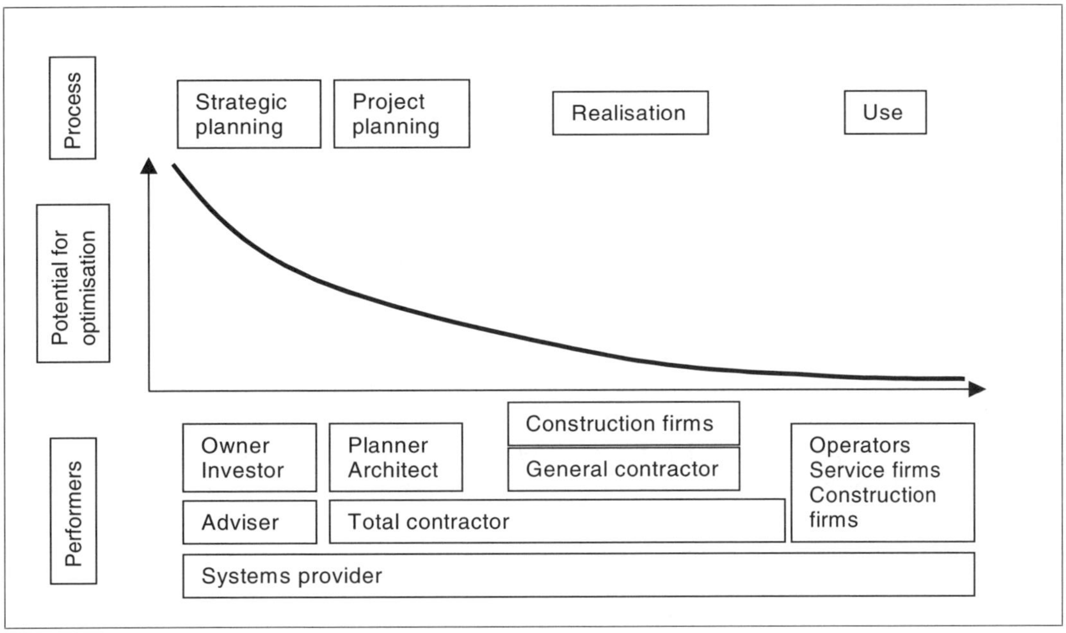

Source: UBS (2000).

Cluster-related policies

Policies and actions at different state levels (local, regional, national, international) influence the construction cluster. Generally, state agencies act as regulators, clients and promoters or brokers. The following paragraphs concentrate on contemporary, Swiss-specific and innovation-relevant issues.

The following two cases provide examples of how *regulatory changes* are intended to bring more market elements into construction-relevant regulations and thus increase the potential for innovation. First, following a liberal tradition, state actions and regulations are called into question on a regular basis. Recently, a compulsory regulatory impact analysis for federal regulations and projects of significant economic impact was introduced. In addition, for certain issues, a compatibility test for SMEs has to be conducted. This general simplification of new and existing rules has beneficial effects on regulations and projects specifically relevant to the construction cluster. Second, a Swiss specificity – which has consequences for regulation – is the fact that around 70% of the population lives in rented apartments or houses. For political reasons, rents are coupled with mortgage rates, which implies that rents reflect costs rather than market conditions. Ongoing discussions favour the replacement of this mechanism with an inflation-coupled index which would allow more market-driven adaptations of prices and a further segmentation of the market that would feed back to firms and their products.

Spatial planning and traffic policies set rules for various players but also show a demand effect through the building of infrastructure, especially large-scale projects such as, for example, railways (Bahn2000, NEAT). In general, the quality and innovativeness of state demand is determined by *procurement* rules (Gauch and Stöckli, 1999). In terms of volume, the cantons and communities outperform the federal state in public procurement. Due to decentralised legislation, many differing procurement rules apply, although the (Swiss) common market law as well as an agreement between cantons do have an integrating effect. Nevertheless, there are some drawbacks for transparency and costs since, for example, a thorough search of procurement projects has to be done by looking through two dozen different procurement publications. Recently, federal procurement relevant for the construction cluster is co-ordinated by a joint committee of the federal building and real estate organs. Concerning the selection process, the federal procurement law sets out certain selection criteria, obviously price, quality, etc., but also including sustainability. However, innovation is not explicitly mentioned. Federal procurers are said to select traditional, market-proven solutions rather than innovative, experimental ones.

In terms of procurement methods, again a traditional picture appears. It is commonly believed that a good procurer has to determine very many details, thus functional procurement is not common. Nevertheless, there are some signs that procurement methods are changing, especially at the cantonal level but also at the federal level where the introduction of electronic procurement for standard goods is now being discussed. Furthermore, there have been some private and public attempts (labels, technological innovations, computer tools, conferences, etc.) to incorporate ecological requirements in construction activities ("sustainable construction") since it is obvious that the construction cluster has a vast potential for ecological improvements (50% of construction investment is spent for new buildings; 60% of total Swiss energy consumption is for construction and operation of buildings). Although a number of initiatives exist, the possibilities for combining ecological requirements and public procurement to bring about innovation have yet to be fully exploited.

With some initiatives, the federal state takes on the role as a *broker*. From an innovation perspective, the CTI funding mechanism and the Effi-Bau initiative are relevant for the construction cluster. The CTI (Commission on Technology and Innovation) has been the funding mechanism for applied R&D for more than 50 years, sponsoring the costs of non-profit partners in joint industry-science R&D projects up to a maximum of 50% of total costs. In 1986-2000, CHF 24 million were spent on projects directly related to construction, inducing private R&D expenditure of an additional CHF 36 million. In terms of public funding, the ETHs, with around CHF 14.7 million of public funds benefited most, followed by the technical colleges (*Fachhochschulen*), with CHF 3.5 million. Looking at the thematic priorities, CHF 12 million were spent for construction technology, CHF 4.2 million for material science and CHF 3.3 million for information technology (software).

An example of a fully-fledged cluster policy is the recent federal initiative Effi-Bau (the acronym stands for efficiency potential of the Swiss construction business).[5] The severe recession of the 1990s was the catalyst for the action. Starting from an initial general impression that the construction cluster with its over-segmentation was being challenged by the increasing complexity and additional requirements of construction processes, a reorientation as well as the application of new production and management concepts such as integrated construction planning and production were seen as possible solutions which would enable SMEs to survive the recession. The idea of the initiative was to obtain a comprehensive judgement on the state of the construction cluster and to communicate the result, to locate efficiency potentials within the cluster and to concretise them within certain projects, and to apply and diffuse the knowledge gained.

The initiative encompassed four phases, starting in 1995 with an interdisciplinary steering committee defining the overall goals of the initiative and launching three parallel analyses to ensure a comprehensive review of the state and problems of the construction cluster and to locate efficiency potentials.

In the second phase (1995-96), these analyses were realised. Paying attention to feedback loops, different hearings with representatives of the construction cluster were carried out. Not surprisingly, the different analyses highlighted more or less the same weaknesses but with different weightings. Most weaknesses and even recommendations for improvements were widely recognised but had not yet been implemented. The overall message for improving the construction cluster can be summarised as follows: increase the competence of the cluster actors while providing them with an optimal playing field. Three main objectives were deduced from this vision:

- *Increase the competence of the clients (owners or users).* Emphasis was put on unprofessional owners (those not backed by an advisor) who are responsible for a substantial part of construction spending.

- *Increase the competence of the suppliers.* The aim was to better adjust the activities of planning and construction firms to market needs, especially by supplying integrated services and products as well as better project management.

- *Improve framework conditions*, focusing on further education, procedures for the obtention of permits, and procurement.

In the third phase (1997-78), various development and pilot projects were launched and carried out (for a total volume of less than CHF 3 million). Out of 200 proposals (within the three main objectives), 16 were selected and realised. Projects had to deliver concrete products which could be used in practice and were tested in real cases. Examples of the achievements include: a computer program to determine the value of a building; a manual to calculate the expected costs for strategic planning; the application of the concept of a virtual enterprise, some recommendations and model contracts; the analysis of the new public procurement law and some recommendations.

The fourth phase (1998) was dedicated to diffusing the end results using different methods (publications, seminars, etc.). This process was mainly carried out by the organisations which supported the initiative.

Effi-Bau is considered a success since it revitalised the cluster and brought together different actors. Most important was the initiation of dialogue, networking and broader co-operation. Concerning the weaknesses of the construction cluster and its innovation potential, few novelties were identified. However, some aspects were developed further and, with its reference to practice, it helped to point to certain perspectives and action fields for industry. Also important was the fact that weaknesses were explicitly mentioned and discussed.

Summary and conclusions

This chapter has described the construction business and its related activities using a value-chain (cluster) approach focused on dynamics and innovation. The construction cluster is made up of established businesses under heavy strain from recession and price competition. Firm behaviour and market segmentation is traditional, leaving insufficient room for demand criteria. Today's strengths were built up over the years and are often based on existing economic structures or other Swiss specificities (*e.g.* tunnelling, construction chemicals, polymeric materials).

Despite the maturity of the cluster, innovative segments do exist, especially in the peripheral sectors where some European and international players are present. From the perspective of technological innovations, the construction business scores fairly well thanks to an adequate innovation rate and fairly strong patenting activity. The strong interface between the core construction businesses and suppliers paves the way for the diffusion of supplier-generated innovations in the construction business. On the other side, customers are not sufficiently demanding or innovative to have a dynamic impetus on the cluster. Furthermore, cluster actors face an (inherent) organisational problem due to the traditional project-oriented business with sub-optimal co-ordination, knowledge management and learning. Even though the value chain in the Swiss construction business is heavily fragmented (mostly between planning and production), some private and public initiatives (SMART, Effi-Bau) highlighted the usefulness and appropriateness of a cluster perspective integrating the different value chains. Thus, innovation and growth in the cluster is dependent on the ability to organise the necessary horizontal and vertical co-ordination.

Policies related to the construction cluster represent many societal issues, such as the importance of the cluster for employment, the importance of rents in the living costs of the majority of citizens, the importance of construction for spatial development, etc. It is clear that the role of the state is comprehensive, particularly when public procurement makes up an important part of demand. The orientation of these policies towards a cluster perspective is difficult but beneficial, especially when the aim is to target innovation while respecting other goals. In general, government and other actors should strive to enable change, promote dialogue, support cultural and behavioural shifts, etc. A cluster policy needs to identify the strengths and weaknesses of the relevant cluster in order to bolster the former and remedy the latter. In the case of the Swiss construction cluster, this will entail: *i)* improving innovation conditions for suppliers as well as supporting the interface between suppliers and construction firms, since supplier-driven innovations constitute a dynamic element in the cluster; and *ii)* fostering competence building on the demand side in order to create additional dynamism.

Annex

DEVELOPMENT OF EMPLOYMENT IN THE CONSTRUCTION CLUSTER

Annex Table 1. **Persons employed in different (potential) segments of the construction cluster**

NACE	Description	1985	1991	1995	1998
	Tertiary sector	2 038 073	2 475 922	2 436 011	2 447 225
	Secondary sector	1 239 385	1 284 981	1 112 804	1 024 203
45	Construction	348 831	382 732	339 454	293 893
	Core business				
45.2	Building of complete constructions or parts thereof; civil engineering			190 585	156 701
45.21	General construction of buildings and civil engineering works			127 610	95 385
45.22	Erection of roof covering and frames			27 652	27 851
45.23	Construction of highways, roads, airfields and sport facilities			16 930	15 433
45.24	Construction of water projects			358	118
45.25	Other construction work involving special trades			18 035	17 914
	Supporting industries				
45.1	Site preparation			3 250	3 644
45.3	Building installation			89 252	81 262
45.31	Installation of electrical wiring and fittings			40 672	36 524
45.32	Insulation work activities			3 730	3 154
45.33	Plumbing			41 445	36 824
45.34	Other building installation			3 405	4 760
45.4	Building completion			56 144	52 079
45.41	Plastering			8 314	6 218
45.42	Joinery installation			4 587	4 311
45.43	Floor and wall covering			11 232	11 617
45.44	Painting and glazing			30 368	28 421
45.45	Other building completion			1 643	1 512
	Potential suppliers (selection)				
14	Other mining and quarrying	6 168	7 178	5 968	5 187
14.1	Quarrying of stone			1 217	1 013
14.2	Quarrying of sand and clay			4 406	3 445
20	Manufacture of wood and of products of wood and cork, except furniture; manufacture of articles of straw and plaiting materials	54 520	52 573	43 731	38 328
26	Manufacture of other non-metallic mineral products	26 235	26 397	21 734	19 902
26.1	Manufacture of glass and glass products			4 241	4 267
26.3	Manufacture of ceramic tiles and flags			430	20
26.4	Manufacture of bricks, tiles and construction products, in baked clay			1 444	1 071
26.5	Manufacture of cement, lime and plaster			1 298	1 059
26.6	Manufacture of articles of concrete, plaster and cement			6 809	6 849
28	Manufacture of fabricated metal products, except machinery and equipment	81 792	90 446	90 692	83 108
45.5	Renting of construction or demolition equipment with operator			223	207

Annex Table 1. **Persons employed in different (potential) segments of the construction cluster** *(cont'd.)*

NACE	Description	1985	1991	1995	1998
	Potential related services (selection)				
74	Other business activities	180 911	256 902	258 359	266 449
74.2	Architectural and engineering activities and related technical consultancy			83 225	73 864
74.20A	Architects			34 930	31 625
74.20B	Interior designers			1 128	1 111
74.20C	Construction engineers			19 240	17 541
74.20D	Other engineering firms			16 183	11 485
74.20E	Surveyors			2 978	3 116
74.20F	Town and country planning firms			845	793
74.20G	Landscape gardeners			1 374	636
74.20H	Other technical consultancy & planning firms			6 547	7 557
70	Real estate activities	15 215	18 778	17 292	18 492
70.1	Real estate activities with own property			1 576	811
70.2	Letting of own property			1 790	1 637
70.3	Real estate activities on a fee or contract basis			13 926	16 044
70.31	Real estate agencies			3 024	3 880
70.32	Management of real estate on a fee or contract basis			10 902	12 164

Source: STATWEB 2001 (http://www.statweb.admin.ch/).

NOTES

1. A cluster is a set of organisations either characterised by their interdependence (buyer-supplier, input-output or value chain relationships) or their commonalties (geographical location, the generation or use of a common technology, the use of common distribution channels, or a common labour pool). See, for example, Bergman and Feser (1999); DeBresson (1996).

2. The so-called "method of maxima" yields the most important user-supplier relationships.

3. Site preparation (NACE 45.1) can also be included here.

4. The specialisation index RPA_{ij} is defined as: $100 \tanh \ln ((PAT_{ij}/\Sigma_i PAT_{ij})/(\Sigma j PAT_{ij}/\Sigma_{ij} PAT_{ij}))$, where PAT_{ij} is the number of patents of country i in technology field j.

5. The basis for this federal initiative can be found in the message to the Parliament (Schweizerischer Bundesrat, 1994), as well as in the relevant legal documents. Further information on the initiative can be found at http://www.effi-bau.ch.

REFERENCES

Arvanitis, S. *et al.* (1998), Innovationstätigkeit in der Schweizer Industrie. Teil I: Industrie. Teil II: Bauwirtschaft und Dienstleistungssektor. Eine Analyse der Ergebnisse der Innovationserhebung 1996. Strukturberichterstattung, Studienreihe, herausgegeben vom Bundesamt für Wirtschaft und Arbeit BWA.

Bauforum2000 (2000), *Nachlese Bauforum 2000*, Reporting der Session 14, http://www.bauforum2000.ch/deutsch/nachlese.html.

Bergman, E.M. and E.J. Feser (1999), "Industrial and Regional Clusters: Concepts and Comparative Applications", http://www.rri.wvu.edu/WebBook/Bergman-Feser/contents.htm.

Berwert, A. and L. Mira. (2000), *Cluster-Skizzen für die Bereiche Bau – Agro-Food – ICT: Vorabklärungen im Auftrag des Schweizerischen Wissenschaftsrates / Technologie- und Innovationspolitik*, Rütter & Partner, http://www.cest.ch.

DeBresson, C. (1996), *Economic Interdependence and Innovative Activity: Input-output Analysis*, Edward Elgar, Cheltenham.

Diepold (1995), "Analyseprojekt Effizienzpotentiale der Schweizer Bauwirtschaft. Zwischenbericht Phase 2.1. Gesamtanalyse und Modellierung", 30 September, Bundesamt für Konjunkturfragen, Zurich/Bern.

Gauch, P. and H. Stöckli (1999), *Thesen zum neuen Vergaberecht des Bundes*, Institut für schweizerisches und internationales Baurecht, Freiburg Schweiz.

Girmscheid, G. (1997), *Restrukturierung von Bauunternehmungen – Chance für die Zukunft? Einführungsvorlesung*, Institut für Bauplanung und Baubetrieb, ETH Zurich.

Girmscheid, G. (2000), "Innovationen – Schlüssel für eine kundenorientierte Bauwirtschaft", paper presented at BauForum 2000, http://www.ibb.baug.ethz.ch/7.x/bauforum2000/vortrag_girmscheid/sld001.htm.

Girmscheid, G. and E. Hofmann (2000), "Industrielles Bauen – Fertigungstechnologie oder Managementkonzept?", *Bauingenieur Band 75*, September, http://link.springer.de/link/service/journals/94194/papers/0075009/00750586.pdf.

OECD (1999), *Managing National Innovation Systems*, OECD, Paris.

Peeters, L., M. Tiri and A. Berwert (2001), "Identification of Techno-Economic Clusters using Input-Output Data: Application to Switzerland and Flanders", Chapter 12 of this volume.

Schmoch, U., A. Jungmittag and R. Rangnow (2000), *Innovationsstandort Schweiz. Studie für das Bundesamt für Berufsbildung und Technologie*, Frauenhofer-Institut für Systemtechnik und Innovationsforschung, Karlsruhe, February.

Schweizerischer Bundesrat (1994), *Botschaft über die Finanzierung der Tätigkeit der Kommission zur Förderung der wissenschaftlichen Forschung (KWF) im nationalen und europäischen Rahmen (EUREKA) in den Jahren 1996-1999*, Bern, 28 November.

STATWEB (2000), "STATWEB Schweiz. Die statistische Online-Datenbank", Bundesamt für Statistik, Neuenburg, http://www.statweb.admin.ch/.

UBS (2000), "Outlook Bauwirtschaft", http://www.ubs.com/g/cc/outlook/construction.html.

Wettstein, A. (2000), "Innovation aus Sicht des Unternehmens", paper presented at BauForum 2000, http://www.ibb.baug.ethz.ch/7.x/bauforum2000/vortrag_wettstein/sld001.htm.

PART III

APPLIED ANALYSES AND CLUSTER-BASED INNOVATION

Chapter 12

IDENTIFICATION OF TECHNO-ECONOMIC CLUSTERS USING INPUT-OUTPUT DATA: APPLICATION TO FLANDERS AND SWITZERLAND

by

Ludo Peeters, LUC, ITEO-DAM, **Marc Tiri,** LUC, ITEO-DAM, and
Adrian Berwert, Rütter & Partner, Concert Research, Switzerland*

Introduction

Since the appearance of Porter's work on the competitive advantage of nations (Porter, 1990), cluster analysis has become a popular instrument in determining the innovativeness and competitive power of national and regional economies. This chapter reports the results of a workable methodology for identifying aggregate techno-economic clusters – so-called "mega-clusters" – in the regional or national economies of Flanders and Switzerland, using readily available input-output (I/O) data. Clusters are identified, based on existing trade linkages (Hauknes, 1999; Bergman *et al.*, 1999). Further, the underlying assumption of thc approach adopted here is that economic (supplier-user) linkages between industries – as reflected in the I/O tables (intermediary flows of goods and services between industries) – are the main "carriers" of technology diffusion in an economy, through interactive learning processes (Lundvall, 1992; DeBresson *et al.*, 1994; Edquist, 1997). In this sense, economic clusters are part of national innovation systems on a smaller scale, acting as relatively independent "units" of innovation diffusion. The analysis here is closely related to that of Roelandt *et al.* (1997) and van der Gaag (1995), who investigated clusters at sector level.

The clusters that emerge from this I/O approach are made up of *industries* that are closely connected, and not necessarily *companies* that develop innovations through co-operation. In this respect, the notion of "clusters" used in the present analysis deviates from Porter's notion of clusters. Specifically, Porter (1998, p. 199) defines a cluster as "a geographically proximate group of interconnected *companies* [emphasis added] and associated institutions in a particular field, linked by commonalities and complementarities", while the present study focuses on *industries* rather than companies. Moreover, Porter uses a qualitative approach where we will use a quantitative approach. In addition, aspects such as location factors (proximity) and innovativeness are lost. On the other hand, the method presented here covers the whole economy, thus allowing empirical results to be compared across regions and/or industries.

---

* The authors are grateful to Dr. Theo Roelandt and to Drs. Hessel and Verbeek of the Dutch Ministry of Economic Affairs, for their co-operation in developing the methodological basis of the cluster algorithm. The Swiss part of the study was supported by the Centre for Science and Technology (CEST), attached to the Swiss Science and Technology Council. The usual disclaimer applies.

This chapter contains three sections. The first gives a brief account of the I/O-based algorithm used for the identification of economic clusters. The second section presents the empirical results of the cluster analysis at the macro level, as applied to Flanders and Switzerland, respectively. In addition, the economic and technological profiles of the clusters are described, following investigation of a number of important indicators. The final section summarises the major findings of the study and formulates some policy considerations.

Methodology of cluster identification: the "Method of Maxima"

The cluster analysis presented here is conducted at the macro level. In other words, linkages within and between sector aggregations – designated here as "mega-clusters" – are considered. By using the I/O methodology, the analysis focuses on the intermediate supplier-user linkages. In addition to flows of capital goods, the intermediate supplier-user linkages are viewed as the most relevant type of linkages to define techno-economic clusters, because they follow a pattern that is strongly correlated with the inter-sectoral exchange of information, common R&D and innovation activities, and the diffusion of embodied technology. Capital flows are not covered by the analysis due to insufficient data availability. In addition, our attention is limited to domestic flows of intermediate goods and services, leaving the influence of linkages with foreign suppliers and users open for further investigation.

The method used, the "Method of Maxima" (M-method), classifies sectors according to substantial mutual dependency through their deliveries and purchases in the same economic cluster, and is applied in two consecutive phases. The first phase examines the forward linkages, *i.e.* we look primarily for deliveries that are important from the suppliers' point of view; the second phase investigates the backward linkages, *i.e.* we identify those deliveries that are particularly important from the customers' point of view (for further detail, see Annex). The M-method uses cut-off points (in terms of percentage of total output or total input). When these threshold values are reached – and a sector hence exhibits strong linkages with another sector – both sectors will be attributed to the same cluster. Careful analysis of these cut-off values is necessary as too-low values cause excessive aggregation. On the other hand, setting the threshold values too high could result in many very small clusters and runs the risk that some sectors may not be assigned to any cluster at all.

In the final phase of the M-Method, where the results of the two steps described above are combined, two alternative approaches can be adopted. The first, which is applied to the Swiss I/O data, implies a "strict" cluster delineation that is fully determined by the (objective) outcome of the algorithm. Obviously, when using such an orthodox approach, an appropriate choice of weights is of crucial importance. The second approach, that applied to the Flemish I/O data, allows for some "degree of freedom", in the sense that it is explicitly acknowledged that any final cluster delineation is difficult and involves some prior (subjective) judgement by the researcher. This implies that any particular cluster identification that emerges from the strict application of the algorithm can only be considered as a "starting point" for the final cluster analysis. Hereby, additional clustering criteria like "functional dependency" of the sectors are taken into account.

Empirical results for Flanders and Switzerland

Identification of Flemish mega-clusters

Applying the M-method to the Flemish I/O table for 1995 led to the identification of five mega-clusters (Figure 1): "Agro-food" (AF), "Construction and Metal" (CM), "Chemicals" (CH), "Transportation and Distribution" (TD), and "Services" (SV).

Some overlap between these mega-clusters is apparent and is the result of the existence of linkages between sectors of different clusters. For example, there is an important overlap between the SV cluster and the other mega-clusters. In general, the identified mega-clusters are built on one or two core sectors, around which there exists a network of supplying and using sectors. The members of a mega-cluster exhibit strong preferential linkages with each other, but in a number of cases, there are also important linkages with sectors belonging to other mega-clusters. Further, it can be seen that the various clusters show important differences in size, shape and the number of linkages between the constituent sectors. The largest and most centralised cluster of the Flemish economy is the CM cluster, which has important links with other clusters. The AF cluster, on the other hand, has only weak linkages with the other mega-clusters and is primarily self-sufficient.

Figure 1. **Overview of the five Flemish mega-clusters (AF, CM, CH, TD and SV)**

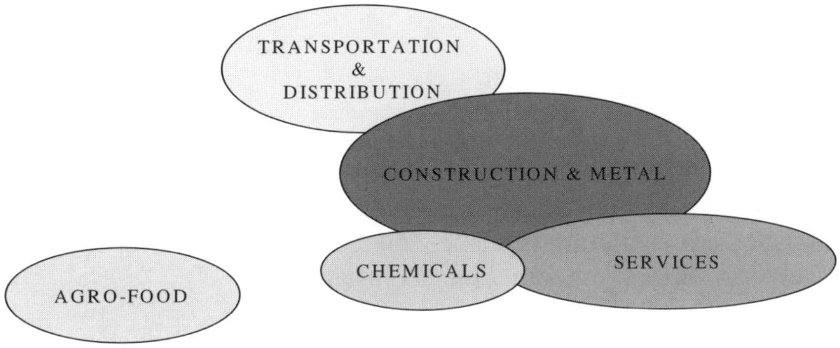

Box 1. **Interpretation of cluster figures for Flanders**

a's most important customer is b [all conditions (see Annex) are met]

a's most important supplier is b [all conditions (see Annex) are met]

Economic significance (Value added)

High R&D intensity

Low R&D intensity

High share of innovative and/or new strongly improved products

Low share of innovative and/or strongly improved products

The Agro-food(AF) cluster

The AF cluster is a "compact" mega-cluster, almost entirely made up of sectors involved in the food production system (Figure 2). The various sectors belonging to the AF cluster have strong mutual linkages, yet only weak linkages with sectors from other clusters. Furthermore, the "Agricultural, forestry and fishery products" sector [1] and the "Meats and meat products" sector [20] are obvious and mutually dependent core sectors of the AF cluster. The (small) sector of "Tobacco products" [24] does not meet the predetermined threshold values. Yet, it was assigned to the AF cluster based on the fact that a very large part of its purchases originate from this cluster and because of the nature of the products involved.

Figure 2. **Structure of the Agro-food cluster**

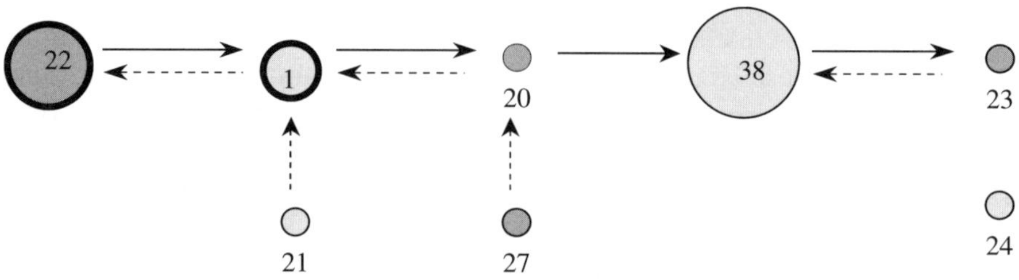

Key: [1] Agricultural, forestry and fishery products; [20] Meats and meat products; [21] Milk and dairy products; [22] Other food products; [23] Beverages; [24] Tobacco products; [27] Leathers, leather and skin products, footwear; [38] Lodging and catering services.

The Construction and Metal (CM) cluster

The CM cluster is a large mega-cluster, dominated by the "Building and construction" sector [34] and the metal-related sectors of "Iron and steel products" [5] and "Metal products" [14]. Around those core sectors, a network of supplying and using sectors is built (Figure 3). The CM cluster is very closely linked with the other mega-clusters. Hence, the CM cluster is an "open" mega-cluster, keeping up important linkages with sectors from other mega-clusters.

Figure 3. **Structure of the Construction and Metal cluster**

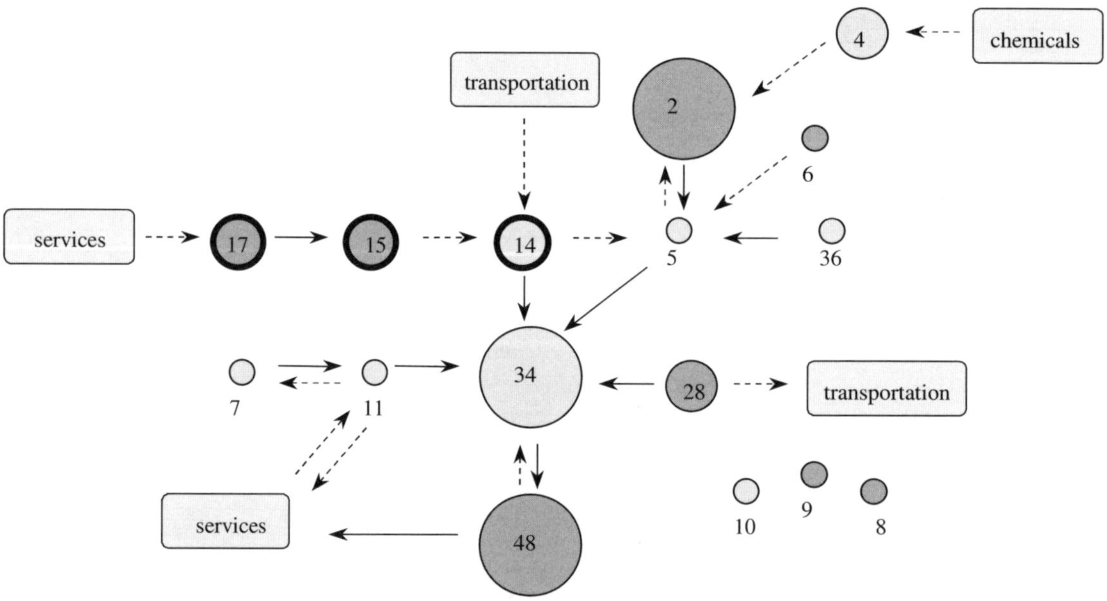

Key: [2] Energy (excluding electricity); [4] Electricity; [5] Iron and steel products; [6] Non-ferrous metals; [7] Cement, lime, plaster; [8] Glass; [9] Earthenware and ceramic products; [10] Other minerals and derived (non-metallic) products; [11] Construction materials; [14] Metal products; [15] Agricultural and industrial machinery; [17] Electrical equipment; [28] Timber and wooden furniture; [34] Building and construction; [36] Recycling and repair services; [48] Renting of immovable goods.

The Chemicals (CH) cluster

The CH cluster is a small cluster, consisting only of four sectors (Figure 4). In this cluster, the sectors of "Chemical products" [12] and "Plastics" [32] take a central role. Both sectors exhibit strong mutual linkages. In addition, the "Chemical products" sector has strong linkages with the CM and SV clusters. The other two sectors belonging to this mega-cluster are the sectors "Rubber products" [31] and "Other manufactured products" [33].

Figure 4. **Structure of the Chemicals cluster**

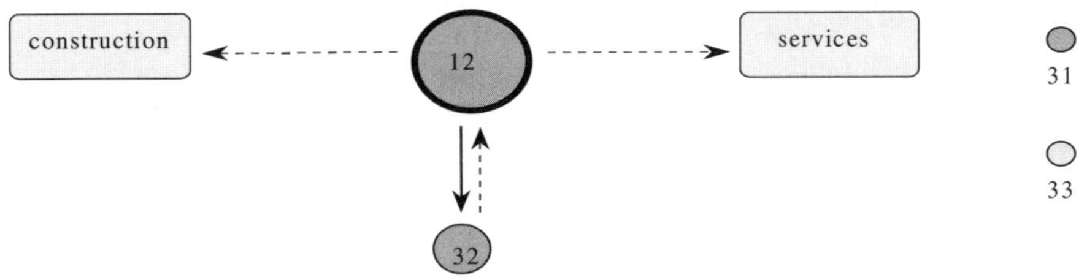

Key: [12] Chemical products; [31] Rubber products; [32] Plastics; [33] Other manufactured products.

The Transportation and Distribution (TD) cluster

The TD cluster is a large and heterogeneous cluster, where the various composing sectors exhibit strong mutual linkages (Figure 5). It is a "bipolar" sub-cluster, in the sense that two sub-systems can be detected: *i)* a sub-system built on sectors related to the "Motor vehicles" sector [18], including the "Car repair" [35] and "Wholesale and retail trade" [37] sectors; and *ii)* a sub-system built around the "Textiles" industry, including the sectors of "Synthetic fibres" [13], "Confection" [25] and "Other textiles" [26]. Further, the TD cluster comprises many transport services sectors that are split up into several "sub-sectors". This leads to a fragmentation of the supplier-user linkages. However, they are attributed to the TD cluster based on considerations of obvious functional dependency.

Figure 5. **Structure of the Transportation and Distribution cluster**

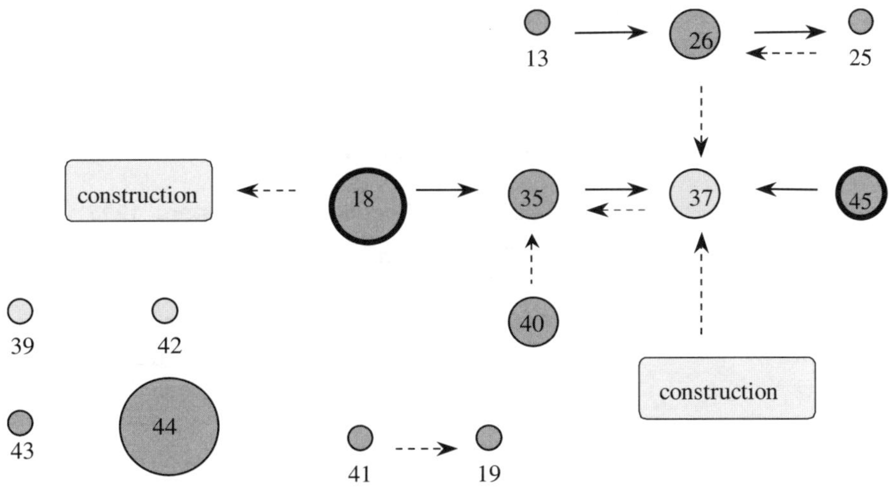

Key: [13] Synthetic fibres; [18] Motor vehicles; [19] Other transport equipment; [25] Confection; [26] Other textiles; [35] Car repair services; [37] Wholesale and retail trade; [39] Railway transport services; [40] Road transport services; [41] Inland navigation; [42] Maritime navigation; [43] Air transport services; [44] Auxiliary transport services; [45] Communication services.

The Services (SV) cluster

The SV cluster contains both commercial services, such as "Banking & insurance" [46] and "Business services" [47], and non-commercial services, such as "Health care" [49], "Public services" [50], and "Education and research" [51] (Figure 6). The central sector "Business services" exhibits strong linkages with the "Banking and insurance" sector, the "Water" sector [3], and the sector of "Paper goods and products of printing" [30]. This latter sector in turn has strong linkages with the "Pulp, paper, board" sector [29]. Several constituent sectors have important connections with other clusters due to the very nature of services, but are categorised here given their functional association with this cluster and the rather important backward linkages.

Figure 6. **Structure of the Services cluster**

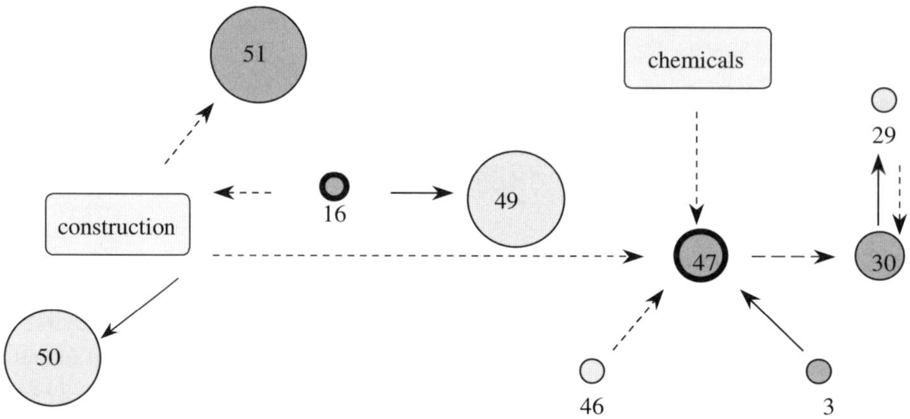

Key: [3] Water; [16] Office machines; [29] Pulp, paper, board; [30] Paper goods, products of printing; [46] Banking and insurance; [47] Business services; [49] Health care; [50] Public services; [51] Education and research.

Economic and technological profiles of the Flemish mega-clusters

The economic profiles of the mega-clusters are summarised in Table 1 which shows production value, gross value added, employment, imports and final demand for the various clusters. Relatively speaking, the TD cluster is the most important cluster, in terms of gross value added, closely followed by the SV and CM clusters. The SV cluster is by far the most important cluster in terms of employment, accounting for 50% of total employment in Flanders.

Table 1. **Economic profiles of the Flemish mega-clusters, 1995**

Percentages

Mega-cluster	Production	Gross value added	Employment	Imports	Final demand
AF	13.5	8.7	7.7	15.2	13.0
CM	32.7	27.9	15.0	38.7	32.4
CH	8.5	5.4	2.4	15.3	9.1
TD	25.5	29.7	25.8	23.8	26.1
SV	19.8	28.3	49.2	7.1	19.4
Total	**100**	**100**	**100**	**100**	**100**

Insight into the domestic (Flemish) transaction matrix for the identified mega-clusters is presented in Table 2. The intermediary deliveries (outputs) are to be read horizontally; the intermediate purchases (inputs) vertically. It can be seen that, for example, for the AF cluster, 92.2% of the deliveries of the constituent sectors occur to sectors within this cluster. At the same time, 61.4% of its purchases originate from sectors within this cluster. These high percentages of intra-cluster deliveries and purchases are apparent in all clusters and are a direct consequence of the algorithm, which attributes sectors having strong (mutual) linkages in the same cluster. Further, the so-called "diffusion indices" are shown. The diffusion index of a cluster is calculated as the (natural) log of the ratio of the row-total over the column-total. A negative value of the diffusion index indicates that the cluster is a net-user of intermediary products and services; a positive value means that a cluster is a net-supplier of intermediary products and services.

Table 2. **The Flemish I/O table at the mega-cluster level, 1995**

	AF	CM	CH	TD	SV	Total interm. deliveries
AF	**254.2**	5.3	3.2	3.4	9.5	275.6
	92.2 %	1.9 %	1.2 %	1.2 %	3.4 %	100.0 %
	61.4 %	*0.8 %*	*1.2 %*	*1.2 %*	*3.4 %*	
CM	29.0	**406.4**	30.1	77.1	57.0	599.6
	4.8 %	**67.8 %**	5.0 %	12.9 %	9.5 %	100.0 %
	7.0 %	***62.7 %***	*24.0 %*	*25.4 %*	*20.8 %*	
CH	17.2	25.3	**40.8**	12.8	12.1	108.2
	15.9 %	23.4 %	**37.7 %**	11.8 %	11.2 %	100.0 %
	4.2 %	*3.9 %*	***32.5 %***	*4.2 %*	*4.4 %*	
TD	69.4	114.3	24.9	**155.8**	43.5	407.9
	17.0 %	28.0 %	6.1 %	**38.2 %**	10.7 %	100.0 %
	16.8 %	*17.6 %*	*19.9 %*	***51.3 %***	*15.9 %*	
SV	44.3	97.0	26.4	54.5	**152.1**	374.3
	11.8 %	25.9 %	7.1 %	14.6 %	**40.6 %**	100.0 %
	10.7 %	15.0 %	*21.1 %*	*18.0 %*	***55.5 %***	
Total interm. purchases	414.1	648.3	125.4	303.6	274.2	1 765.6
	100.0 %	*100.0 %*	*100.0 %*	*100.0 %*	*100.0 %*	
Diffusion index	-0.177	-0.034	-0.064	0.128	0.135	

In defining the technological profile of the mega-clusters, the R&D intensities of the mega-clusters are calculated as weighted averages of the R&D percentages of the constituent sectors, based on the results of the most recent Flemish R&D survey. The R&D percentage of a sector is defined as the ratio of the internal R&D expenditure of the sector over the total value of production (excluding VAT). Several important differences with respect to R&D efforts are noticeable between the various economic sectors in Flanders; specifically, the R&D percentages range from 0% for certain sectors to 8.4% for the "Electrical equipment" sector. These differences are present in the mega-clusters as well (Table 3). For the mega-clusters, the AF cluster exhibits the lowest R&D percentage (0.1%). This figure contrasts sharply with the R&D efforts of the CH cluster (2.6%). Moreover, the CH cluster is responsible for 39% of the total R&D expenditures in Flanders. Also, the R&D intensity of the CM cluster is higher than the average percentage (0.65%); the CM cluster is responsible for 48% of the total Flemish R&D expenditures. The internal R&D efforts of the AF, TD and SV clusters are relatively low. Yet, the technological positions of the TD and SV sectors are possibly influenced to an important extent by "externally" (*i.e.* from other clusters) purchased intermediary products and services and the corresponding embodied technology.

Table 3. **Technological profiles of the mega-clusters**

Mega-cluster	R&D % (1993)	Innovation % (strongly improved products 1997)	Innovation % (new products 1997)
AF	0.1	12.9	8.3
CM	0.8	13.9	14.4
CH	2.6	9.9	11.6
TD	0.3	14.6	11.2
SV	0.4	14.6	12.0
Weighted average	0.6	13.7	12.1

In the preceding paragraph, the focus was on the "input side" of the innovation process. A high level of R&D expenditures does not necessarily mean a strongly innovative output, although a high correlation between innovation input and innovation output can be expected. Using the results of the latest (1997) Flemish survey on technology diffusion, the average innovation intensity for each cluster was calculated, distinguishing between (incrementally) "strongly improved products" and (radically) "new products" (that were non-existent two years ago). The calculations are based on the weighted innovation intensities per sector, where "intensity" is defined as the share of the strongly improved and new products in the total production assortment. From these results, the low value of the AF cluster stands out: only 8.3% of the total assortment is designated as "new". This is not surprising given the low R&D intensity of the AF cluster. It appears that the SV cluster is by far the most innovative in terms of innovative output. Combining these cluster maps with sector data on innovation allows the key transmitting sectors of innovation and the potentially benefiting sectors within each cluster to be identified. In Figures 2 to 6, the key R&D performing sectors (in terms of R&D intensity), as well as the key innovative sectors are shown.

From this, it becomes clear that there is no strict relationship between R&D efforts and innovation output. Sometimes a sector which invests heavily in R&D scores low on innovation output. On the other hand, moderate innovation efforts sometimes generate high innovation output. This is also observable at cluster level. For example, in the AF cluster, the Agricultural, forestry and fishery products sector scores high on R&D investment (0.13%), but very low on innovation output. This indicates that R&D figures only tell part of the story. In addition to identifying these key transmitting sectors of innovation (high share of new and/or improved products), it would be worthwhile in future research to investigate in more detail how the innovation performance of the cluster is influenced by the degree of interconnectedness – and the possibility of knowledge transfer – between the sectors within the clusters.

Identification of Swiss mega-clusters

Based on the Swiss I/O table of 1995, the orthodox approach of the "Method of Maxima" led to the identification of five pure forward clusters and seven pure backward clusters (Figure 7). Merging the two types of clusters gives five mega-clusters: *i)* Agro-food (AF); *ii)* Service-related Industries (SRI); *iii)* Construction-Services-Metals/Machinery-Electrical Equipment [CS(ME)]; *iv)* Metals/ Machinery-Electrical Equipment-Chemicals (MEC); and *v)* Textiles (TX). All of the mega-clusters comprise internal backward or forward sub-clusters. Further, the mega-clusters are not strictly delineated and exhibit substantial overlapping.

Figure 7. **Overview of the five Swiss mega-clusters (AF, SRI, CS(ME), MEC and TX)**

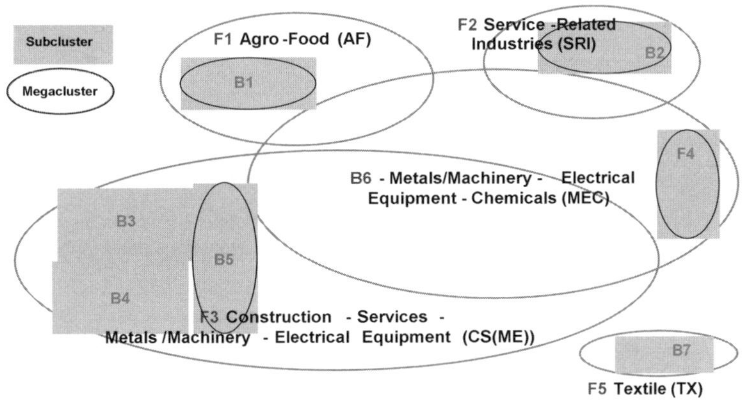

Box 2. **Interpretation of cluster figures for Switzerland**

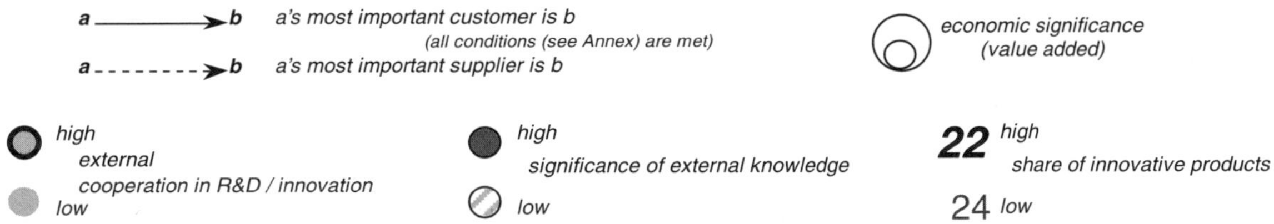

In addition to the main user and supplier linkages, each of the following cluster diagrams indicates the economic importance (value added) of the sectors within the cluster. In each case, there is also an innovation and diffusion oriented characterisation with regard to: *i)* high/low external co-operation in R&D and innovation; *ii)* high/low significance of external knowledge; and *iii)* high/low share of innovative products.

The Agro-food (AF) cluster

The AF cluster (Food processing and related activities) is entirely formed by forward linkages. Closer inspection shows that the sector "Hotel and food services" [24] is the most important user of the sectors "Alimentary products" [3], "Beverages" [4] and "Tobacco" [5].[1] Additionally, there are two important backward linkages within the AF cluster. The linkages between the sectors [1], [3] and [24] result in a backward chaining cluster B1. With the exception of sector [22] "Wholesale trade", which is the most important user of sector [24] within the mega-cluster, the cluster is characterised by its internal economic linkages. Sector [22], characterised by a relatively high share of innovative products, has important backward linkages within mega-cluster 4. However, the forward chain between sectors [22] and [24], which is clearly shown in the I/O coefficients, needs further economic interpretation.[2] The AF cluster exhibits a low external co-operation in R&D/innovation for [4] and [24] as well as a low significance of external knowledge for [24].

Figure 8. **The Agro-food cluster**

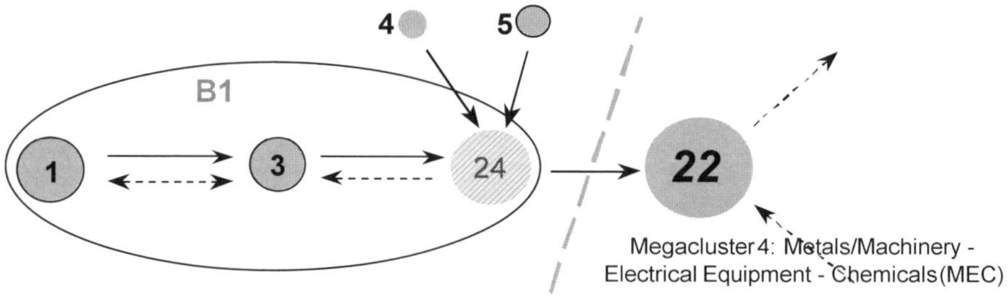

Key: [1] Agriculture; [3] Alimentary products; [4] Beverages; [5] Tobacco; [22] Wholesale trade; [24] Hotel and food services.

The Service-related Industries (SRI) cluster

The SRI cluster is a compact cluster with strong mutual dependencies formed entirely by significant forward linkages in a linear value-added chain. It consists of four sectors: "Paper industry" [10], "Graphical industry" [11], "Education, R&D and general leisure" [33] and a less homogeneous and more economically important sector "Business and personal services (including construction-related services)" [32]. Strong backward linkages can be identified between [10] and [11], as well as between [11] and [33] which result in the backward chain B2. In particular, sector [11] is characterised by low external co-operation in R&D/innovation and low significance of external knowledge. Sector [32], which has a relatively high share of innovative products, is an important transmitting sector between the SRI and the MEC clusters.

Figure 9. **The Service-related Industries cluster**

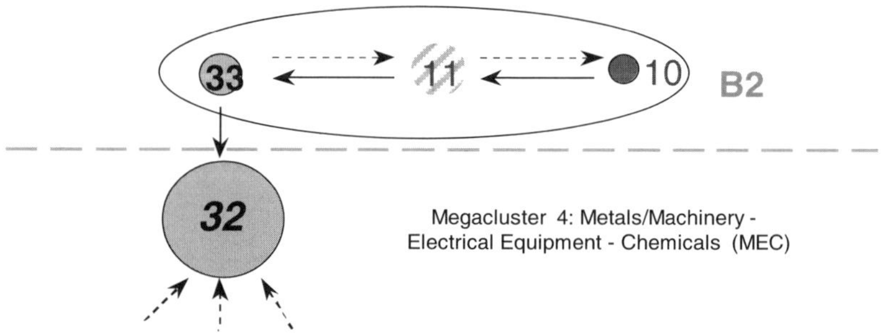

Key: [10] Paper; [11] Graphical industry; [32] Business and personal services; [33] Education, R&D and general leisure.

The Construction – Services – (Metals/Machinery – Electrical Equipment) CS(ME) cluster

The CS(ME) cluster is the largest cluster in the Swiss economy, consisting of 14 sectors. Within this forward cluster, three important backward sub-clusters have been identified. Sub-clusters B3 and B4 belong to the construction sector ("Construction" [20] and "Building [21]) and related activities. These backward clusters have a strong forward link with a third backward cluster B5 "Services". Within this sub-cluster B5, the sector "Real estate" [31] requires special attention due to its central position and the numerous linkages with several forward and backward clusters. Another important feature is related to the large overlapping section with mega-cluster 4.

Figure 10. **Construction - Services - (Metals/Machinery – Electrical Equipment) [CS(ME)] cluster**

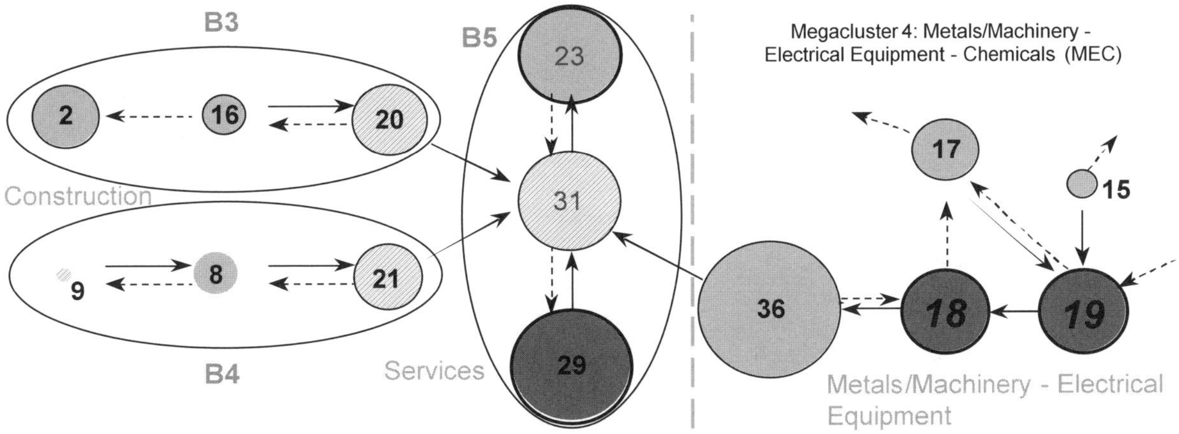

Key: [2] Electricity, gas and water; [8] Wood and furniture; [9] Wood industry; [15] Rubber and plastics; [16] Non-ferrous metals; [17] Metals; [18] Machine building; [19] Electrical equipment and watch industry; [20] Construction (main activities); [21] Building (finishing and completion); [23] Retail trade; [29] Banking; [31] Real estate; [36] Government.

Due to the fact that these two mega-clusters are only linked by the forward chain between the sectors “Government” [36] and “Real estate”, a separation of the mega-cluster 4 related part (*i.e.* the ME part of the CS(ME) cluster with the ME part of the MEC cluster) may be more relevant and may lead to a clearer distinction between these two mega-clusters. It is interesting to note that the economically important financial sector in Switzerland, “Banking” [29], has strong and mutual linkages with “Real estate”, and is characterised by high external co-operation and use of external knowledge sources.

The Metals/Machinery – Electrical Equipment – Chemicals (MEC) cluster

The MEC cluster, consisting of 11 sectors, comes out as the second largest cluster. Contrary to the other mega-clusters, the MEC cluster is the result of strong backward linkages. The identified cluster is, to a large degree, constructed around two core sectors, “Machine building” [18] and “Electrical equipment and watch industry” [19], both of which have several forward and backward linkages. Other important mutual backward/forward linkages can be observed between sector [18] and the “Government” sector [36] as well as between sector [19] and the sector “Metals” [17]. Further, a strong backward linkage was identified with the forward sub-cluster F4 “Health care”. In addition, within the MEC cluster, the three sectors “Chemicals” [13], [18] and [19] play a central role in that they exhibit high external co-operation in R&D and innovation-related activities, high significance of external knowledge, as well as a high share of innovative products. Finally, if we argue that mega-cluster 3 should be separated into two different clusters, then mega-cluster 4 would become the most important cluster for the Swiss economy.

Figure 11. **The Electrical Equipment - Metals/Machinery - Chemicals cluster**

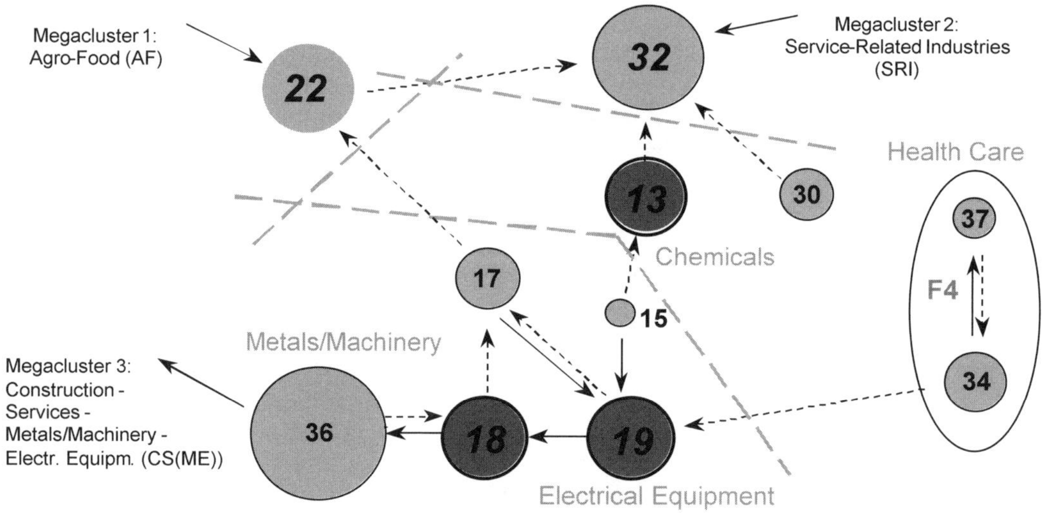

Key: [13] Chemicals; [15] Rubber and plastics; [17] Metals; [18] Machine building; [19] Electrical equipment and watch industry; [22] Wholesale trade; [30] Insurance; [32] Business and personal services; [34] Health care; [36] Government; [37] Social insurance.

The Textiles (TX) cluster

Compared to the other clusters, the TX cluster is a very compact cluster, consisting only of the sectors "Textile industry" [6] and "Confection" [7], between which strong mutual linkages exist. From this analytical perspective, this mega-cluster displays an "introverted" dependency. Despite the relatively small economic significance, the sector "Textile industry" [6] shows high external co-operation in R&D combined with a relatively high share of innovative products.

Figure 12. **The Textiles cluster**

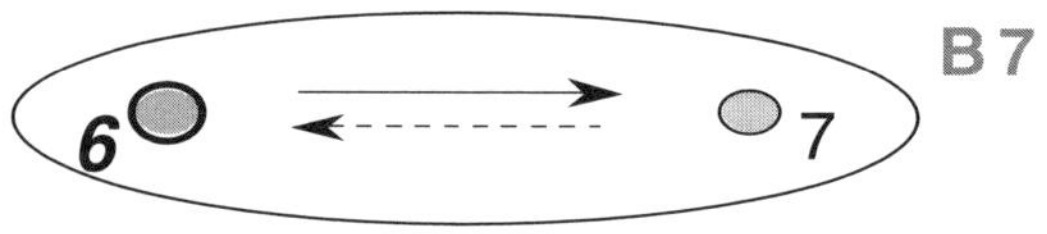

Key: [6] Textile industry; [7] Confection.

Free-floaters

Finally, the quantitative analysis leads to more isolated sectors which do not reach the forward or backward threshold values. Following the orthodox application of the algorithm, they have not been assigned to a specific cluster. In particular, this relates to the following sectors: "Leather and shoe industries" [12], "Oil refineries" [14], "Railroad transportation" [25], "Road transportation and related services" [26], "Waterway transportation" [27], "Communication services" [28] and "Non-profit organisations and household services" [35].

Economic and technological profile of the Swiss mega-clusters

As pointed out in the cluster-related figures above, substantial differences can be found between the various sectors within the clusters in terms of economic significance. At the level of the identified mega-clusters, Table 4 provides an overview of the economic profile for 1995. Among the different mega-clusters, the MEC cluster is the most important in terms of: (1) gross production, (2) gross value added, (3) employment and (6) final demand.[3]

Table 4. **Economic profile of the mega-clusters, 1995**

Cluster	Gross production[1] (as a % of total) (1)	Gross value added[2] (as % of total) (2)	Gross value added per employee[3] (in CHF) (3)	Employment (full-time equiv., as a % of total) (4)	% of imported intermediate consumption[4] (5)	(6) Final demand[5] (as a % of total) (6)
AF	14.9%	(12.0%)	71 297	(18.5%)	9.4%	(14.2%)
SRI	10.4%	(10.6%)	96 373	(12.1%)	5.4%	(4.7%)
CS of CS(ME)	41.1%	(39.7%)	103 033	(42.3%)	10.4%	(56.0%)
ME of CS(ME)	23.3%	(25.3%)	110 823	(25.4%)	2.9%	(21.1%)
MEC	50.7%	(49.6%)	110 741	(49.1%)	8.4%	(56.4%)
TE	0.9%	(0.6%)	68 592	(1.0%)	30.8%	(2.2%)
Total Swiss economy	100% (CHF 640 375 million)	100% (CHF 352 620 million)	100% (CHF 109 69)	100% (3 214 423 FTE)	14.1%	100% (CHF 433 084 million)

1. I/O table 1995 (results per sector).
2. I/O table 1995 (results per good) and conversion to sectors with results from production account 1994.
3. Full-time equivalents.
4. Estimation (Schnewlin I/O table 1990).
5. I/O table 1995 (results per good).

Most of the clusters are quite heterogeneous and consist of industry and service-related sectors. This is shown specifically in (3) gross value added per employee and per branch. This explains the considerably higher numbers in (3) in the construction services cluster with some "high-value-added branches" (Banks, Electricity and Real estate). It should also be noted that there is a comparatively high percentage of imported intermediate consumption in the TX cluster.

Table 5 illustrates the flows of intermediate deliveries and supplies as well as the diffusion index for the Swiss mega-clusters and the rest of the economy.[4] If we look at the diffusion index, the AF cluster is a significant net user of intermediate products and services. On the other hand, the SRI cluster can be interpreted as a net supplier. As expected from the cluster algorithm, the grey-shaded diagonal elements yield the highest values and percentages. Although the level of "internal linkages" is also a function of the size of the mega-cluster, the high values of the linkages within the "Agro-food" cluster are the result of the very strong and mutual forward and backward linkages between the "Agriculture" and "Alimentary products" sectors. Outside of the diagonal elements, important user and supplier linkages can be found between SRI and MEC as well as between MEC and TX.

Table 5. **Transaction matrix on the aggregation level of the mega-clusters and rest of economy, 1995**

Million CHF

	AF	SRI	CS of CS(ME)	MEC	TX	Rest of Swiss economy	Total intermediate deliveries
AF	19 057.8	185.4	1 123.0	4 835.4	165.2	1 531.1	26 898.3
	70.9%	0.7%	4.2%	18.0%	0.6%	5.7%	100%
	50.5%	**1.6%**	**1.7%**	**3.3%**	**4.9%**	**8.0%**	
SRI	1 411.5	5 916.8	2 338.5	6 293.6	51.7	962.6	16 974.7
	8.3%	34.9%	13.8%	37.1%	0.3%	5.7%	100%
	3.7%	**52.2%**	**3.5%**	**4.2%**	**1.5%**	**5.0%**	
CS of CS(ME)	5 754.4	1 064.3	38 724.1	21 151.2	3 63.8	3 428.19	70 486.0
	8.2%	1.5%	54.9%	30.0%	0.5%	4.9%	100%
	15.2%	**9.4%**	**58.1%**	**14.2%**	**10.7%**	**17.9%**	
MEC	8 719.7	3 69.5	20 345.3	104 789.4	1384.5	5 277.1	143 785.6
	6.1%	2.3%	14.1%	72.9%	1.0%	3.7%	100%
	23.1%	**28.9%**	**30.5%**	**70.5%**	**40.7%**	**27.5%**	
TX	262.3	25.6	197.5	1 111.2	1 234.1	290.5	3 121.2
	8.4%	0.8%	6.3%	35.6%	39.5%	9.3%	100%
	0.7%	**0.2%**	**0.3%**	**0.7%**	**36.3%**	**1.5%**	
Rest of Swiss economy	2 548.5	870.5	3 965.1	10 543.6	202.9	7 676.7	25 807.4
	9.9%	3.4%	15.4%	40.9%	0.8%	29.7%	100%
	6.8%	**7.7%**	**5.9%**	**7.1%**	**6.0%**	**40.1%**	
Total interm. supplies	37 754.3	11 332.2	66 693.7	148 724.3	3 02.7	19 166.2	287 073.2
	100%	**100%**	**100%**	**100%**	**100%**	**100%**	
Diffusion index	-0.339	0.404	0.056	-0.034	-0.086	0.296	

Source: Based on Antille (1999).

Tables 6 and 7 clearly point out different innovation styles and ways of acquiring knowledge and types of knowledge.

The measurement of the innovation activity within clusters using indicators provides an approximate identification of some important aspects of innovation intensity that cannot be measured directly. This measurement can start at various phases of the innovation process. The indicators chosen in Table 6 can be classified into input-related indicators (1,2), output-related indicators (3,4,5) and market-oriented indicators (6,7). Indicator (8) shows R&D-relevant co-operation activities.[5] The results indicate significant differences among the mega-clusters. In principle, the AF and SRI clusters are significantly lower in innovation intensity and performance compared to the other mega-clusters. Taking a more detailed look at individual sectors and sub-sectors, it is apparent that the sectors "Chemicals", "Machinery" and "Electronics/Instruments/Electrical Equipment" come out well above average (all are in MEC). The significance of "Paper industry" (SRI) in the area of process innovation and "Textile industry" (TE) in the area of product innovation must also be pointed out. The service sectors and their sub-sectors show an above-average innovation intensity, in particular, innovation performance, especially in a part of the sector "Computer and research" (SRI), less pronounced in the sectors "Banking" (CSME) and "Insurance" (MEC).

The co-operation intensities, with respect to R&D, show substantial variations among the mega-clusters. Taking a more detailed look, sectors [13], [18], and [19] (MEC) show the highest co-operation intensities (MEC). Again, [10] (SRI) and [6] (TE) are above average. Within service-related sectors, above-average intensities can be found in "Computer and research (SRI), [32] (CSME) and [30] (MEC).

Table 6. **Innovation profile of the Swiss mega-clusters**

Mega-cluster sectors	AF (3,4,22,24)	SRI (10,11,32)	CS of CS(ME) (8,9,16,20,21; 23,29,31)	ME of CS(ME) (15,17,18,19)	MEC (13,15,17,18, 19,22,30,32)	TX (6,7)
(1) % of firms with domestic R&D activities	46.0%	57.9%	57.1%	**73.0%**	**73.2%**	**73.5%**
(2) % of personnel in R&D (only industry related sectors within cluster)	2.6%	1.5%	2.4%	**6.5%**	**8.9%**	3.6%
(3) % of firms with product innovation	69.1%	69.3%	58.5%	**76.5%**	**78.8%**	**87.4%**
(4) % of firms with process innovation	65.0%	65.0%	67.2%	**81.2%**	**80.0%**	70.7%
(5) % of firms with patents pending	8.2%	8.5%	27.1%	**45.7%**	**44.6%**	21.2&
(6) % of firms with innovative products and services	**19.9%**	**21.3%**	8.8%	17.1%	**29.0%**	**20.4%**
(7) % of firms with worldwide new products (only industry related sectors)	1.46%	0.69%	3.84%	**6.28%**	**7.16%**	**7.43%**
(8) % of firms with co-operative activities in R&D and innovation	30.8%	56.5%	64.6%	**73.0%**	66.9%	**70.7%**

Source: Arvanitis *et al.* (1998) and own calculations.

Table 7 illustrates the importance of external knowledge sources – user or market oriented knowledge, supplier-oriented knowledge, knowledge from the same sector/company and science-based knowledge – for innovation activities. The analysis shows a general focus on user-oriented knowledge, which is, for example, very important for the beverage, rubber and plastics industries. Emphasis on supplier-oriented knowledge can be found in the alimentary products or the watch industry. Science-based knowledge (universities, research institutes, etc.) is, for example, acquired especially by the chemicals industry. It is interesting that the confection industry relies strongly on different kinds of external knowledge sources. The last column in Table 7 provides information on the importance of exports for the different sectors.

From the strict quantitative analysis of Swiss I/O data, some interesting results arise for discussion, although the analysis only applies to an I/O table with 37 sectors. The linkage between the overlapping mega-clusters is found in sectors which also have above-average importance from the economic or innovation-relevant point of view: Wholesale trade [22], Real estate [31] and Government [36]. A special role is played here by Business and personal services [32]. Within this heterogeneous sector are situated the so-called KIBS (knowledge-intensive business services) companies which are of growing importance with regard to both information transfer and knowledge generation (Reuter, forthcoming).

The quantitative input-output analysis identifies three key sectors, all of which are situated within the MEC clusters – Chemicals [13], Machine building [18] and Electrical equipment and watch industry [19]. They are characterised by: *i)* strong networking by forward and backward linkages; *ii)* high innovation relevance with regard to innovation intensity and performance; and *iii)* above-average economic importance.

Table 7. **Importance of external knowledge and export orientation**

Cluster		Sector	Importance of external knowledge					
			User linkages	Supplier linkages	Same sector[1]	Same company[1]	Universities / research institutes/ consultants	**Export orientation**
AF	3	Alimentary Products		High				
	4	Beverage	Very high			Very high		
	22	Wholesale Trade					High	
	24	Hotel and Food Service	High					
SRI	10	Paper Industry	High	Very high				High
	11	Graphical Industry	High					
	32a	Renting of Machinery and Equipment						
	32b	Legal and Business Consultancy						
	32c	Architectural and Engineering Activities					High	High
	32d	Computer and Related Activities	High					
	32e	Other Business Activities, Social Activities						
CS	8	Woods and Furniture		High				High
	9	Wood Industry	High	High				
	16	Non-ferrous Metals				High		
	20	Construction						
	21	Building (Finishing and Completing)						
	23	Retail Trade						
	29	Banking	High				High	
	31	Real Estate						
MEC	13	Chemicals			Very high	Very high	Very high	Very high
	15	Rubber and Plastics	Very high				High	High
	17	Metals	High					
	18a	Machine Building	High		High	High	High	Very high
	18b	Office Machinery and Motor Vehicles	Very high					Very high
	19a	Electrical Equipment	Very high		High	High	High	Very high
	19b	Watch Industry		High	Very high			
	19c	Other Manufacturing	High					
	22	Wholesale Trade						
	30	Insurance					High	
	32a	Renting of Machinery and Equipment						
	32b	Legal and Business Consultancy						
	32c	Architectural and Engineering Activities					High	High
	32d	Computer and Related Activities	High					
	32e	Other Business Activities, Social Activities						
TX	6	Textile Industry	High					Very high
	7	Confection Industry	Very high	High	High			High

1. Only industry-related sectors.

Very high   High

Source: Own calculations, based on Arvanitis *et al.* (1998).

The innovation and diffusion behaviour of the Swiss cluster can be regarded in general as more implementation- and application-oriented than science-based (Hotz-Hart *et al.*, 2001). Important impulses for innovation thus arise in the "strategic triangle" between focal firms and their users and suppliers of knowledge/technology. They are largely the result of market relations. In the case of product innovations, the user provides the main information, while in the case of process innovations, the supplier is generally the main source (Eisinger, 1996). The focus of the analysis on national user-supplier linkages – except for the indication of the percentages of imported intermediate consumption in Table 4 and the export orientation in Table 7 – should be seen against the background of a strong trend towards growth abroad (direct investments, R&D outlay), with a simultaneous loss of dynamics in the national economy and a downsizing of the technology portfolio (SWR, 1999).

Concluding remarks and policy recommendations

Method

This chapter has presented an workable methodology for identifying aggregate techno-economic clusters – so-called "mega-clusters" – in (regional) economies, using input-output (I/O) data. The I/O method proposed here, which is known as the "Method of Maxima", was applied to the economies of Flanders (Belgium) and Switzerland. Through this application, we have attempted to demonstrate the usefulness of the method proposed and to introduce an (international) comparative element into the analysis. The underlying assumption of our approach is, basically, that economic (supplier-user) linkages between industries – as reflected in the I/O tables – are the main "carriers" of technology diffusion in an economy.

However, the quantitative user-supplier linkages analysed represent only one important part of a national innovation system. Not shown here, in particular, are the linkages to the knowledge system (universities, research institutes), the informal knowledge and innovation networks. Although some qualifications are in order concerning the general applicability of the proposed methodology – and the fact that the method may, of course, not capture all the facets of regional or national innovation systems (due to, for example, the impact of data aggregation, the "too narrow" focus on domestic supplier-user linkages, the existence of technology diffusion even without economic interdependencies, etc.), the applied method is a good starting point for cluster analysis in an international comparative context.

Results

The results of the study revealed some structural similarities between Flanders and Switzerland, but also highlighted significant differences. The results showed clearly that each country has its own collection of clusters and specialisations and that the individual clusters identified – even at the aggregate level of the present analysis – have different characteristics and play a distinctive role in the economy. In addition, inspection of the profiles of the various clusters revealed the heterogeneity of the economic activities, in terms of size, "connectedness", R&D intensity, share of innovative products, etc.

One should be cautious, however, when comparing results across countries. Being fully aware of the fact that alternative methods may lead to different outcomes (we used two "versions" of the M-method, for Flanders and Switzerland, respectively), it should be clear that no uniform or "best" methodology exists from a theoretical point of view, and that any method should take into account the specificities of each country – in terms of the availability and aggregation of the I/O data, the general economic situation and/or the "openness" of the countries involved, as well as the specific purposes of the cluster analysis. Recently, some research efforts have been directed towards comparing different methods (*e.g.* Hoen, 2000), but general conclusive results concerning the appropriateness and robustness of the various methods for different countries are still lacking. Furthermore, we believe that it is not possible or at least not straightforward to test the robustness of one particular method for various countries, given the diversity of (predetermined) national and/or regional policy goals or options and priorities in each country.

Policy implications

In view of the importance of interdependent actors in modern innovation theory and the relation between the interconnectedness of economic agents and the production and diffusion of specialised knowledge (DeBresson, 1996), an overview of the strong linkages within regional and national economies is useful for the purpose of policy design. Therefore, we extended the cluster identification analysis with quantitative and qualitative innovation and knowledge-based indicators. From this we learned, among other things, that although it is easy to accept the idea that the core sectors or clusters characterised by high levels of R&D expenditures should be favoured when designing industrial and technological policies, high levels of R&D expenditure are not necessarily a prerequisite for innovativeness, and *vice versa* – as is the case, for example, for the Flemish services cluster (which shows a low level of R&D and a high level of innovativeness). This is, of course, an important issue from a policy point of view.

Given the existence – or non-existence – of intensively linked clusters, the policy maker has several options available (which can possibly be combined). The first option is to focus on the core clusters and to foster or maintain growth by stimulating the dynamic character of these core clusters (and constituent sectors). In so doing, the optimal leverage of invested funds can be obtained and large parts of the economy can benefit from the measures taken. Alternatively, one might choose to boost those sectors or clusters that display only weak techno-economic linkages with the rest of the economy. Such a policy may trigger new (endogenous) developments that may lead – in the long-run – to positive effects on the rest of the economy. Regardless of which particular policy option is chosen, the question of the appropriateness of direct (and selective) state intervention vs. market-oriented policy (*e.g.* improvement of market conditions or reduction of systemic failures) remains.

The cluster maps presented in this chapter, which point at the strong and privileged innovation potentials and/or weak or missing links in the economy, can act as a solid starting point for designing such policies.

Annex

M-METHOD FOR CLUSTER IDENTIFICATION

In the M-method, the diagonal elements of the I/O matrix (z_{ii}) are initially set to zero, in order to emphasise the inter-sectoral flows (for further detail, see Peeters and Tiri, 1999). Subsequently, the algorithm consists of two phases.

Phase 1: analysing forward linkages

The analysis of the forward linkages is conducted in two steps. The analysis begins with a "horizontal" or "row-wise" reading of the matrix of domestic intermediary flows (step 1). For each supplying sector i, the most important delivery (*i.e.* the highest absolute value of z_{ij} in row i), z_{ik}, is selected and divided by the total of row i (excluding the corresponding diagonal element). If this ratio is larger than a predetermined threshold value q_r, then the buying sector k is viewed as closely related with the supplying sector i. In other words, it can be concluded that there exists a strong forward linkage between supplying-sector i and using-sector k. The latter is called the (single) "best" user of sector i. Repeating this test for all supplying sectors i in the economy yields a binary [0,1] matrix, containing a "1" in the cells indicating strong forward linkages, and a "0" in the remaining cells.

Next, the matrix of the domestic intermediary flows is read "vertically" or "column-wise" (step 2). For each (single) "best" using-sector k buying from sector i, identified in step 1, the corresponding z_{ik} is divided by the total of column k (excluding the corresponding diagonal element z_{kk}). If this ratio is larger than a second predetermined threshold value q_c, then the most significant delivery from the supplier's point of view is also a significant delivery from the user's point of view. Repeating this test for all the sectors k identified in step 1 yields a new binary [0,1] matrix, containing a "1" in the cells indicating strong user linkages, and a "0" in the remaining cells.

Finally, the two binary matrices are "merged" or summed, showing several cells containing a value of 2. The algorithm allows for the identification of a number of strictly delimited chains of forward linkages, which represent the final forward-linked economic clusters.

Phase 2: analysing backward linkages

Along similar lines, the backward linkages are analysed in a second phase. For each using-sector j in the I/O table, the most important supplier k is again identified in two steps, starting with a vertical or column-wise reading of the matrix of intermediary flows, followed by a horizontal or row-wise reading. Applying similar tests as in Phase 1 for all the supplying-sectors k and using-sectors j, yields two new binary [0,1] matrices, which, when merged or summed, allow for the identification of a second set of strictly delimited chains of backward linkages, which represent the final backward-linked economic clusters.

NOTES

1. Based on the chosen threshold values, at least 20% of the deliveries of these sectors go to "Hotel and food services". At the same time, each of these deliveries represents at least 5% of the input of "Hotel and food services".

2. For an overview of the Agro-food and Construction clusters, see Berwert and Mira (2000).

3. Due to substantial overlapping, the ME-related part of the CS(ME)cluster was assigned to mega-cluster 4.

4. To avoid double counting, the "Wholesale trade" sector was calculated only in mega-cluster 4 and not in mega-cluster 1; and the "Business and personal services" sector was calculated only in mega-cluster 4 and not in mega-cluster 2.

5. Table 6 show some of the results of the 1996 Swiss Innovation Survey, which was analysed according to sectors and aggregated to mega-clusters using weighted averages. For an elaborate analysis of innovation indicators in terms of innovation intensity and performance on a sectoral and sub-sectoral level, see Arvanitis *et al.* (1998).

REFERENCES

Antille, G. (1999), "Tableaux entrées-sorties 1995 pour la Suisse", *Laboratoire d économie appliquée*, Université de Genève, Geneva.

Arvanitis, S., L. Donzé *et. al.* (1998), *Innovationstätigkeiten in der Schweizer Wirtschaft. Teil I: Industrie. Teil II: Bauwirtschaft und Dienstleistungen – Eine Analyse der Ergebnisse der Innovationserhebung 1996*, Reihe Strukturberichterstattung, Bundesamt für Wirtschaft und Arbeit BWA, Bern.

Bergman, E. and E. Feser (1999), "Industry Clusters: A Methodology and Framework for National and Regional Development Policy in the United States", in OECD, *Boosting Innovation: The Cluster Approach*, OECD, Paris.

Berwert, A. and L. Mira (2000), *Cluster-Skizzen für die Bereiche Bau – Agro-Food – ICT: Vorabklärungen im Auftrag des Schweizerischen Wissenschafts- und Technologierates (SWTR) / Technologie- und Innovationspolitik (TIP),* Schweizerischer Wissenscharfts- und Technologierat, Bern.

DeBresson, C., G, Sirilli, X. Hu and F.K. Luk (1994), "Structure and Location of Innovative Activity in the Italian Economy, 1981-85", *Economic Systems Research* 6, pp. 135-158.

DeBresson, C. (1996), *Economic Interdependence and Innovative Activity: An Input-Output Analysis*, Cheltenham, UK, 483p.

Edquist, C. (ed.) (1997), *Systems of Innovation, Technologies, Institutions and Organisations*, Pinter, London.

Eisinger, A. (1996), "Unternehmerisches Innovationsverhalten am Standort Schweiz", in Bundesamt für Konjunkturfragen (ed.), *Modernisierung am Technik-Standort Schweiz*, vdf, Zürich.

Hauknes, J. (1999), "Norwegian Input-Output Clusters and Innovation Patterns", in OECD, *Boosting Innovation: The Cluster Approach*, OECD, Paris, pp. 61-90.

Hoen, A. (2000), "Clusters: Welke Clusters?", *Economische Statistische Berichten*, Vol. 85, No. 4277, pp. 859-861.

Hotz-Hart, B., A. Reuter and P. Vock (2001), *Innovationen: Wirtschaft und Politik im globalen Wettbewerb*, Peter Lang Verlag, Bern.

Lundvall, B.-Å. (1992), *National Systems of Innovation: Toward a Theory of Innovation and Interactive Learning*, Pinter, London.

Peeters, L. and M. en Tiri (1999), *Identificatie van techno-economische clusters in Vlaanderen op basis van input-outputgegevens voor 1995*, IWT-Studie No. 27, IWT-Observatorium, Brussels.

Porter, M.E. (1990), *The Competitive Advantage of Nations*, The Free Press, New York.

Porter, M.E. (1998), *The Competitive Advantage of Nations: With a New Introduction*, MacMillan.

Reuter, A. (forthcoming), *KIBS-Firmen: Brückenorganisationen im schweizerischen Innovationssystem? Eine Untersuchung zur Bedeutung von wissensbasierten und unternehmensorientierten Dienstleistern für die Innovationsprozesse ihrer Kunden,* Zentrum für Wissenschafts- und Technologiestudien (CEST), Bern.

Roelandt, T.J.A. (1986), *Vervlechtingsconglomeraten en sectorstructuurbeleid*, EUR-EOV, Rotterdam.

Roelandt, T.J.A. (1997), P. den Hertog and D. Jacobs (1997), "Nederlandse clusters in beeld", *ESB*, 12-2-1997, pp. 124-128.

Schnewlin, M. (1995), *Input-Output Tabelle Schweiz 1990. Weiterführende quantitative Analysen*, KOF Konjunkturforschungsstelle der ETH Zürich, unveröffentlichtes Excelfile, Zürich.

SWR (1999), *Die technologische Wettbewerbsfähigkeit der Schweiz – Indikatoren, Bewertungen, Diskussion. Fakten&Bewertungen*, 2/99, Schweizerischer Wissenschaftsrat, Bern.

van der Gaag, H. (1995), "Clusters economisch en technologisch bekeken: een verkenning van statistische identificatiemogelijkheden", *TNO-Rapport* STB/95/039, Apeldoorn.

Chapter 13

INDUSTRIAL CLUSTERS IN THE FINNISH ECONOMY

by

Sakari Luukkainen
VTT Group for Technology Studies, Finland

Introduction

Traditional technology policy focuses on managing the decline of industries or on targeting future growth industries and technologies. Modern policies emphasise the importance of technology diffusion and networking. This approach is called the "cluster approach" (OECD, 1999). One advantage of the cluster approach is that it focuses on the existing strengths of an economy and takes into account the benefits to be gained from connecting different economic activities and forms of knowledge by creating new combinations (Jacobs, 1997). Also, according to Porter (1998), successful new industries and clusters often grow out of established ones.

Synergies between domestic production sectors are especially important for a small country like Finland, which has limited R&D resources. Spillovers benefiting other firms can represent up to one-third of the original R&D costs (Irwin and Klenow, 1994). Clusters provide important tools for analysing this kind of technology diffusion and for understanding interconnections among sectors. Any change in the production in the core sector of one cluster is linked with the performance of a number of other sectors.

A firm can obtain the necessary knowledge either directly by performing its own R&D or indirectly by purchasing disembodied technologies developed elsewhere. Indirect ways of obtaining knowledge, *i.e.* through technology diffusion, include the acquisition of the technology embodied in intermediate goods and disembodied spillovers. Trade flows presented in the national input/output (I/O) statistics can be used as proxies of embodied technology diffusion among domestic production sectors. In spite of the growing importance of imported technology and globalisation, and the fact that small economies benefit more from foreign R&D than do big ones (Coe and Helpman, 1995), the focus in this study is only on domestic trade-flow-based networking in clusters.

Between 1991 and 1998, the share of private firms' R&D expenditures in Finnish GDP increased from 1.2% to 2.0% while public financing decreased from 10% to 7% (Statistics Finland, 1999c). This is an indication that the importance of the traditional innovator's risk-reduction-based technology policy is decreasing as the share of private risk financing of firms increases. The aim of venture capital financing is to increase the value of the innovating company; spillovers to other companies are not considered as important. The market failures created by R&D spillovers are one of the main justifications for government policy: clusters should be the major targets of technology policy because, left to themselves, markets fail to facilitate efficient networking.

However, fostering networking is a demanding task. For instance, the Finnish forestry sector's value chain is so complex that it proved extremely difficult to ensure that equal consideration was given to all parts of the value chain in the same publicly funded cluster programme. In these programmes, co-operation occurred with naturally close partners and little innovative R&D-oriented new collaboration was initiated with private companies (Pentikäinen, 2000).

There would appear to be a need for more efficient, bottom-up cluster policy. This kind of policy can be realised by financing R&D projects that stimulate the networked economy. Government intervention is rationalised when alternative projects, for instance the expansion of existing production facilities, are more profitable for a firm than R&D investment and when the associated spillovers are high. Concrete technology policy occurs in the selection of publicly financed R&D projects. In this kind of decision making, it is crucial to understand the mechanisms of technology diffusion in order to evaluate the economic impacts of the R&D investment in advance.

In this study, the focus is on technology diffusion as measured by I/O statistics. The work complements earlier, more qualitative Finnish cluster studies based on the Porter framework, in which firms that hold a remarkably high international market share are at the core of the cluster (Hernesniemi *et al.*, 1996; Rouvinen and P. Ylä-Anttila, 1999). The general objective of this study is to produce knowledge to support the design of efficient cluster-oriented technology policy. The Finnish industrial clusters are identified using quantitative value-chain analysis, and their economic profiles are analysed. In addition, a framework is presented that describes how to evaluate and select government-supported R&D projects for carrying out cluster policy in practice.

Industrial sectors and clusters: composition and dynamics

Sectoral composition of clusters

In this chapter, technology diffusion in the Finnish economy is analysed using I/O statistics provided by Statistics Finland. Identification tables for 1995 were constructed, consisting of 68 production sectors using the NACE system of classification. The goal was to discover the trade flows that describe the linkages between the sectors.

The cluster structures were retrieved by editing out insignificant links that contained proportionally little trade. Trade flows that were under 8% were treated as insignificant, which significantly reduced the number of the original links (4 624), while maintaining the clear structure of the value chains. The significant links were classified at three levels: forward and backward linkages, whose intermediate trade was over 20%, between 14% and 20%, and between 8% and 14% of total deliveries or supplies of the receiving or supplying sector. The diagonal lines of the matrix, *i.e.* intra-trade values, were set to zero in the calculations. Some sectors had significant links with several other clusters. The weight of a sector belonging to a cluster was calculated as a ratio of the sum of significant deliveries and supplies inside the cluster and total deliveries and supplies of this sector. If the ratio is 1, the sector has significant links only to one cluster.

Five clusters (foodstuffs, ICT, metals, construction, forestry) were identified using significant trade linkages. The cluster maps and related production sectors are shown in Annexes 1-5. The production sectors that did not clearly belong to any specific cluster were placed in a separate group, "other". The minimisation of trade flows between clusters was quite successful and the resulting clusters explained a major part of the intermediate trade that can be seen from the figures in the diagonal line of Table 1. In these calculations, each production sector was placed in the cluster where its weight was most significant.

Table1. **Intra, inter and extra deliveries and supplies of the clusters**

Percentages

	Foodstuffs	ICT	Metals	Construction	Forestry	Other	***Total deliveries***
Foodstuffs							
Deliveries	**74.5**[1]	1.6	2.8	0.9	6.6	13.7	***11.7***[2]
Supplies	**58.7**[3]	3.4	2.1	0.7	3.1	6.2	
ICT							
Deliveries	2.5	**38.7**	7.5	2.2	10.4	38.7	***6.9***
Supplies	1.1	**49.3**	3.4	1.1	2.9	10.2	
Metals							
Deliveries	3.4	3.7	**70.8**	9.4	7.6	5.0	***12.8***
Supplies	3.0	8.8	**59.5**	8.7	3.9	2.5	
Construction							
Deliveries	7.6	1.8	4.6	**63.4**	6.1	16.5	***10.5***
Supplies	5.4	3.6	3.2	**48.1**	2.6	6.7	
Forestry							
Deliveries	6.3	1.7	4.6	8.8	**61.0**	17.5	***27.1***
Supplies	11.5	8.5	8.2	17.2	**66.9**	18.2	
Other							
Deliveries	9.7	4.6	11.5	10.8	16.3	**47.0**	***31.1***
Supplies	20.4	26.5	23.5	24.2	20.6	**56.2**	
Total supplies	***14.8***[4]	***5.4***	***15.2***	***13.9***	***24.7***	***26.0***	***100.0***

1. Relative share of cluster's total deliveries.
2. Relative share of cluster's deliveries in total deliveries.
3. Relative share of cluster's total supplies.
4. Relative share of cluster's supplies in total supplies.

Source: Statistics Finland (1999b).

In earlier studies based on the Porter framework, ten clusters were identified: forestry, base metals, telecommunications, energy, well-being, environmental, transportation, chemicals, construction and foodstuffs (Hernesniemi *et al.*, 1996). However, according to the analysis based on trade flows, the chemical industry has links with the foodstuffs, construction and forestry clusters. Also, energy-related production sectors belong to the ICT, metals and forestry clusters; well-being-related sectors to the ICT cluster; base metals and transportation to the metals cluster; and environmental-related sectors to the forestry, construction and metals clusters.

Networking in the clusters contributes to growth of GDP via two mechanisms. It increases economies of scale through outsourcing and stimulates technical change through technology diffusion among production sectors. It has been estimated that the main source of embodied technology is the machinery and equipment production sector, with the addition of a few service sectors (Vuori, 1997). According to the meso level cluster maps (Annexes 1-6), the core production sectors that are important network builders and in which production is growing fast are electronics [36], machinery [33] and pulp and paper [20].

Economic profiles of the clusters

The Finnish economy is very clustered: 75% of value added was produced in the five clusters identified for study. In addition to most public activities, the textiles and leather industries did not clearly belong to any particular cluster, and these sectors have faced the most severe decline. The largest cluster, forestry, generated almost 40% of exports (Figure 1). The forestry cluster plays a remarkable role in the trade balance; its production is mainly based on domestic raw materials. The metals cluster was the main importer. The foodstuffs cluster employs the highest share of the workforce. The construction and foodstuffs clusters are domestic oriented, with very low imports and exports.

Figure1. **Relative economic profiles of the clusters, 1995**

Percentages

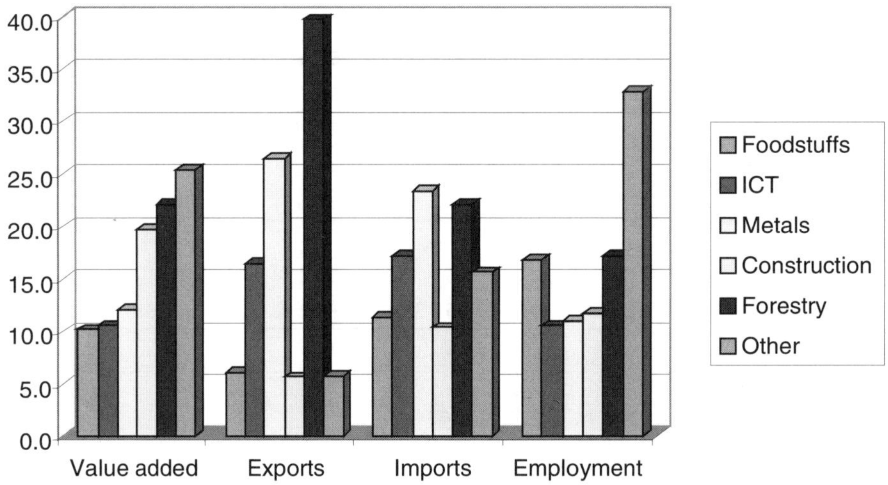

Source: Statistics Finland (1999b).

The relative share of the clusters in production between 1990-98 is shown in Table 2. It was assumed that the weight of a cluster in a single production sector remains constant over time. The share of the metals, construction and forestry clusters and non-cluster-related sectors has remained fairly stable, while growth in ICT has offset the decline of the foodstuffs cluster.

Table 2. **Relative share of the clusters in production, 1980-98**

Percentages

	1980	1990	1998
Foodstuffs	13.7	12.7	9.4
ICT	8.5	10.0	13.2
Metal	11.8	11.7	11.9
Construction	19.5	20.7	21.5
Forest	26.6	22.3	21.9
Other	19.9	22.7	22.1

Source: Statistics Finland (1999a, 1999b).

In the 1980s, growth was driven in equal measure by the pulp and paper, machinery and electronics industries, whereas in the 1990s, it was mainly driven by electronics and, especially, by firms belonging to the ICT cluster (see Paija, Chapter 2 of this volume). The Porter (1990) cluster framework based industrial structures have clearly influenced the excellent profitability and fast growth of this sector. The telecommunications industry has benefited from both highly technology-oriented domestic customers and subcontracting. Liberal telecommunications policy has fostered favourable domestic market conditions in the run up to international development, thus providing Finnish companies with a lead position on export markets (Mäenpää and Luukkainen, 1994).

As industrial networks become more complex, there is a need for efficient communication and information transfer channels between the firm and the new or externalised entities. Knowledge-intensive business services (KIBS) act as a knowledge routing mechanism in the networks. KIBS provides a platform for connectivity and receptivity among sectors. Just as in-house R&D is the basis for firms' knowledge and performance, KIBS provides essential information and technologies to other sectors. As an intermediate input, KIBS shows statistically significant linkages with industrial production (Luukkainen and Niininen, 2000).

Figure 2. **R&D and KIBS intensities of the clusters**

Percentages

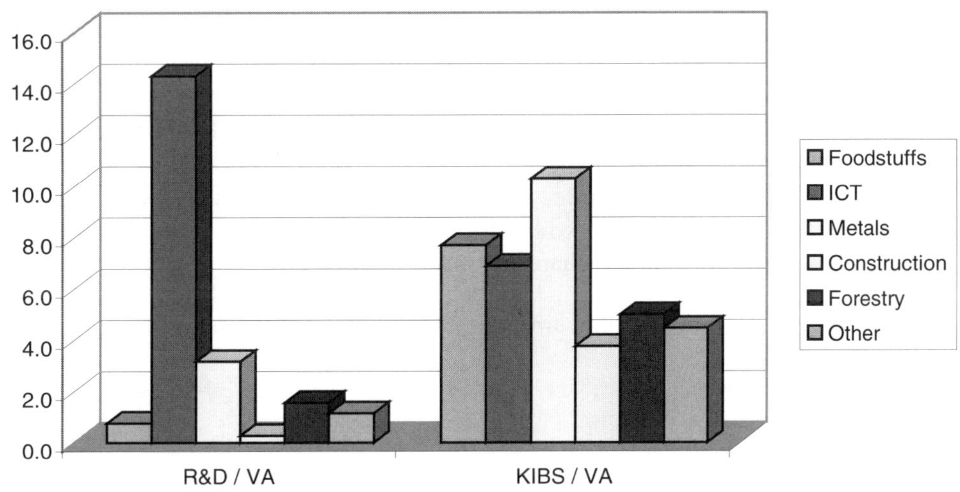

Source: Statistics Finland (1999b, 1999c).

The R&D intensities describe the structural change of Finnish industries towards high-technology products, although it is the ICT cluster that conducts the major part of the R&D (Figure 2). However, the ICT cluster is less KIBS-intensive. The metals cluster uses the highest level of business services and is R&D-intensive. Although the foodstuffs cluster has a very low R&D intensity, it has the second highest KIBS intensity. Both dimensions are at a very low level in the construction cluster. Surprisingly, the KIBS intensity of the forestry cluster is also quite low. It can be seen from the cluster maps that KIBS have strong linkages with several industrial production sectors, notably in the foodstuffs, ICT and metals clusters.

Market failure in the innovation value chain network

As indicated above, market failure caused by R&D spillovers is one of the main justifications for government intervention. Since markets alone fail to facilitate efficient networking, clusters should be the major targets of future technology policy. Technology diffusion policies should therefore focus on the relationship between the innovator rate of return (RoR) and the social RoR as an indicator of market failure and under-investment in R&D (Tassey, 1997).

The innovator RoR describes the anticipated benefits which will accrue to the innovating firm, while the social RoR indicates how profitable an R&D investment will be for the whole value chain through technology diffusion. In order to be able to evaluate projects of different sizes, there is a need for relative measures of their prospects. The usual method for calculating such indicators is the internal rate of return.

Because firms' decisions are based on their expected rate of return, they will not invest in projects that are desirable from the point of view of the economy as a whole. However, maximising the social return is complicated, due to the risk that government support will displace private funds. Governments should fund private R&D projects that have a high social rate of return and that would otherwise be under-funded or delayed.

The example projects A, B, C and D presented in Figure 3 all lie above the 45° line and thus generate spillovers (sRoR includes RoR). However, their characteristics vary significantly, with implications for the firm's investment behaviour and for the financing structure of the projects. The vertical lines describe the ranges of expected social RoR, and the horizontal lines the ranges of expected innovator RoR by the projects. Risk is the probability of the R&D investment not fulfilling the requirements, *i.e.* the hurdle rates of the government or innovating firm. The farther the project's profitability from the innovator hurdle rate, the greater the government's share of costs. In low-risk projects, financial instruments, such as tax reductions, low-interest or subordinated loans, can also be provided in addition to small percentage shares of direct funding.

Figure 3. **Government R&D project selection**

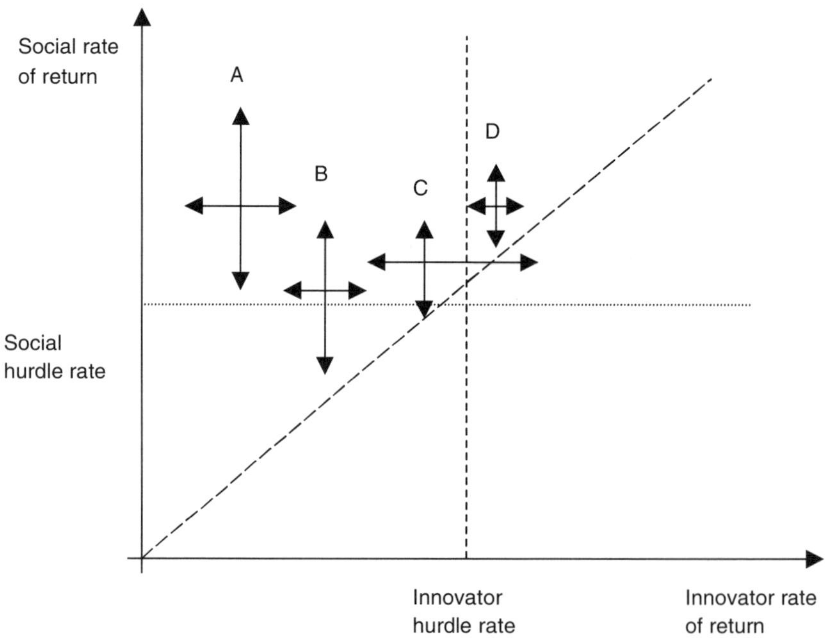

Source: Modified from Tassey (1997).

Project D is most likely to be funded by private funds because from the innovator's point of view the hurdle rate will be exceeded at no risk. Project D represents a typical project ordered from the R&D department by a business unit of a technology firm. Its risk and spillover effects are low, which explains why the firm is able to finance the project from its own cash flow.

Project C comprises greater risk as the worst estimates are below firms' requirements, although it could turn out to be more profitable than Project D. In addition, Project C is not an interesting case for government support because of the low level of spillover effects generated. This project thus illustrates a situation in which venture capital financing would be the most desirable alternative, since the investment is too risky, and perhaps also too large, for the shareholders alone, but it has the potential to gain very high returns.

Project B should be funded neither by government nor by the private sector because its profits are estimated to be below the requirements of both. The government should select projects with the highest social rate of return and which will not be funded by the private sector in the absence of government support.

Project A is the most desirable of the four projects and should be the prime candidate for public funding. The government share of funding should be set at a level where the innovator hurdle rate will be exceeded.

RoR can be calculated by using innovator input data (innovator RoR) and data from the whole value chain (social RoR). Typically, the spillover effects are enjoyed by the subcontractors and customers of the innovator. In addition, horizontally linked firms can also benefit, *e.g.* through the creation of a new market for a supplementary product. According to earlier studies, on average, innovator RoR accounts for 30% and social RoR for 50% of private R&D investments (Tassey, 1997).

In practice, in the start-up phase of R&D projects, only rough estimates of the rate of return exist. It is therefore useful to calculate a range of estimates using different business scenarios. Figure 3 presented profitability estimates for project proposals based on the estimates made during the project proposal phase. It is equally important to follow what happens over time, since estimates of risk will change over the duration of the project. Typically, government support for a project is divided into several components which are paid out as progress on the project is monitored.

Continuous evaluation of the profitability indicators enables decisions to be taken as to whether to "stop" or "continue" the funding. The project should be terminated if the estimates fall below the original ones (and the funds should be targeted to alternative projects); otherwise the project should be funded to the point where the innovator rate of return exceeds the hurdle rate. Thus, the management process creates the possibility of targeting public funds to a project portfolio in which the *spillover effects are continuously maximised and loss of funds minimised.*

Typically, the most important spillover effects are generated by profitable large companies operating in the growing production sectors that have wide network connections in the economy. Government support to such firms is justified because of their leading role in the clusters.

Conclusions

The increasing share of venture capital in R&D financing indicates that the importance of the traditional innovator's risk-reduction-based technology policy is decreasing. The objective of venture capital financing is, however, to increase the value of the innovating company; spillovers to other companies are therefore not as important. The market failure created by R&D spillovers is one of the main justifications for government policies. For this reason, clusters should be the major targets of future economically driven technology policy, since markets alone fail to facilitate efficient networking.

There is a need for reactive bottom-up cluster-oriented technology policy. This kind of policy can be realised by financing R&D projects that stimulate the networked economy. Cluster policy finance instruments should be designed in such a way that they encourage innovators to generate spillovers. Governments should fund the private projects that have the highest spillover effects and that would otherwise be under-funded or postponed.

The Finnish manufacturing sectors have faced rapid structural changes over the last decade. The size of the telecommunications industry more than quadrupled in the 1990s; it is currently the largest single production sector. Synergies in the value chain have clearly influenced this cluster's excellent profitability and fast growth. The telecommunications industry has benefited especially from knowledgeable domestic customers and from subcontracting. The liberal Finnish telecommunications policy has fostered market conditions which have been favourable to international development and have placed Finnish companies in a lead position in export markets.

The trade-flow-based clusters identified in this study provide a meso-level view of how different sectors in the Finnish economy are interlinked, and indicate the main paths of domestic technology flows at an aggregated level. Despite the fact that the telecommunications-related ICT cluster has been the main growth determinant, its size and impact on GDP is still lower than in more traditional clusters. The influence of the forestry cluster is dominant, both in terms of its positive trade balance and its share of production.

Clusters should be seen as a technology policy tools which allow industrial structures to be analysed and developed. In this kind of decision making, it is crucial to understand the mechanisms of technology diffusion in order to evaluate the economic impacts of R&D investments in advance. Core industries that are important network builders and in which production is growing rapidly include electronics, machinery, and pulp and paper. In addition, knowledge-intensive business services play a very important role in the clusters as an external knowledge transfer mechanism. It can be assumed that the R&D projects of companies belonging to these sectors would generate the largest spillover effects in the Finnish economy.

Annex 1

THE FOODSTUFFS CLUSTER

	Description	NACE classification	Weight[1]	VA[2]
1	Agriculture and related service activities	011-014	1	8 930
3	Hunting and fishing	015, 05	1	837
7	Production, processing, preserving of meat, fish, prods thereof	151, 152	1	2 790
8	Processing and preserving of fruit and vegetables	153, 154	1	1 004
9	Manufacture of dairy products	155	0.95	1 732
10	Man. of grain mill products, starches and starch products	156	0.79	352
11	Manufacture of prepared animal feeds	157	1	674
12	Manufacture of other food products	158	1	3 938
13	Manufacture of beverages and tobacco products	159, 16	1	1 811
20	Manufacture of pulp, paper and paperboard	211	0.01	289
21	Manufacture of articles of paper and paperboard	212	0.16	286
24	Manufacture of basic chemicals	241	0.17	851
48	Wholesale and retail trade	50,51,52	0.18	8 957
49	Hotels and restaurants	55	1	7 706
50	Land transport; transport via pipelines	60	0.11	2 058
52	Air transport	62	1	3 005
54	Activ. of travel agencies and tour operators; tourist assist. act.	633	1	773
60	Other real estate activities	701,7022,7031	0.5	3 250
61	Business activities	71-74	0.05	1 139
67	Recreational, cultural and sporting activities	92	0.1	858
	Total			**51 240**

1. Describes how strong trade links a sector has to the cluster.
2. Weighted value added, FIM 1 000 000, 1995.

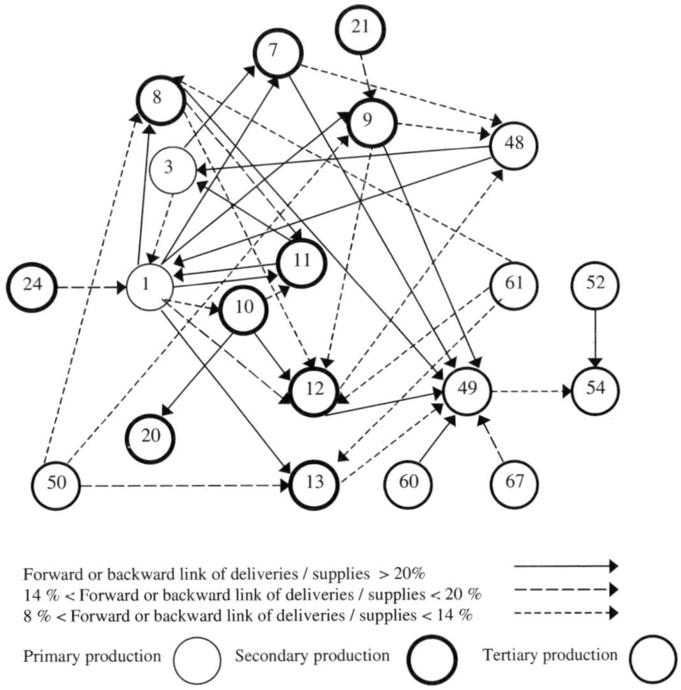

Annex 2

THE ICT CLUSTER

	Description	NACE classification	Weight	VA
22	Publishing and printing	22	0.09	760
27	Manufacture of plastic products	252	0.23	650
32	Manufacture of fabricated metal products	28	0.02	158
33	Manufacture of machinery and equipment n.e.c.	29	0.02	275
34	Manufacture of office machinery and computers	30	1	886
35	Manufacture of electrical machinery and apparatus n.e.c.	31	0.6	2 578
36	Man. of radio, television and communication equipment and apparatus	32	1	9 360
37	Manufacture of medical and precision products	33	0.72	1 728
47	Civil engineering	4502	0.06	458
48	Wholesale and retail trade	50,51,52	0.26	12 764
50	Land transport; transport via pipelines	60	0.02	412
56	Post and telecommunications	641,642	1	11 048
57	Financial intermediation and insurance	65,66,67	0.02	335
61	Business activities	71-74	0.4	9 683
62	Public administration and defence; compulsory social security	75	0.05	1 342
63	Education	80	0.01	267
67	Recreational, cultural and sporting activities	92	0.08	717
	Total			**53 420**

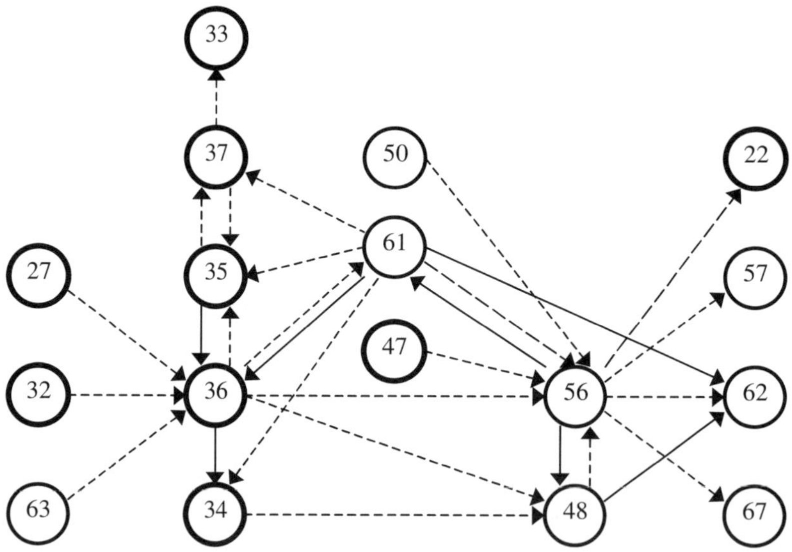

Forward or backward link of deliveries / supplies > 20%
14 % < Forward or backward link of deliveries / supplies < 20 %
8 % < Forward or backward link of deliveries / supplies < 14 %

Annex 3

THE METALS CLUSTER

	Description	NACE classification	Weight	VA
5	Mining of metal ores	13	1	184
6	Other mining and quarrying	14	0.45	440
20	Manufacture of pulp, paper and paperboard	211	0.02	578
30	Manufacture of basic iron and steel and ferro-alloys	271	1	4 967
31	Manufacture of basic metals n.e.c.	272-275	1	2 709
32	Manufacture of fabricated metal products	28	0.79	5 044
33	Manufacture of machinery and equipment n.e.c.	29	0.84	12 645
35	Manufacture of electrical machinery and apparatus n.e.c.	31	0.4	1 701
37	Manufacture of medical and precision products	33	0.1	230
38	Manufacture of motor vehicles, trailers and semi-trailers	34	1	1 508
39	Building and repairing of ships and boats	351	1	2 784
40	Manufacture of other transport equipment n.e.c.	352-355	1	1 177
43	Recycling	37	1	116
44	Electricity, gas, steam and water supply	40	0.07	748
46	Building of complete constructions or parts thereof, service activities	4501,4509	0.07	1 126
47	Civil engineering	4502	0.94	7 783
48	Wholesale and retail trade	50,51,52	0.01	224
50	Land transport; transport via pipelines	60	0.28	5 352
53	Road and railway maintenance	630	1	3 768
61	Business activities	71-74	0.3	7 405
62	Public administration and defence; compulsory social security	75	0.02	565
	Total			**61 054**

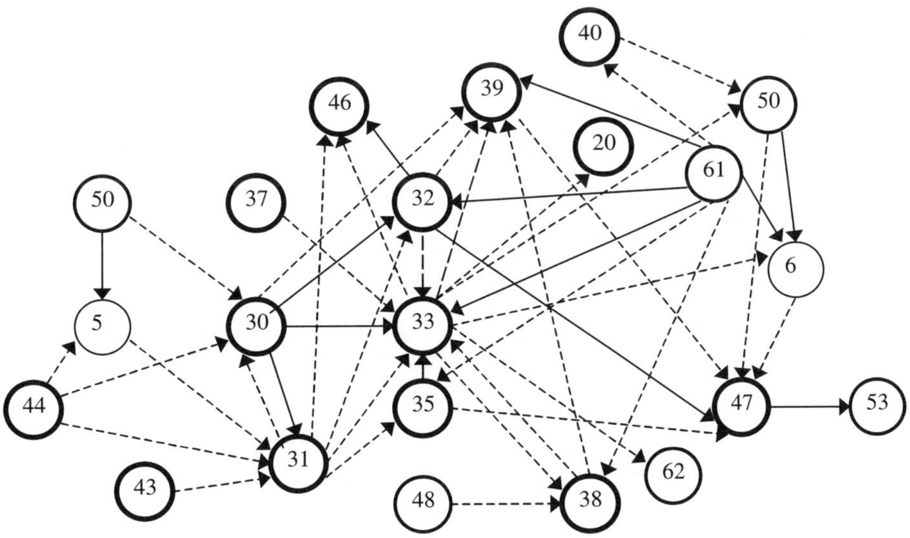

Forward or backward link of deliveries / supplies > 20%
14 % < Forward or backward link of deliveries / supplies < 20 %
8 % < Forward or backward link of deliveries / supplies < 14 %

Annex 4

THE CONSTRUCTION CLUSTER

	Description	NACE classification	Weight	VA
6	Other mining and quarrying	14	0.22	220
17	Saw-milling and planing of wood, impregnation of wood	201	0.11	365
18	Manufacturing of veneer sheets, plywood, laminate, particle board	202	0.08	114
19	Manufacture of other products of wood	203,204,205	0.63	1 162
24	Manufacture of basic chemicals	241	0.1	510
25	Manufacture of other chemical products and man made fibres	241-247	0.49	1 545
27	Manufacture of plastic products	252	0.77	2 176
28	Manufacture of glass and glass prods, non-refractory ceramic goods	261,262	1	1 183
29	Manufacture of ceramic tiles and bricks, cement, articles of concrete	263-268	1	2 324
32	Manufacture of fabricated metal products	28	0.19	1 214
37	Manufacture of medical and precision products	33	0.19	451
41	Manufacture of furniture	361	0.39	799
44	Electricity, gas, steam and water supply	40	0.21	2 430
45	Collection, purification and distribution of water	41	1	1 495
46	Building of complete constructions or parts thereof, service activities	4501,4509	0.69	10 475
48	Wholesale and retail trade	50,51,52	0.38	18 810
57	Financial intermediation and insurance	65,66,67	0.02	446
58	Letting and operation of dwellings	7021	1	43 805
59	Management of real estate on a fee or contract basis	7032	1	3 859
60	Other real estate activities	701,7022,7031	0.5	3 250
61	Business activities	71-74	0.12	2 848
65	Sewage and refuse disposal, sanitation	90	0.38	992
	Total			**100 474**

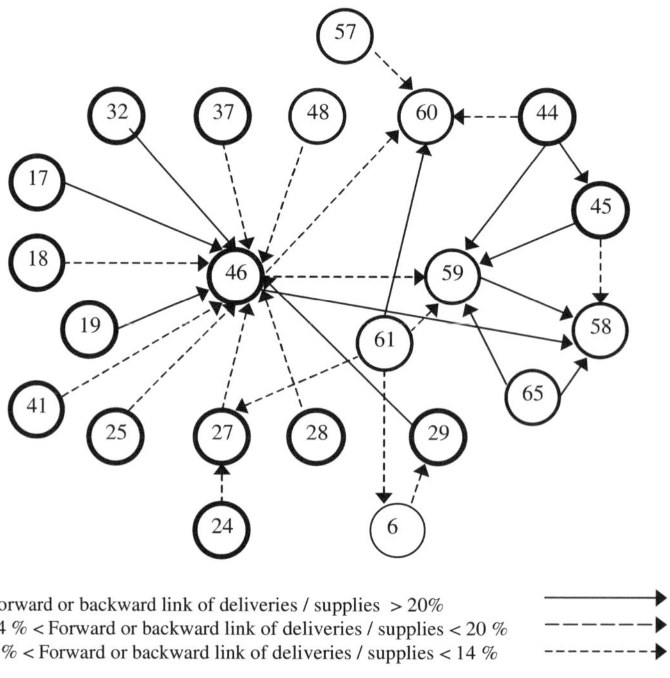

Annex 5

THE FORESTRY CLUSTER

	Description	NACE classification	Weight	VA
2	Forestry, logging and related service activities	02	1	13 044
4	Mining and quarrying of energy producing materials	10	1	634
6	Other mining and quarrying	14	0.33	320
9	Manufacture of dairy products	155	0.05	83
10	Manufacture of grain mill products, starches and starch products	156	0.21	93
17	Sawmilling and planing of wood, impregnation of wood	201	0.89	3 008
18	Manufacturing of veneer sheets, plywood, laminate, particle board	202	0.92	1 383
19	Manufacture of other products of wood	203,204,205	0.37	672
20	Manufacture of pulp, paper and paperboard	211	0.96	22 540
21	Manufacture of articles of paper and paperboard	212	0.84	1 471
22	Publishing and printing	22	0.91	7 598
24	Manufacture of basic chemicals	241	0.73	3 743
25	Manufacture of other chemical products and man made fibres	241-247	0.51	1 613
33	Manufacture of machinery and equipment n.e.c.	29	0.15	2 199
41	Manufacture of furniture	361	0.61	1 255
44	Electricity, gas, steam and water supply	40	0.72	8 223
46	Building of complete constructions or parts thereof, service activ.	4501,4509	0.23	3 535
48	Wholesale and retail trade	50,51,52	0.18	8 957
50	Land transport; transport via pipelines	60	0.59	11 116
51	Water transport	61	1	3 156
55	Other supporting transport and auxiliary activities	631,632,634	0.12	764
61	Business activities	71-74	0.14	3 304
63	Education	80	0.17	4 544
67	Recreational, cultural and sporting activities	92	0.82	7 177
	Total			**110 432**

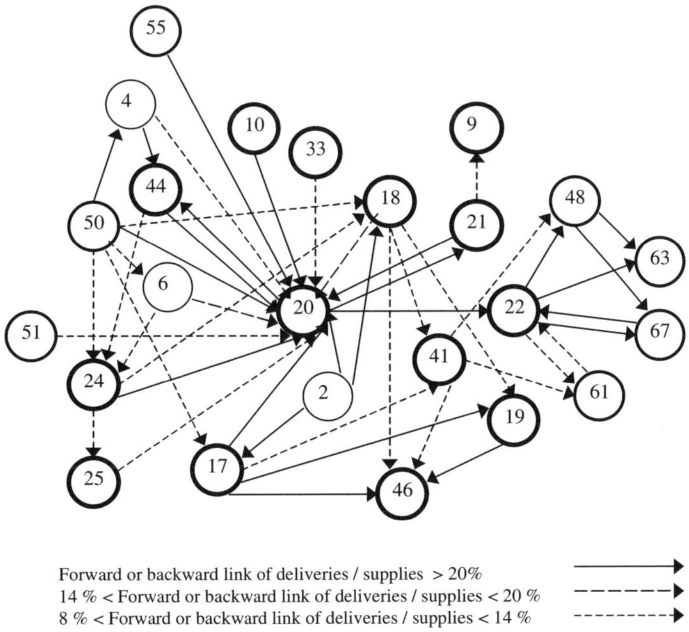

Annex 6

NON-CLUSTER-RELATED PRODUCTION SECTORS

	Description	NACE classification	Weight	VA
14	Manufacture of textiles	17	1	1 458
15	Manufacture of wearing apparel; dressing and dyeing of fur	18	1	1 370
16	Manufacture of leather and leather products	19	1	567
23	Manufacture of refined petroleum products, coke, nuclear fuel	23	1	1 457
26	Manufacture of rubber products	251	1	740
42	Manufacturing n.e.c.	362-366	1	940
55	Other supporting transport and auxiliary activities	631,632,634	0.88	5 600
57	Financial intermediation and insurance	65,66,67	0.96	17 817
62	Public administration and defence; compulsory social security	75	0.93	24 938
63	Education	80	0.82	21 919
64	Health and social work	85	1	42 190
65	Sewage and refuse disposal, sanitation	90	0.62	1 618
66	Activities of membership organisations n.e.c.	91	1	5 266
68	Other service activities	93	1	2 466
	Total			**128 346**

REFERENCES

Coe, D. and E. Helpman (1995), "International R&D Spillovers", *European Economic Review*, Vol. 39, No. 5, pp. 859-888.

Hernesniemi, H., M. Lammi and P. Ylä-Anttila (1996), "Advantage Finland: The Future of Finnish Industries", Etla Series B 113.

Irwin, D. and P. Klenow (1994), "Learning-by-Doing: Spillovers in the Semiconductor Industry", *Journal of Political Economy*, December.

Jacobs, D. (1997), "Knowledge-intensive Innovation: The Potential of the Cluster Approach", *The IPTS Report, Technology and Competitiveness*, No. 16, July.

Luukkainen, S. and P. Niininen (2000), "Technology Intensive Services and National Competitiveness", VTT Group for Technology Studies Working Papers No. 46.

Mäenpää, K. and S. Luukkainen (1994), "From Telecommunications Technology to Multimedia Communication: The Competitiveness of the Telecommunications Cluster", Etla Series B 96.

OECD (1999), *Boosting Innovation: The Cluster Approach*, OECD, Paris.

Paija, L. (2001), "ICT Clusters – The Engine of Knowledge-Driven Growth in Finland", Chapter 2 of this volume.

Pentikäinen, T. (2000), "Economic Evaluation of the Finnish Cluster Programmes", VTT Group for Technology Studies Working Papers No. 50.

Porter, M.E. (1990), *The Competitive Advantage of Nations*, The Free Press, New York.

Porter, M.E. (1998), "Clusters and the New Economics of Competition", *Harvard Business Review*, November-December.

Rouvinen, P. and P. Ylä-Anttila (1999), "Finnish Cluster Studies and New Industrial Policy Making", in OECD (1999), *Boosting Innovation: The Cluster Approach*, pp. 361-380.

Tassey, G. (1997), *The Economics of R&D Policy*, Quorum Books.

Statistics Finland (1999a), *National Accounts Time Series 1975-1998.*

Statistics Finland (1999b), *Input-Output 1995.*

Statistics Finland (1999c), *Research and Development in Finland 1998.*

Vuori, S. (1997), "Technology Sources and Competitiveness – An Analysis of Finnish Industries", Etla Series B 138, Helsinki.

Chapter 14

KNOWLEDGE TRANSFER IN AGGLOMERATIONS: A REGIONAL APPROACH TO JAPANESE MANUFACTURING CLUSTERS

by

Kinji Gonda, Waseda University and NISTEP, and **Fumihiko Kakizaki,** NISTEP

Introduction

To stimulate and accelerate regional innovation, our group at NISTEP has been conducting studies on regional systems of innovation. This has taken place in the context of the framework of science and technology policy (S&T) and the national system of innovation of Japan (Collins *et al.*, 1998; Gonda, 1998). Since the responsibilities of regional governments in S&T have been strengthened by the Science and Technology Basic Law enforced in 1995,[1] a new framework for regional S&T policy has been launched. Regional systems of innovation and clusters are seen as major contributors to national competitiveness (Porter, 1998; Council on Competitiveness, 1999).

From our research on industrial location, industries typically locate by agglomeration and/or migration (Gonda *et al.*, 1999). The spatial mobility of industries depends on the size of firms as well as industrial sectors. Industries with manufacturing processes that are dependent on tacit knowledge, such as design and/or know-how, tend to co-locate in agglomerations. These results lead us to conclude that clustering mechanisms might be more complex and dynamic than the Marshallian model would lead us to expect (Marshall, 1920).

The purpose of this chapter is to describe the current cluster approach in our research on regional innovation systems, and provide some insight into the results of this activity. It should be noted that our present approach to cluster studies deviates from that of the majority of the chapters in this volume: we do not treat clusters as value chains, but rather see them as geographical or spatial agglomerations. Firms and industries may migrate (disperse) or agglomerate in accordance with the opportunities available to exploit key resources. Of particular interest is the degree to which knowledge represents a key resource leading to distinct agglomeration patterns. In the next section, using our *regional* cluster approach, we present the findings of the study on spatial mobility and agglomerating tendencies of manufacturing industries, while the following section presents patterns of knowledge linkages among Japanese small and medium-sized enterprises. The overall hypothesis which binds the two parts together is that industries that are dependent on tacit or new, emerging knowledge will agglomerate in regions which offer conditions for knowledge exchange, and that this pattern is observable in the knowledge transfer links between firms and non-business organisations in particular regions.

Industrial locations and spatial mobility

Some industries typically agglomerate spatially to specific regions. On the other hand, a "hollowing out" of industries has occurred through dispersal away from losing regions towards gaining regions or areas. Arguments about industrial location and regional economies have focused on the problem of inter-regional, comparative advantage between developed and developing regions. Industries have tended to move spatially to maintain their competitiveness. The prosperity or decline of industrial zones through agglomeration or dispersal can be easily measured in Japan and other countries. In this context, the first approach of our research on regions was to develop quantitative indicators of spatial agglomeration and/or dispersal of manufacturing industries across regions (Gonda *et al.*, 1999).

The Ministry of International Trade and Industry (MITI) provides detailed statistical data on manufacturing industry (Census of Manufactures). We have used four key economic variables: number of firms (NOF); number of employees (NOE); value of shipments; and value-added production. We re-tabulated these variables by 23 industries based on the Japanese Standard Industrial Classification (JSIC) at the two-digit level, and by 47 prefectures (proxy for regions) for 1980-94 (15 years). The tabulation regarding prefectures mainly provided information on the industrial location process across prefectures over the study period.

This measurement helps to illustrate industrial concentration in specific areas over the period studied and also reflects industrial policies. It was expected that a high degree of concentration would be observed around the highly populated regions. Thus, we analysed deviations from the average industrial distribution across regions. The Gini co-efficient indicates the deviation or inequality compared with average values (Krugman, 1991; Fujita *et al.*, 1999), but it is more difficult to measure the deviation from an average distribution of industries across space. To analyse the spatial agglomeration and/or migration, we have to consider geographical distribution in detail.

We introduced a method to evaluate the difference in these distributions. If each industrial parameter, such as NOF or NOE of the industries studied, showed the same distribution over regions, we would find isometric dispersion. In addition, we could also observe agglomeration as a relatively high percentage of specific industries in certain regions combined with a low percentage of the same industries in the remaining regions. Therefore, we obtained the Index of Industrial Location (IIL) as follows:

The Index of Industrial Location (IIL) for industry *r* is:

$$\text{IIL} = (1/2) \sum |(A_{ri} / A_r) - (A_{ni} / A_n)|,$$

Where summation is taken over all regions *i*.
A_{ri} = number of firms (or employees, etc.) in industry *r* of region *i*.
A_r = total number of firms (or employees, etc.) in industry *r*.
A_{ni} = total number of firms (or employees, etc.) from all industry in region *i*.
A_n = total population of firms (or employees, etc.) in all regions.

The value ranges from zero to one. An IIL value of 0.0 means isometric dispersion, and a value of 1.0 indicates complete agglomeration. The IIL values were obtained over the 23 manufacturing industries and over the 15 years. Generally, an increase in the IIL value indicates a location process of an industry towards spatial concentration; a decrease, on the other hand, indicates out-migration.[2]

The IIL values show highly complicated patterns. Comparing the changes in NOEs among the industries from 1980-94, an increase in NOE values (indicating industries with growth) were observed in eight industries (foods, apparel, pulp and paper, chemicals, tannery, metal fabrication, general machinery, electric machinery, and transportation machinery). Among these eight growing industries, a decrease in the IIL values was found in the metal fabrication, general machinery, and electric machinery industries, while these values increased significantly in apparel and in transportation machinery (Figure 1).

Figure 1. **Examples of increase or decrease of the IIL values *vs.* industrial parameters**

A. Case of increasing IIL values

ILL (NOE). 30 Electrical machinery

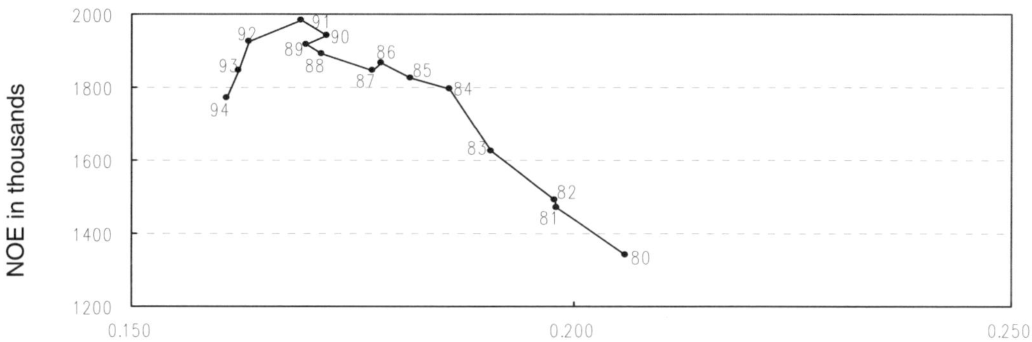

B. Case of decreasing IIL values

ILL (NOE). 15 Apparel

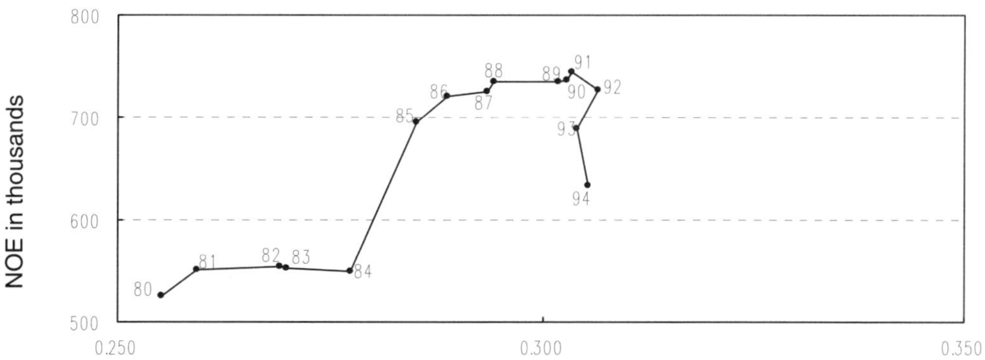

On the other hand, we also observed that two types of industries exhibited spatially distinct patterns of decline as shown in Figure 2. Textiles is typical of an industry declining in size through agglomeration. The steel industry shows a more complicated pattern of spatial mobility in the process of decline (Figure 3). This indicates that some industries locate through spatial agglomeration through growth processes, while others agglomerate in the process of decline. The major findings from this analysis using IILs are as follows:

- Industries have their own individual characteristics of spatial mobility, showing agglomeration or dispersal.
- Industries agglomerate and/or disperse spatially not only during their growth but also in processes of decline.
- The spatial mobility of industries is quantitatively obtained by the IILs.
- Industries locating through agglomeration, in the processes of either growing or declining, can be categorised as industries whose key technologies depend on implicit or tacit knowledge such as design or know-how. Dispersing industries are less dependent on the exchange of tacit knowledge.

Figure 2. **Example of a declining process**

ILL (NOE). 14 Textiles

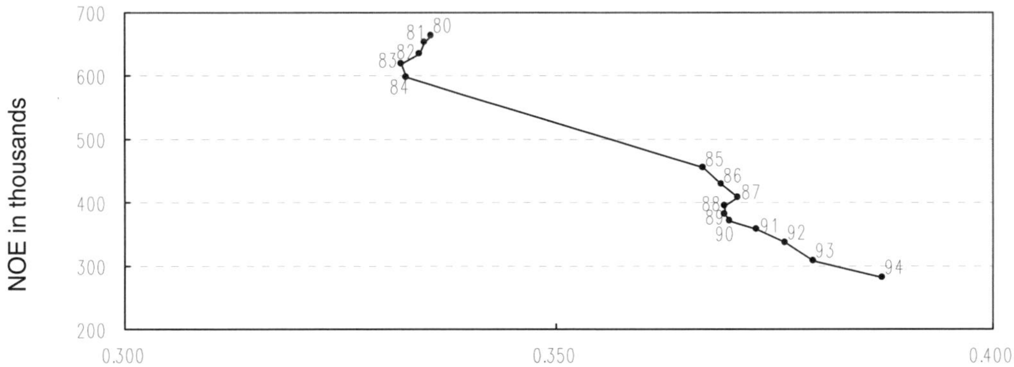

Figure 3. **Example of a complicated change**

ILL (NOE). 26 Steel

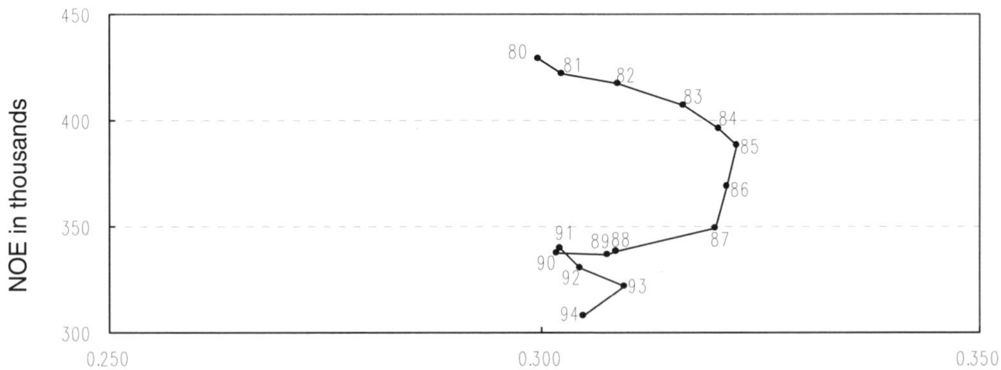

Discussions on the spatial agglomeration of industries have focused on the external economies in regions (proximity, spillover effects). In the classical theories on industrial location, the theoretical framework was based on intra-regional economic development. Discussions therefore centred on whether or not such regions can provide geographical and/or regional comparative advantage, as represented in least-cost theory and maximum-demand theory. The basic theory of industrial location, as described by Paul Krugman (1991) in *Geography and Trade*, has focused on the relations between locations and productivity of manufacturing industries, but not on the productivity of knowledge as the basis for the emergence of regional innovation.

According to the product and regional life-cycle theory, industries have characteristics that induce their dispersal over growth and maturation processes (Norton and Rees, 1979). This is based on the premise that products evolve through three stages in their life cycles, *i.e.* an innovation stage, a growth stage, and a standardisation stage. In the final stage, the theory predicts the hollowing out of industries from developed regions; that is, production may shift to low-cost regions.

From a series of theoretical studies by Gonda and Shamizu (1999), this process was explained in terms of "clustering as an emerging space". This work attempts to explain the non-physical mechanisms for the formation of spatial clusters. Generally, the key technology dominating competitiveness of, for example, the apparel and motor vehicle industries, is found in design technology and/or manufacturing know-how. These key technologies or knowledge types are very often difficult to express in procedures and operational manuals as explicit knowledge. Emerging or new knowledge is often implicit or tacit, such as design of new concepts and/or manufacturing know-how, and researchers and engineers are required to discuss and interact with colleagues as well as customers, suppliers and even competitors through person-to-person contacts (Gonda, 2001). This explains why service-intensive industries agglomerate to specific regions to generate comparative advantage. On the other hand, technology-intensive industries also migrate across nations. The key technologies of the manufacturing processes of these industries are less dependent on implicit or tacit knowledge, and rely more on explicit knowledge, the latter being easier to transfer through manuals and even through the patents that protect intellectual property rights. In addition, explicit knowledge is increasingly transferred globally by reliable computer networks. This explains why industries migrate away from industrial districts in order to seek competitive advantage through manufacturing productivity or cost advantages.

Knowledge transfer in clusters

We have described the method to measure geographical agglomeration and/or migration of industries. The next step is to assess the factors that enable cluster formation and strengthen existing clusters, as well as indicate what kind of policy could be desirable for emerging and existing clusters (Boekholt and Thuriaux, 1999). According to Alfred Marshall's theory of industrial districts, external economies are the essential factor for cluster formation. On the other hand, in terms of new or emerging knowledge, Saxenian (2000) refers to the limits of external economies and emphasises the importance of networks in a region. She pointed out that "Silicon Valley's regional advantage lies not in its entry onto a technological pathway, but in an institutional environment that supports continuous innovation and collective learning."

We have studied the industrial location and business activities of Japanese small and medium-sized enterprises (SMEs) engaged in research and technological development. These enterprises are the most important actors of regional innovation systems. A survey was carried out of 5 000 firms, including manufacturing industries and software and services industry across Japan. The response ratio was about 24% (1 230 firms). Four industrial sectors were taken into account: traditional and mature (textiles and apparel), high-technology (electric and machinery), software and services, and the rest of manufacturing industries. The analysis was conducted as follows:

- All sample firms were classified into self-brand firms (SBF with ownership of self-brand products and/or technologies) that produce *final goods* at the end of a value chain, and non-self-brand firms (NSBF), often defined as subcontracting SMEs that produce *intermediate goods*.

- Sources of technology and know-how, and external communication of the individual SBFs and NSBFs with other organisations or institutions were measured by comparing the importance of external linkages and relations.

The proportion of final goods producers (SBF ratio) is about 44%, which implies that the majority of SMEs produce intermediate goods. The SBF ratios by sector and region offer interesting insights into the different locational preferences of sectors across regions. We adopted seven geographical regions according to the usual Japanese classification: the Kanto, Chubu and Kansai districts include large urban areas such as Tokyo, Nagoya and Osaka, respectively. The urban regions are characterised by relatively low SBF ratios in the software and services industries, unlike the other four peripheral regions where the equivalent SBF ratios were markedly higher. A relatively low SBF ratio for textiles and apparel was found in the districts of Chubu, Hokuriku and Chu/Shikoku, all of which are known to be traditional agglomerating areas of clothing firms. The software and services industry tends to be located in the urban areas, implying that this industry is demand-driven, while the textiles and apparel industries preferred to take advantage of a manufacturing environment. Significant regional preferences were not observed among electric and machinery industries.

The IIL[3] values discussed above prove useful here. Since the SBF ratio is larger than that for NSBF (Table 1), this indicates that the SBF were relatively more concentrated in specific regions or geographical clusters. Examining the IIL values by industrial sector, the textiles and apparel industry was found to have the largest IIL value, indicating high agglomeration (Table 2). This shows that the firms in this industrial category were primarily locating by forming clusters in specific regions, and developing their own self-brand products as observed by the higher SBF ratio. Those districts are shown in Figure 4.

Table.1. **IIL values for SBFs and NSBFs**

	SBFs	NSBFs
IIL values	0.341	0.275

Table 2. **The IIL values of the industrial groups**

	Textiles and apparel	Electric and machinery	Software and services	Others
IIL values	0.648	0.556	0.520	0.419

Table 3. **IIL values of the industrial groups, broken down into SBFs and NSBFs**

	Textiles and apparel	Electric and machinery	Software and services	Others
SBFs	0.711	0.699	0.546	0.510
NSBFs	0.706	0.626	0.616	0.472

Note: The IIL values of SBFs and NSBFs by industry were calculated using the distributions of SBFs and NSBFs, respectively.

Figure 4. **SBF ratio of three industrial sectors by seven districts**

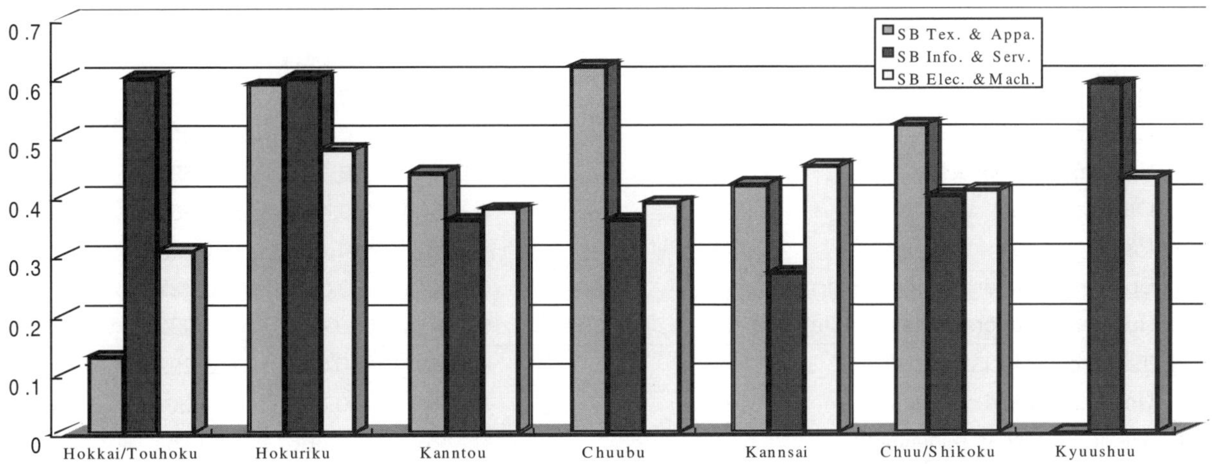

Note: Tokyo, Nagoya and Osaka are located in the Kanto, Chubu and Kansai districts, respectively.

Generally, subcontractors (NBSFs) that manufacture intermediate goods tend to locate in rural areas to a greater extent than do final producers, as shown in Table 3. However, in the case of the software and services industries, final goods producers tend to locate in rural areas, while the intermediate goods producing subcontractors (NSBF) were more often found in urban areas. The difference found among the industries indicates that industrial clusters may be categorised into two types: manufacturing clusters and customer-oriented clusters.

Formal and tacit knowledge transfer in regional clusters

The survey questions also focused on detecting knowledge-related causes for locating in given regions and for possible future location decisions in order to isolate the dominating factors behind the formation of clusters. If the formation of clusters was dependent on external economies, the outsourcing attitude of firms, *i.e.* transfer of technologies and/or know-how, would reflect the benefits of locating in clusters. We considered 13 types of external knowledge resources as shown in Table 4, and asked firms to rate their importance (on a scale of 1-5, with 1 for least important to 5 for most important).

Table 4 presents the results for three industries. In addition to distinguishing between final goods (SBF) and intermediate goods (NSBF) producers, we also separated firms which carried out research and technology development (RTD) from those which did not, and classified them respectively. RTD activity was measured according to the percentage of RTD expenses in net sales. As very few SBF firms were categorised as non-RTD final goods producers, the table contains only the three relevant sub-categories.

Table 4. **External knowledge and know-how: the textiles and apparel industries**

Source of technical knowledge and know-how	SBF RTD	NSBF non-RTD	NSBF RTD	OSI category
1. Recruitment	0.870	0.667	0.637	BAI
2. Customers	1.061	0.623	1.260	BAI
3. Equipment suppliers	0.284	0.191	0.521	BAI
4. Other suppliers	0.758	0.448	0.867	BAI
5. Patent providers	-0.288	-1.091	-0.581	BAI
6. Business partners (same sector)	0.574	0.338	0.729	BAI
7. Business partners (other sectors)	0.294	-0.471	0.022	BAI
8. Industrial associations	-0.045	-0.373	0.292	RII
9. Universities and colleges	-0.456	-0.866	-0.304	RII
10. Regional research institutes	-0.188	-0.866	-0.261	RII
11. Consultants	-0.309	-0.706	-0.234	RII
12. Trade fairs	0.103	-0.362	0.553	RII
13. Publications	0.014	-0.412	0.298	RII

Note: "+2" = most important; "-2" = least important.

Technical knowledge and know-how are sourced very heavily from other firms in an extended value chain of customers, suppliers, business partners, and so forth, while other non-business organisations typically considered part of national (or regional) innovation systems are seldom, if ever, involved (consistently negative values).[4] This proves true for all three types of classified firms. To generate a more precise profile of external linkages, we defined an "Outsourcing Index (OSI)" that consists of "business activity infrastructure (BAI)" and "research and information infrastructure (RII)", using the 13 categories shown in Table 4. BAI is the average unweighted value of categories from recruitment to associations (Questions 1-7). RII refers to ranging categories from universities and colleges to publications (Questions 9-13). Figure 5 shows the OSI-BAI and OSI-RII for several industrial groups.

Intermediate goods industries (NSBF) which do not undertake RTD show very weak relationships to external resources of all kinds, including business activity infrastructure. By contrast, RTD-intensive firms show relatively strong dependence on the BAI. Dependence upon the RII in final goods (SBF) industries, such as electric and machinery and other manufacturing industries, is very limited. The results suggest that internal characteristics of clusters vary across the different industries as well as by stage of production (final *vs.* intermediate goods producers).

According to Marshallian external economies, informal mechanisms of technology transfer are a key factor responsible for spatial agglomeration – *geographical clustering*. Strong value-chain linkages between buyers and suppliers seem to support this view. The full results of our research as summarised in Figure 5 indicate, however, that *formal* research and information infrastructure does not function as an institutional framework for knowledge and technology transfer, at least for Japanese SMEs. Only SBFs which conduct RTD use the RII for their technology acquisition. This suggests that the role of policy should be reconsidered in order to accelerate RTD in SMEs and to strengthen the competitiveness of clusters through enhancing and building upon existing business linkages within them.

Figure 5. **Outsourcing indices among three different industrial groups**

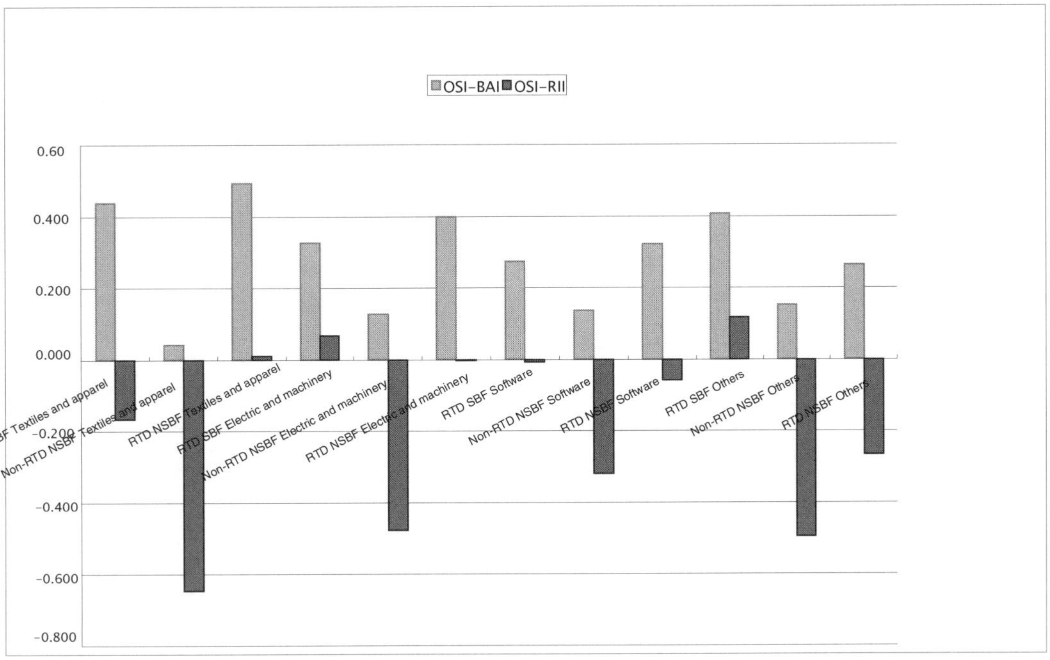

To better analyse the mechanism of implicit or tacit knowledge transfer, we also collected data on the intensity and frequency of external linkages among firms. Table 5 shows the results for the electric and machinery industries for similar categories as in Table 4. Strong relationships with business partners, such as customers, suppliers and competitors were again found for all types of firms in the value chain, but particularly for intermediate goods producers with strong RTD activities. In addition, strong interaction with financial institutions was observed, which is entirely logical for typically capital-starved SMEs. These figures show that clusters are formed relative to the patterns of connectivity among firms and organisations.

Table 5. **Intensity/frequency of firm's interaction through external linkages: the electric and machinery industry**

External linkage	SBF RTD	NSBF non-RTD	NSBF RTD	LI
1. Customers	1.375	1.256	1.409	BP
2. Suppliers	0.875	0.741	0.848	
3. Competitors	0.625	0.212	0.439	
4. Industrial associations	-0.143	-0.155	0.015	BI
5. Financial institutions	0.500	0.756	0.985	
6. Service providers	-0.143	-0.447	-0.031	
7. Training institutions	-0.167	-0.235	0.078	
8. Universities and collages	0.333	-0.541	0.094	RI
9. Public Institutions	-0.143	-0.318	0.168	
10. Regional research institutes	0.286	-0.424	0.109	

Note: "+2" = most important and "-2" = least important.

Similar results were obtained for the other three industrial groups. Based on these results, we defined "Cluster Indices (CI)", consisting of individual "business partnership (BP)", "business infrastructure (BI)", and "research and infrastructure (RI)" indices. BP is the average values for the categories from customers to competitors (Questions 1-3), (no weighting to obtain values). Similarly, BI is the average for the categories from industrial associations to training institutions (Questions 4-7). RI is based on the average values for the categories for academic and public institutions (Questions 8-10). Figure 6 shows CI-BP, CI-BI and CI-RI for all of the industrial groups.

Figure 6. **Cluster Indices of the industrial groups**

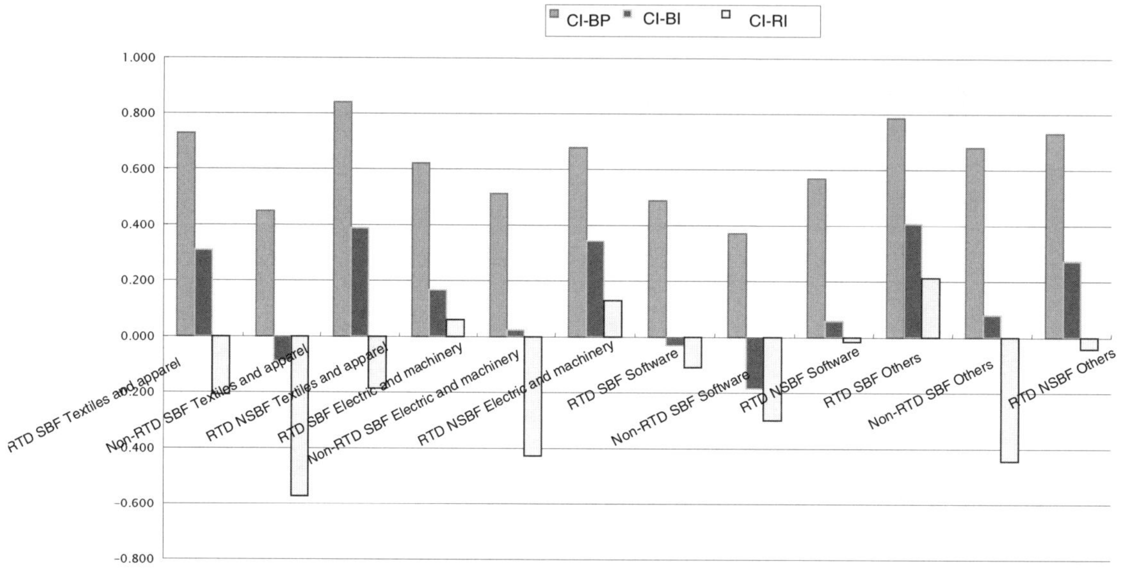

We observe much higher intensity in the relationships between business partners across all industrial groups. On the other hand, there are lower but marked differences in the linkages with business infrastructure and research infrastructure. In the textiles and apparel group, RTD-intensive firms show relatively intense relationships with business infrastructure; however, its non-RTD firms are poorly linked. In addition, the research infrastructure was deemed less important throughout this industrial group. In the electric and machinery group, business partnership was also of great importance; in particular, RTD-intensive firms indicated a relatively positive attitude towards the research infrastructure. The software and services industrial group shows positive signs for business partnership, as well as with business and research infrastructure. As far as we can learn from the results on cluster formation, business partnership is generally the most important factor, confirming results from studies in other countries on innovation behaviour. On the other hand, significant differences were found among the industrial groups in their links with business infrastructure and research infrastructure.

Overall findings and policy conclusions

To summarise the main findings of our analysis: first, the IIL value of the textiles and apparel industrial group indicated a high degree of geographical agglomeration. The Cluster Index, with large positive values of business partnership and business infrastructure in the RTD/SBF and non-RTD/NSBF, suggests that these industries utilised the spillover advantages offered by spatial industrial agglomeration. On the other hand, the software and services industrial group showed relatively smaller IIL values compared to textiles and apparel. The Cluster Index suggests an almost exclusive importance of business partnership for knowledge transfer. It also suggests that intermediate producers of software and services primarily favour location in urban areas. The electric and machinery industrial group shows its location propensity in between these two industrial groups. In particular, RTD-intensive firms tend to relate to business partnership and business infrastructure, as do textiles and apparel. Electric and machinery RTD firms were considered to be performing their business activities in regions offering cost-related comparative advantage.

While the OSI measures technology or knowledge transfer contact with outside firms, non-physical or informal mechanisms could be measured through the intensity or frequency of daily and personal communication links. Although the comparative advantage of clusters, as geographical (spatial) agglomeration, can be quantitatively explained by product links and/or technology transfer among firms, it might be generally more difficult to evaluate the knowledge transfer through relationships among persons. Still, "proximity" could be understood in terms of not only the actual distance between the actors but also the informal mechanisms, reflecting this kind of knowledge transfer as "an adhesive factor" in cluster formation.

The evidence shown in this chapter indicates that the industrial reallocation policy of the Japanese Government since the 1970s has been successful in regulating the over-concentration of industries in specific regions and in developing new industrial zones through the migration of Japanese manufacturing industries. However, the geographical agglomeration developed through new industrial zones may not have strengthened regional innovation through formal institutions. These agglomerations appear to have maintained the competitiveness of Japanese industries by generating regional comparative advantages through traditional Marshallian externalities. The Techno Polis policy which was implemented in 1983, focused on the intra-regional economic development and was designed within the framework of classical industrial location policy, where the importance of regional systems of innovation was not stressed and clusters were therefore not explicitly formed through policy in the regions.

The Japanese Government's regional industrial policy has a long history dating back to the 1860s, and has mainly focused on technology transfer from the developed countries and modernisation of regional SMEs. Some of the regional technology institutes, the so-called *Kosetsushi*, have histories going back more than a hundred years and have functioned as institutions for technology transfer. Typically, institutions for technology transfer have been established in the regions, although those supplying broader knowledge transfer have not been implemented at the regional level. The results of our research on clusters seem to reflect the fact that SMEs have acknowledged the necessity of innovating on a continuous basis in order to survive in the global market. However, much remains to be done to strengthen business links and knowledge transfer in clusters which increasingly depend on new and tacit knowledge.

As a concluding remark, we suggest that the key implication for cluster-based policy is to strengthen value-chain mechanisms for knowledge transfer in clusters and to establish workable regional innovation systems for the creation and distribution of new knowledge in relevant geographical areas.

NOTES

1. "The Responsibility of Local Government", Article 4: Necessary Consideration to be Given by the National and Local Governments in Formulating Policies; Article 5: The Science and Technology Basic Law, Law No.130, effective 15 November 1995.

2. Two other indicators were obtained using the similar definitions: the Indexes for Conversion of Regional Industrial Structure and the Coefficient of Regional Industrial Concentration. If we look at the change of industrial structures in specific regions and their competitiveness, these indicators will be able to provide relevant information. The results are available in Gonda *et al.* (1999).

3. The IIL values were obtained using the Japanese five-digit standard address code with the number of firms in the four industrial sectors.

4. This confirms in Japan what is rapidly becoming common wisdom in the United States (Cooke *et al.*, 2000; Bergman and Feser, 2001).

REFERENCES

Bergman, E.W. and E. Feser (2001), "Innovation System Effects on Technological Diffusion in a Regional Value Chain", *European Planning Studies*, Vol. 9, No. 5.

Boekholt, Patries and Ben Thuriaux (1999), "Public Policies to Facilitate Cluster: Background, Rationale and Policy Practices in International Perspective", in OECD, *Boosting Innovation: The Cluster Approach*, pp. 381-412.

Collins, S. K. Gonda, F. Kakizaki and K. Kato (1998), "Analyzing the Spatial Allocation of Industrial Resources", International Workshop on Cross Regional Science and Technology Collaboration in Asia, NISTEP, Kobe, Japan, 17-18 March.

Cooke, P., P. Bocholt and F. Toedtling (eds) (2000), "The Governance of Innovation in Europe", Pinter, London.

Council on Competitiveness (1999), *The New Challenge of America's Prosperity: Findings from Innovation Index*, Washington, DC.

Fujita, Masahira, Paul Krugman, and Anthony J. Venables (1999), "The Spatial Economy: Cities, Regions, and International Trades", MIT Press.

Gonda, K. (1998), "Information Society and Emerging Space in Terms of Regional Innovation", 4th RESTPOR '98 Conference, Chapel Hill, North Carolina, United States, 21-24 November.

Gonda, K. (2001, forthcoming), "Dynamics of Clustering and Emerging Knowledge in Terms of Regional Innovation", 5th RESTPOR Conference, Proceedings, September 2000.

Gonda, Kinji and Hiroshi Shimizu (1999), "Clustering and the Mechanism of Innovation", 14th Annual Conference, The Japan Society for Science Policy and Research Management, Proceedings, pp. 255-260.

Gonda, Kinji, *et al.* (forthcoming), "Clusters of SME in Japan", to be published as a NISTEP Report.

Gonda, Kinji, Masato Kyuui, and Fumihiko Kakizaki (1999), "Study on Spatial Mobility of Manufacturing Industries and Structural Change of Regional Industries in Japan", *NISTEP Report* No.60, March.

Krugman, Paul (1991), *Geography and Trade*, MIT Press, Cambridge.

Marshall, Alfred (1920), "Principle of Economics", Macmillan, London.

Norton, R.D. and J. Rees (1979), "The Product Cycle and the Spatial Decentralization of American Manufacturing", *Regional Studies* 141, p. 13.

Porter, Michael E. (1998), "Clusters and The New Economics of Competition", *Harvard Business Review*, November-December.

Saxenian, AnnaLee (2000), "Regional Networks and Innovation in Silicon Valley and Route 128", in Zoltan J. Acs (ed.), *Regional Innovation, Knowledge and Global Change*, Pinter, London and New York, pp. 123-138.

Chapter 15

DYNAMICS OF INITIAL CLUSTER FORMATION: THE CASE OF MULTIMEDIA AND CULTURAL CONTENT

by

Michael Peneder
Austrian Institute of Economic Research (WIFO)

> *The new media of the twentieth century have significantly expanded the possibilities of worldwide communications. These new forms of media have intensified the discrepancy between what can be communicated and what is communicated. They have intensified the problem of selection, to which society has reacted on the one hand with organisation and on the other hand with individualisation of selective operations.*
> (Niklas Luhmann, 1997, p. 311)

Introduction

The general purpose of this chapter is to synthesise the available theoretical concepts into a policy-relevant perspective on the potential dynamics of initial cluster formation. The case of multimedia and cultural content serves as an illustrative example to put these ideas into a more concrete form. Its theoretical background, as well as the general implications of an evolutionary approach to *competitiveness policy*, is further developed in Peneder (2001). Empirical evidence on the development of multimedia industries in Austria is reported in Warta *et al.* (1998). The major results of these and other cluster case studies are surveyed in Peneder (1999).

This chapter is structured as follows. The next section provides an eclectic recourse to the theory of evolutionary change, presenting the major connecting principles applied thereafter. A stylised industry life-cycle model is then used to illustrate the potential long-term developments of cultural industries. This is followed by a compilation of institutional components decisive in the formation of a new cluster at the crossroads of multimedia technologies and cultural content. The final section presents the summary and conclusions.

Principles of evolutionary change

The starting point of our considerations is the question of whether and under which prerequisites a completely *new branch of production* can arise from the increasing integration of multimedia technologies and cultural industries. The initial task is to consider the evolutionary fundamentals necessary for the creation of a new specialisation, in the sense of a new type of economic activity. The evolutionary ideas presented are based, among others, upon previous work by Nelson andWinter (1982) and Metcalfe (1998). Luhman (1997) offers an additional social systems perspective.

Human communication generally creates complex fields of interaction. Luhman (1997) describes social systems as sets of reflexive and operationally closed communications which establish their own modes of selection within the system. These only partly reflect the external forces of selection carried into communications via structural feedback between the social system and its outside environment, so as to guarantee at least its physical and economic viability. However, social systems are also characterised by autonomous intentions and purposes which require the establishment of corresponding selective operations created from within the system. Even in this general and abstract model of social interaction – which is arguably a more apt basis for understanding cultural activity than, for example, that provided by purely economic models – the well-known evolutionary trinity of variation, cumulation and selection is the prerequisite for complex dynamics and structural change such as the emergence of a new industry. Consequently, the following exposition is organised along these fundamental distinctions, which we have nevertheless chosen to restate briefly in terms of a deliberately economic interpretation:

- The *principle of competitive selection* is the most familiar to economists and basically states that relative scarcities direct the flow of productive resources. This mechanism works by differential growth (as well as entry and exit as limiting cases), and depends positively on the ability either to use the productive resources efficiently or being more effective in pushing the overall demand constraints. In other words, those firms which are better at adapting to existing (social, economic, or technological) constraints, *compete* more successfully, grow faster, and are able to expand their fields of activity.

- Social systems exhibit a considerable degree of inertia. The *principle of cumulation* holds that economic change can be evolutionary only if the system performs some storage of information, which causes current action to depend on information generated from past experience and action. In other words, some kind of memory is inherent in the system. At the most fundamental level, this storage of information works through the cumulation of information and knowledge which in the past have been advantageous in the process of competitive selection.

- Finally, the *principle of variation* stresses that evolutionary change depends on the existence of diversity in economic actions. Otherwise the process of selection has nothing left to work on. To be sustainable, the principle of variation needs both initial variety and continuing creation of novelty (*i.e.* new information). If the latter is missing, the initial diversity of economic actions will be consumed during the process of selection – a process associated with the related notion of *spontaneous order*.

Industry life cycles and cluster formation

Theoretical models and empirical observations of the dynamics of industrial development in individual branches describe typical elements in the sequence of market growth, market structures and corporate strategy (see, for example, Afuah and Utterback, 1997; Klepper, 1996, 1997; Klepper and Simons, 1997). Real phenomena, however, rarely occur sequentially, but are rather characterised by various overlaps and mutual dependencies. As all the three functional elements continue to operate in each of the developmental stages, the following sequential presentation is therefore highly simplified.

At this general level of observation, however, we can demonstrate how the typical dynamic pattern of an industry life cycle is generated by shifts in the relative importance of the three functional elements of variation, cumulation and competitive selection occurring over the course of time. Consequently, we expect that the potential emergence of a new industry of multimedia and cultural content moves through different broadly defined stages, where outcomes of the prior phases restrict the possible range of operations in the next phase. Figure 1 provides a stylised representation of the kind of life cycle envisaged in the following considerations. It is partly inspired by a similar illustration in Tether and Storey (1997).

Figure 1. **Stylised "life-cycle" representation of industrial evolution**

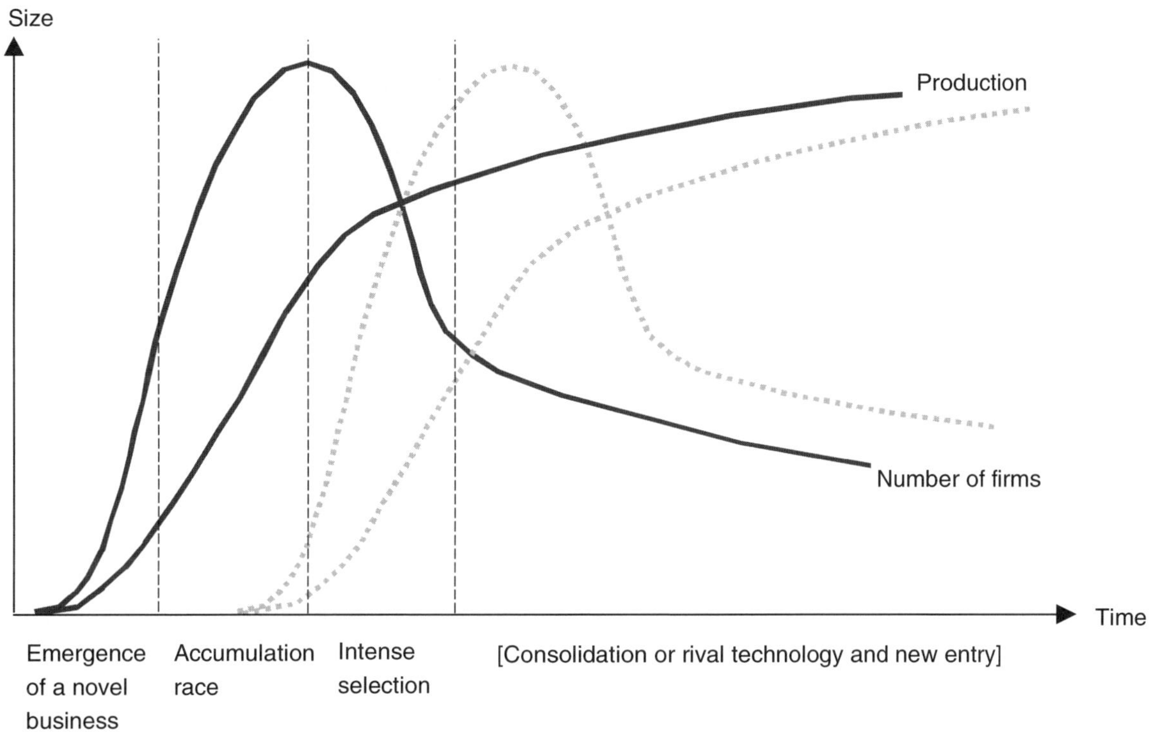

Emergence of a *novel* business

When new branches of industry are in their earliest phases of evolution, particular technological breakthroughs or changes in consumer demand create room for new products and sales opportunities. The substitutional relationships between individual firms are, in the beginning, usually slight, and markets are highly differentiated (Afuah and Utterback, 1997). Since there is still enough room for the expansion of their own business activities, smaller players have the chance to establish themselves in the market during the first phase of development. The greatest competition is most often created by traditional products and services. In the context of multimedia, this realisation is best illustrated by the role of traditional media, such as the printing of books and newspapers, as well as radio and television. Firms whose business revolves around traditional media defend their shares of time and money in consumer budgets with traditional yet more firmly established technologies and marketing strategies.

Innovation and creativity are to a great extent bound to individuals. However, new ideas depend on more than individual achievement; they also require a supportive environment. Organisational structures which facilitate learning help a firm to become more receptive to new knowledge (Nonaka and Takeuchi, 1997). Beyond the level of the firm, supporting factors also include political and social structures which provide ideas for new solutions with sufficient space for diversity. An example is a climate open to new and unusual ideas, which in this case is a special responsibility of social as well as educational policy. In the realm of technological policy, the creation and maintenance of incentives for investments in innovation and high-risk activities with corresponding returns is of greatest priority. The actual realisation of these tasks might depend on promotive activities, as well as, for example, on the functioning of copyright law. In many cases, success in this first phase depends on the early elimination of distortive regulatory restrictions, which are very often in the interest of the supplier of established services and products but disadvantageous to new technologies and young firms.

The diverse cluster of various multimedia-based activities is still at an early stage of development and is predominantly characterised by uncertainty. In such a situation, the stylised model of Afuah and Utterback (1997) differentiates the strategic orientation of the suppliers of new products in dependence of their relative market position over the course of time. With respect to market structure, they argue that competition between existing suppliers during this developmental phase tends to be slight, due to the extremely differentiated products. Traditional products and services therefore continue to create the greatest competition. In the context of multimedia, this realisation refers above all to the position of conventional media, such as books and newspapers, as well as radio and television, which, with older, but more established technologies and marketing forms, attempt to maintain their market shares. On the other hand, demand often consists of highly discriminating users with specific ideas and wielding substantial influence.

In this initial situation, the strategic reaction of the supplier depends on how he/she assesses his/her relative position in the market during the transitional phase in which the most dominant designs are gradually established. Following the terminology of Afuah and Utterback (1997), so-called "strategic leaders" invest specifically in the development and introduction on the market of those applications which have the potential to generate and set new standards. Latecomers, following closely behind the strategic leaders ("fast second"), still have a chance to expand into complementary fields, for example by providing their products in a co-operative endeavour. They will stay alert for an opportunity to take over the leading position themselves. Strategic latecomers ("followers"), who see no possibility for closing in on the strategic leaders in the foreseeable future, must, in contrast, attempt to establish themselves in specialised niches.

The "accumulation race"

Further following this stylised exposition of a sequential industry life cycle, the greatest challenge during the second phase of development is to achieve critical levels of growth and to acquire the competitive skills necessary for lasting success as quickly and efficiently as possible. Klepper and Simons (1997) illustrate that the likelihood of being driven out of the market decreases, the older and larger a firm is. They thereby underline the extent to which accumulation and growth determine the long-term ability to survive. From a political point of view, this growth phase provides an opportunity to positively support accumulative processes through the timely creation of complementary institutional and political structures (for example, through specific educational and training programmes, innovative forms of regulation, etc.).

The *formation of clusters* can be seen as a special form of the spatial accumulation of competitive advantages which extend beyond the individual firm. The general idea is attributed to Alfred Marshall

(1920), who, with the concept of "localised industries", postulated a positive connection between the density of economic activity (agglomeration) and its efficiency. Thus, location-specific competitive advantages may arise due to dense informational structures, a common pool of qualified employees, and the emergence of specialised industries which provide complementary inputs and services. These advantages are therefore the result of reciprocal feedback which generates additional impulses for innovation and growth through specialised inputs and complementary functions along the vertical chain of successive production processes, as well as between the horizontal relationships characterised by intensive competition and/or co-operation. The dynamics are particularly important because of the path dependence and irreversibilities involved in phenomena of industrial location (see Fujita *et al.*, 1999).

However, it is precisely in the field of digital media that the contradiction is often pointed out between local clustering, on the one hand, and the steady decline in spatial restrictions to communications achieved through electronic networks, on the other. If we are to clarify this contradiction, we must differentiate between at least two typical, ideal forms of knowledge: *i) explicit knowledge* which can be passed on in formal, systematically codified language; and *ii) implicit knowledge* which is bound to personal experience, specific to a certain context and therefore difficult to communicate beyond the immediate realm of shared experience (Polanyi, 1958; Nonaka and Takeuchi, 1997). With respect to the formation of local industrial clusters, spatial proximity supports the searching for, imparting and dispersing of implicit knowledge. In contrast, progress in communications technologies has simplified the utilisation of global networks in the expansion of accessible explicit knowledge. In other words, the spatial creation of clusters, as well as of global networking, has supported the development of various forms of knowledge. The primary function of both processes is, however, the accumulation of information which will be advantageous in the process of competitive selection. Without doubt, in many industries the significance of both forms of knowledge is increasing. And most probably, it is growing in a mutually dependent way in the sense that more communication of one kind of information simultaneously requires more operations within the other type of knowledge. As a consequence, there is no contradiction if we assume that local interaction and global networking are becoming more important at the same time.

Phase of intense selection

During the third phase of this stylised industry life-cycle model, certain technological standards typically attain general acceptance and consumers become visibly well informed about available products. The market becomes more transparent and price competition increases in intensity. Suppliers who, during the previous phase of accumulation, were not able to sufficiently expand their competitive advantages, or who reckoned with technological or strategic developments which ultimately did not gain general acceptance, are forced to exit the market. At the same time, profit expectations, and with them the number of new firms entering the market at this mature stage of competition, decrease. The result is often the *consolidation* of the market and an increasing concentration of suppliers (Klepper and Simons, 1997). The competitive challenge might only rise again if technological changes allow new firms to enter the market and establish a novel way of doing that business. However, in the case that the innovations and changes which keep the market moving are not steadfast, the interplay between new information, accumulation and competition increasingly loses momentum. Short-term changes in the balance of power (for example, due to individual errors in management, minor technological changes, etc.), as well as attempts at industrial or technological policy intervention, are unlikely to have a lasting effect on the market under these conditions.

The ultimate question of relevance concerns a proper understanding of the precise nature of the particular kind of services which can be commercialised on an electronic marketplace:

- Can we expect, in analogy to the book market, that future earnings will be raised directly via the contents themselves? This picture presently corresponds most closely to the CD-ROM market.

- Or, will the case be similar to the newspaper and magazine market, as well as to radio and television, where advertising is the primary source of income and content provides only the indirect means to attract consumer attention?

This division into two separate markets for content and advertising is typical of most media and increasingly characterises the development of the multimedia offerings available on line.

The most convincing explanations behind the growing significance of the advertising market are the increasing scarcity of *consumer attention*, which to some considerable extent results from the information overflow created by the dramatic expansion of the same media. As we expect consumer attention to become ever more hard to get, we can safely bet that, in the long run, the potential returns on it will similarly increase. Accordingly, "range" and "ratings" have become decisive factors in the success or failure of competitors in the online multimedia market. Consumer attention, however, is an extremely variable asset which bears a great deal of risk, especially if large sunk investments are involved. In order to be less vulnerable to the fast-moving preferences of customers for content, additional business models, increasing the personal ties to the specific (electronic) location, are regularly required. Under these conditions, those suppliers who succeed early in initiating cumulative processes and in establishing *central "marketplaces"* within the worlds of electronic information, will have the best opportunities to survive and grow.

Principal cluster components of multimedia and cultural content

The creation of clusters is closely tied to the concept of *dynamic complementarity*. Dynamic complementarity refers to the positive impulses for innovation and growth which arise between the various functional components of a cluster, contributing to the accumulation of competitive advantages. What appears to be the most important difference compared to other industries is that cultural services are more affected by non-market institutions with direct accountability to the priorities set by cultural policies.

Figure 2 summarises several important components of a potential cluster at the intersection of multimedia technology and cultural industries. It does not intend to offer a blueprint for "drawing up" a new cluster. The only purpose is to give an illustrative example of the manifold institutions involved. If economic policy wants to design an enabling business environment, institutional factors like these should be taken into account.

Within the production chain, access to the hardware required by information and communications technologies is the first prerequisite for multimedia production. The *hardware for electronic data processing* can be traded easily around the world. Production and manufacturing at locations in close proximity to one another is therefore not an unconditional prerequisite for the emergence of a multimedia cluster. Nevertheless, the example of the United States illustrates that in countries with a strong industrial basis in the information technologies, extended interest groups, including hardware producers and the manufacturers from the field of information technologies, can act very effectively in the realisation of common political interests. Examples are the modernisation of communications infrastructure and improvements in the availability of complementary public services (research and training, as well as public programmes to improve access to multimedia technologies).

More significant to the establishment of an independent production basis is access to an efficient and attractively priced *telecommunications infrastructure*, because it touches on one of the vital nerves of multimedia online applications. It is exactly with respect to this point that the reforms enacted by the European Union in favour of the harmonisation and liberalisation of telecommunications services are stimulating the commercialisation of cultural materials which can be transmitted via multimedia.

Figure 2. **Principal cluster components of multimedia and cultural content**

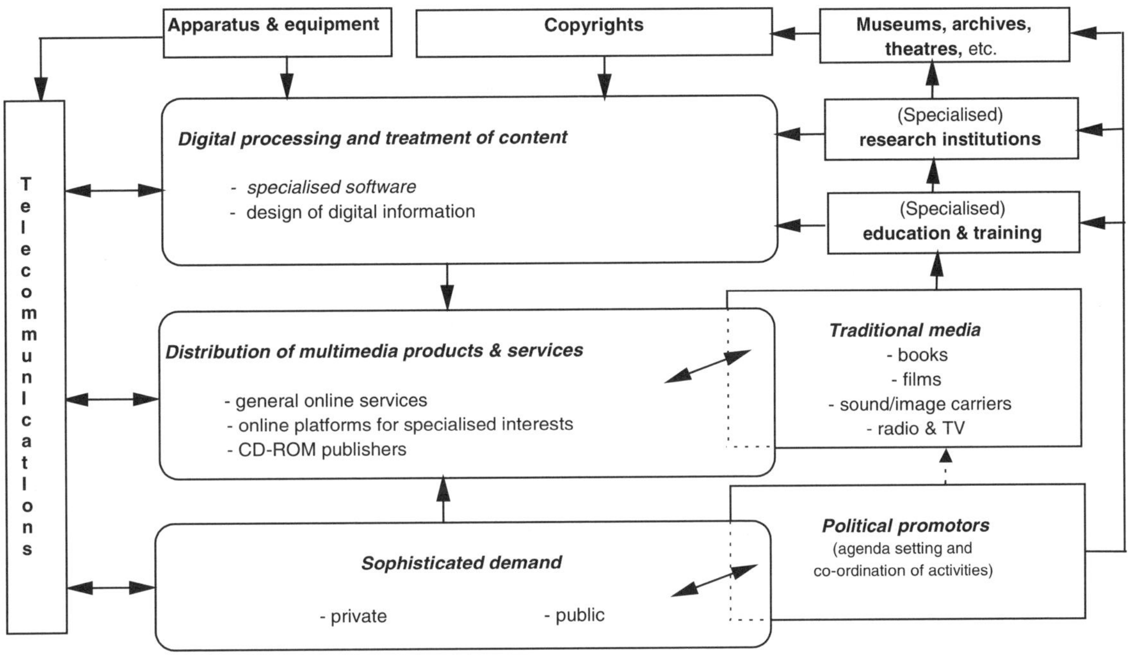

The raw materials most crucial to multimedia production are, nevertheless, the contents themselves. The costs of production and development can only be covered when the product arouses enough attention to translate interest into earnings. Potential sources are, for example, libraries, museums, theatres, musical societies, galleries, as well as archives for audio, visual or literary documents. They are all conveyors of cultural content; their organisation, degree of autonomy and readiness to explore the possibilities of multimedia are of great significance to the content at hand. Legal questions regarding copyright and other property rights are particularly important.

In addition to the equipment and hardware required by information technologies, as well as the telecommunications infrastructure and cultural contents themselves, a further core element of the cluster is the *multimedia processing of cultural contents. Software developers* specialised in multimedia applications can be included in this category, as can *communications designers*, who, outfitted with the appropriate software, undertake the multimedia processing of the content. Since these are highly specialised applications, the spatial proximity of qualified suppliers is a decisive competitive advantage. Supportive policy measures can broaden the basis of available human capital through training and educational programmes, as well as specialised research and developmental facilities.

Following their production, the commercialisation of multimedia contents also depends on efficient and effective *marketing channels* for successful market penetration. Possibilities include specialised CD-ROM publishers, as well as suppliers of online services. An advantageous form of complementary marketing structures could, for example, be the presence of a specialised Internet platform with an international orientation and corresponding marketing strategies, directed towards specific target groups, in the sense of, for example, "virtual city of ..." or a "virtual museum of ..." (some specialised art). Specialised publications or programmes from the field of electronic media could serve as role models, each one opening the advertising market to very specific groups of clients. The appropriate choice of marketing channels is therefore not only a quantitative but even more so a qualitative issue. At least since Porter (1990), it is widely recognised that a firm can gain substantial competitive advantage and lead-time by making use of systematic feedback from informed customers. Ideally, the firm needs to establish regular communications with highly discriminating customers, who are critical, open to new ideas and eager to consume.

In addition to the components mentioned above, the long-run opportunities for local multimedia production depend very much on the strategic reaction and specific timing chosen by the stakeholders of *traditional media*, such as books, audio and visual data carriers, as well as radio and television. The resources and knowledge they convey can become established within the new cluster to an extent which reflects whether and how fast they intend to explore the opportunities of new multimedia applications themselves or to persist as competing systems of communication. High degrees of concentration or even national monopolies of public broadcasting, for example, can delay adaptive responses by masking the competitive pressure from new media.

This is only one example to show why the influence of politics on many of the functional components of the cluster can be considerable. Other examples include research, training and educational facilities, as well as subsidies to book and newspaper publishing and, last but not least, telecommunications regulations. Designing an optimal environment for multimedia clusters therefore requires a great deal of political support, initiative and co-operation. Similar to the way in which an entrepreneur is the dynamic and assertive force behind new ideas, the multimedia cluster might benefit from *political promoters* who are willing and able to identify themselves with this new field and are prepared to push ahead for supportive business conditions.

Summary and conclusions

Evolutionary dynamics illustrate that when several favourable competitive factors converge simultaneously at the right point in time, the groundwork for the emergence of a particular industry at a particular location can be accomplished. If the potential of this situation is recognised and put to good use, the process can mature and self-reinforcing mechanisms develop. As a result, certain locations assume leading roles with respect to different forms of production.

The imaginative phrase of *electronic space* is particularly suggestive in underlining the evolutionary significance of cumulative processes: during the exchange of information, centres of attraction can form within the virtual electronic space while at the same time less frequented peripheries also develop. If we transpose this picture to the commercialisation of information and multimedia content, it can be seen that the unmanageable diversity of the content available can only escape those applications which succeed early enough in becoming centres of attraction. Only if critical thresholds can be surpassed will demand suffice for the emergence of an independent and commercially viable medium. Revenues from advertising or other information-based services can then be invested in long-term improvements and expansions in the information supply, supporting a self-driven cycle of growth and market presence. If the demand for a product does not grow sufficiently rapidly and the product does not succeed in establishing itself as a "central marketplace" for specific interest groups, the growing abundance of information will push it farther and farther to the periphery, where it is highly unlikely that it will be able to maintain a sound economic basis on its own.

What are the functions and responsibilities of economic policy in the light of these perspectives? From an evolutionary standpoint, discretionary intervention and individual policy measures can *potentially* be very effective. However, they can just as easily fail in the event that they are not supported by the prevailing political and economic environment. It should therefore be emphasised that the dynamics of initial cluster formation presented in this chapter imply a particular perspective on innovation policy, which differs in several important respects from the more common policies directed towards strengthening established cluster relationships. In the latter case, typical actions correspond mainly to a process of fine-tuning the existing socio- or techno-economic environment. In this process of mutual dialogue, vested interests are well prepared to pursue their specific wants and needs simply by virtue of their huge informational advantages. In the case of initial cluster formation, however, the particular target group of new entrepreneurs is hardly capable to raise its intents to enter the political agenda. Political promoters therefore have to be strong, independently minded and keen to collect information on their own.

Finally, we must acknowledge that fundamental uncertainty becomes even more important when policy is directed towards the formation of new clusters. Mechanistic approaches, where direct policy interventions are expected to be effective, turn out to be inappropriate. Instead, the systemic nature of innovation policy might best be summarised by the organic *metaphor of gardening*. Gaudin (2001, p. 56) applies this metaphor to explain three essential components of innovation policy: *i)* "preparing the ground", which refers to technical culture and education; *ii)* "weeding", by which he means removing obstacles (*e.g.* entry barriers) and excess domination by vested interests; and finally *iii)* "providing water and nutrients", *i.e.* the point where financial support and public procurement enter the process. In short, cluster-oriented policies bear the important responsibility of cultivating the business environment in such a way that the self-organised dynamics of initial cluster formation are most likely to occur.

REFERENCES

Afuah, A.N. and J.M. Utterback (1997), "Responding to Structural Industry Changes: A Technological Evolution Perspective", *Industrial and Corporate Change*, Vol. 6, No. 1, pp. 183-202.

Fujita, M., P. Krugman and A. Venables (1999), *The Spatial Economy*, MIT Press, Cambridge, Mass.

Gaudin, T. (2001), "Innovation Policy as a Substitute for Failing Economic Policies", in G. Sweeney (ed.), *Innovation, Economic Progress and the Quality of Life*, Edward Elgar, Cheltenham, pp. 54-61.

Klepper, S. (1996), "Entry, Exit, Growth, and Innovation over the Product Life Cycle", *American Economic Review*, Vol. 86, No. 3, pp. 562-583.

Klepper, S. (1997), "Industry Life Cycles", *Industrial and Corporate Change*, Vol. 6, No. 1, pp. 145-181.

Klepper, S. and K.L. Simons (1997), "Technological Extinctions of Industrial Firms: An Inquiry into Their Nature and Causes", *Industrial and Corporate Change*, Vol. 6, No. 2, pp. 379-460.

Knoll, N., M. Peneder and K. Warta (1998), "Multimedia-Produktion in Österreich. Ergebnisse einer Unternehmensbefragung", *WIFO-Monatsberichte* 1, pp. 55-63.

Luhmann, N. (1997), *Die Gesellschaft der Gesellschaft*, Suhrkamp, Frankfurt/Main.

Marshall, A. (1920), *Principles of Economics*, 8th edition, London.

Metcalfe, S. (1998), "The Economic Foundations of Technology Policy: Equilibrium and Evolutionary Perspective", in P. Stoneman (ed.), pp. 409-512.

Nelson, R.R. and S.G. Winter (1982), *An Evolutionary Theory of Economic Change*, Cambridge, MA.

Nonaka, I. and H. Takeuchi (1997), *Die Organisation des Wissens. Wie japanische Unternehmen eine brachliegende Ressource nutzbar machen,* Frankfurt/Main.

OECD (1999), *Boosting Innovation: The Cluster Approach*, OECD, Paris.

Peneder, M. (1999), "Creating a Coherent Design for Cluster Analysis and Related Policies", in OECD (1999), *Boosting Innovation: The Cluster Approach*, OECD, Paris, pp. 339-359.

Peneder, M. (2001), *Entrepreneurial Competition and Industrial Location*, Edward Elgar, Cheltenham.

Porter, M. (1990), *The Competitive Advantage of Nations*, New York.

Tether, B.S. and D.J. Storey (1997), "New Technology-based Firms and Industrial Renewal. Evidence from the United Kingdom", *Wirtschaftspolitische Blätter*, Vol. 44, No. 5, pp. 415-425.

Warta, K., N. Knoll and M. Peneder (1998), *Multimedia, Kultur und Konvergenz: Perspektiven einer Clusterbildung in Österreich*, Vienna.

Chapter 16

TRADE-FLOW-BASED NETWORKS IN THE FINNISH ECONOMY: HETEROGENEOUS PRACTICES

by

Tuomo Pentikäinen
VTT Group for Technology Studies, Finland

Introduction

Networks and clusters have been focal issues in contemporary technology policy. It is generally agreed that governments have a role to play in facilitating appropriate network or cluster conditions and even in implementing focused cluster programmes (Jacobs and de Man, 1996; OECD, 1999). In practice, however, cluster and network policies are highly complex and can entail serious risks. A major complication arises from the fact that these policies are almost always focused on a set of heterogeneous agents. Understanding the natural networking behaviour and incentives for networking of different types of agents is a necessary starting point for successful public initiatives. The second issue is that co-operation is not always fruitful: policies can create needless or destructive or, at the least, costly co-operation. At worst, they can create or bolster market distortions or failures.[1] The third complication originates in the dynamic nature of clusters and networks. Cluster and network policies are uncertain investments that aim at desired future changes in networking practices. The dynamics is an unavoidable dimension of these investments. In this chapter, we develop tools to analyse these three issues, apply the tools thus developed in an empirical Finnish setting, and draw policy conclusions.

Recently, a substantial effort has been devoted to analysing the network and cluster structures of economies as indicated by trade-flow statistics (OECD, 1999).[2] The underlying motivation for these efforts is that trade flows are believed to contain significant information about the network or cluster structure of an economy (Hauknes, 1999). These structures are thought to be important for improving the competitive advantage of firms, sectors and economies (Porter, 1990, 1998; Roelandt *et al.*, 1999).

There are several reasons why these structures may be important. From economic theory, we can find at least three different mechanisms to explain why networking and clustering may have an impact on the competitive advantage of network members. Following the argumentation of Arrow (1994), a key issue is knowledge. Arrow views collaboration, networks and clusters as factors that affect the creation, distribution and diffusion of knowledge. Mohnen (1996) provides an excellent review demonstrating how these interactions take place, how they can be measured and how their impact can be estimated. A general point in this literature is that networks improve the productivity, and hence competitiveness, of network members.

A second mechanism has been studied in the context of network economics. Economides (1996) stresses the central role of network externalities. On the demand side, these externalities arise in cases where the expected utilities of network members increase as the network expands. This is intuitively clear in the case of simple consumption networks – for instance, local telephone or transport operators – where expansion of the network obviously tends to increase the expected utility of end users. Economides (1996) shows that this is a basic feature of several networks, irrespective of whether networking occurs among companies or among consumers, or both. Network externalities can also occur on the supply side. In addition to knowledge externalities, these externalities can occur since networked companies may gain economies of scale, consumers may be more willing to pay higher prices (network effect), transaction costs may decrease, and networked companies may have the possibility to affect prices. These features, as Economides (1996), and especially Williamson (1987), among other industrial economics authors, have shown, are highly dependent on networked companies' negotiation power, competitive situation, strategic behaviour and regulation.

A third issue related to trade-flow-based network analysis concerns vertical integration and the way firms strive for monopolistic or oligopolistic market power. Without going into detail, suffice it to say that the major indications of these kinds of phenomena are changes in market concentration and supernormal profits (Tirole, 1998; Maddigan, 1981; Wolff, 1991).

The common outcome of all these mechanisms is that networks – as long as they do not obtain excessive market power – are likely to increase the effectiveness and productivity of production processes, they are likely to create positive externalities and they may create improved possibilities to handle asymmetric information and moral hazard. However, network members are likely to occupy asymmetric positions in their networks and the diffusion of network benefits is likely to be uneven. Moreover, potential changes in market power need to be considered.

In this chapter, our aim is to study these aspects in the Finnish economy. Our starting point is that networks and clusters are the result of individual agents' private actions, and individual agents face the impacts. For this reason, the most disaggregated level of analysis is desirable. Thus, we analyse networking and its impacts at the sectoral level since this is the lowest level of aggregation at which nation-wide network information is available.

Other researchers (Hernesniemi *et al.*, 1995; Luukkainen, 2001) have analysed the cluster structure of the Finnish economy. These studies give an interesting new insight into sectoral network analysis as they offer information on the systemic environment in which firms and other agents operate. Of course, there is still a great deal of work to be done to understand the connection between different levels of analysis: clusters, sectors and firms. In this chapter, we offer an empirical exploration of these inter-linkages. We complement the sectoral analysis by studying how belonging to specific industrial clusters, as well as sectors' positions in the clusters, have affected the networking behaviour of sectors.

Analytical approach

Our primary objective is to characterise the essence of the networking of the sectors of the Finnish economy. The most significant networking measures have already been emphasised by Leontief (1986). Leontief pointed out that two aspects of inter-sectoral trade links are of special interest. First, inter-sectoral trade as a share of the sector's total factor usage or production reveals how deeply each sector is linked with other sectors. The second point of interest is the concentration of inter-sectoral trade, which reflects the structure of inter-sectoral links; in particular, it is a first-order indicator of market power (Tirole, 1998). These measures are, of course, highly dependent on the specific nature of the sectors. However, as emphasised for example by Rosenberg (1982) and

DeBresson (1996), especially in dynamic settings they contain information on structural dynamics that can be linked with competitiveness.

For our analysis we calculated proportional trade measures, which we called measures of backward and forward dependence, respectively. The backward dependence measure was calculated for each sector as a proportion of domestic inter-sectoral inputs in total factor usage. Forward dependence was calculated as the proportion of domestic inter-sectoral trade in total production. Concentration measures were calculated as Herfindahl-Hirschman indices of inter-sectoral trade[3] (Tirole, 1998).

Table 1. **Calculation of the major networking indicators**

Indicator	Description
Forward dependence	Domestic inter-sectoral supplies/total production
Backward dependence	Domestic inter-sectoral demand/total factor usage
Forward concentration	Herfindahl-Hirschman concentration index of forward inter-sectoral trade
Backward concentration	Herfindahl-Hirschman concentration index of backward inter-sectoral trade

The primary data were Finnish input/output (I/O) statistics for the years 1992 and 1995 (Statistics Finland, 1999a, 1999b). For 1995, data were available for 68 sectors, while for 1992, data were available for 64 sectors. In order to facilitate dynamic analysis, we had to take changes in sector classification into account. After aggregation, 51 sectors were eligible for dynamic analysis.

The sectors were classified to five industrial clusters [forest, information and communication technologies (ICT), metal, construction and foodstuff clusters], or to non-cluster sectors. Classification was adopted from Luukkainen (2001), based on an analysis of significant trade links. In this approach, each trade link for each sector was considered separately, and was termed "significant" if it presented a higher portion of that sector's trade relative to a pre-specified threshold. Such sectors formed a cluster with significant trade flows between one another. Of course, the links were often asymmetric. Moreover, Luukkainen emphasised that some sectors belonged very strongly to a particular cluster, while others belonged weakly to one or several clusters. He operationalised this fact by calculating cluster weights that reflected how strongly sectors belonged to specific clusters. The weights varied from 0 to 1. In this chapter, we address only such sectors of the corresponding clusters that had a cluster weight of more than 0.5 (Annex Table A1).

Furthermore, we classified all sectors belonging to clusters as being demand-oriented, supply-oriented or mixed. Demand-oriented sectors had only backward-significant links within a cluster. Supply-oriented sectors had only forward-significant links; while mixed sectors had both forward- and backward-significant links. These classifications were based on Luukkainen's (2001) cluster maps.

Group comparisons were based on the non-parametric Kruskal-Wallis test. Correlation analyses were based on the Pearson product moment correlation. All analyses were performed by the SAS 8.01 package.

Results

Finnish sectors were deemed to be dependent on their domestic suppliers since, on average, 63% of all factor usage came as an intermediate input from other domestic sectors. Forward dependence in value-adding chains was lower, although around 40% of all supplies were sold to other domestic sectors. Internal dependence – *i.e.* intra-sector trade – was on average below 10% of total factor demand or supply. Variation was high, reflecting the different characteristics of the sectors. Mean dependence values are reported in Table 2, and accurate values for each sector are documented in Annex Table A1.

Table 2. **Dependence and concentration values for all sectors for five clusters and by the position of the sector**

	All observations	Clusters					Position in cluster		
		Forest	ICT	Metals	Cons-truction	Food	Supply	Demand	Mixed
Forward dependence	0.40	0.49	0.27	0.32	0.57	0.41	0.57	0.29	0.38
Backward dependence	0.63	0.65	0.40	0.58	0.72	0.73	0.63	0.70	0.63
Forward concentration	0.23	0.20	0.16	0.35	0.35	0.25	0.34	0.21	0.24
Backward concentration	0.16	0.20	0.10	0.20	0.15	0.22	0.15	0.31	0.17

Concentration indicators revealed that, on average, concentration of Finnish sectors was not high. Interestingly, backward concentration was lower than forward concentration, and the difference was statistically significant. However, both concentration index values were only around 0.2, which does not suggest the prevalence of excessive market power. Naturally, sectors differed radically in their concentration. Some sectors were concentrated to only a very small extent (*i.e.* their inter-sectoral trade was almost symmetrically spread among all sectors), while others were very tightly connected as buyers or suppliers of one or only a few sectors (Table 2 and Annex Table A1).

According to the non-parametric Kruskal-Wallis test, there were significant differences between the dependence measures of different clusters. The major finding was that all dependence measures as well as the forward concentration value were fairly low for the ICT cluster, while in the case of the construction cluster, the opposite result held, *i.e.* sectors belonging to the construction cluster were highly dependent both backwards and forwards, and concentration was high. As far as dependence and concentration were concerned, the forest, foodstuff and metal clusters resembled the construction cluster, although dependence and concentration were lower (Table 2).

A second way to analyse the interaction of cluster structure and sector-specific networking was to separate out from each cluster those sectors that were supply-oriented, those that were demand-oriented, and those that were both supply- and demand-oriented. The natural expectation was that sectors that are predominantly supply-oriented would have higher forward dependence compared to demand-oriented ones; and indeed, this was the observed phenomenon. However, the differences between supply- and demand-oriented sectors were so small that they failed to have statistical significance (Table 2).

Correlation analysis revealed several interesting patterns. First, there was significant correlation between concentration and dependence measures. Forward dependence and concentration, and especially backward dependence and concentration, were positively and statistically significantly correlated. This indicates that those sectors that tended to be highly dependent on their domestic trade partners also tended to have highly concentrated trade with them (Table 3).

Another interesting result was that there were positive correlations between backward and forward dependence, as well as with backward and forward concentration. This would suggest that sectors that buy a large share of their inputs from other domestic sectors also tend to be more active suppliers to other domestic sectors. Similarly, sectors that had a concentrated backward trade strategy also tended to have concentrated forward trade. However, correlations were not high (around 0.3) and significance values were rather low ($p<0.05$), so that strong conclusions should be avoided in this issue (Table 3).

In dynamic considerations, dependence and concentration were rather stable over time. Between the two time periods (1992 and 1995), both dependence and concentration were highly and very significantly correlated (correlation of 0.7-0.9). However, proportional change in dependence was negatively correlated with initial value of dependence. This indicates that there was catching-up – sectors that were initially highly dependent tended to decrease their dependence values, while sectors with low initial dependence tended to increase their dependence. In terms of concentration values, no similar patterns were detected (Table 3).

We tested the correlation of dependence and concentration measures with production and value added in order to see whether the size of the sector was connected with networking. Indeed, large sectors tended to have lower forward dependence and concentration, *i.e.* they tended to be less dependent on domestic sales and their sales tended to be less concentrated. Correlation was rather low (around 0.3), but still significant at the 5% significance level. In case of backward dependence and backward concentration, no similar effects were found.

Table 3. **Correlation analysis of dependence and concentration of the Finnish economy, 1992 and 1995**

	forward dependence 1992	forward dependence 1995	% change in fwd dependence	backward dependence 1992	backward dependence 1995	% change in bwd dependence	forward concentration 1992	forward concentration 1995	% change in fwd concent.	backward concentration 1992	backward concentration 1995	% change in bwd concent.
forward dependence 1992	1.0	**0.83** ***	**-0.54** ***	**0.31** *	0.26	-0.14	**0.30** *	**0.41** **	0.05	-0.10	-0.02	0.09
forward dependence 1995	**0.83** ***	1.0	-0.12	0.26	**0.29** *	0.05	0.25	**0.38** **	0.18	-0.07	0.00	0.10
% change in fwd dependence	**-0.54** ***	-0.12	1.0	-0.06	0.12	**0.31** *	0.00	0.06	0.24	0.21	0.17	0.05
backward dependence 1992	**0.31** *	0.26	-0.06	1.0	**0.88** ***	**-0.44** **	**0.29** *	**0.35** *	0.01	**0.47** ***	**0.49** ***	0.02
backward dependence 1995	0.26	**0.29** *	0.12	**0.88** ***	1.0	-0.01	**0.35** *	**0.40** **	0.10	**0.45** **	**0.55** ***	0.20
% change in bwd dependence	-0.14	0.05	**0.31** *	**-0.44** **	-0.01	1.0	-0.02	0.01	0.20	-0.12	-0.01	**0.29** *
forward concentration 1992	**0.30** *	0.25	0.00	**0.29** *	**0.35** *	-0.02	1.0	**0.74** ***	-0.24	0.08	0.13	0.11
forward concentration 1995	**0.41** **	**0.38** **	0.06	**0.35** *	**0.40** **	0.01	**0.74** ***	1.0	**0.37** **	0.08	0.14	0.16
% change in fwd concentration	0.05	0.18	0.24	0.01	0.10	0.20	-0.24	**0.37** **	1.0	-0.01	-0.06	-0.10
backward concentration 1992	-0.10	-0.07	0.21	**0.47** ***	**0.45** **	-0.12	0.08	0.08	-0.01	1.0	**0.89** ***	-0.19
backward concentration 1995	-0.02	0.00	0.17	**0.49** ***	**0.55** ***	-0.01	0.13	0.14	-0.06	**0.89** ***	1.0	0.23
% change in bwd concentration	0.09	0.10	0.05	0.02	0.20	**0.29** *	0.11	0.16	-0.10	-0.19	0.23	1.0

* Indicates significance value of $p<0.05$; ** indicates $p<0.01$; *** indicates $p<0.001$.

Conclusions

In this chapter, we studied networking in the Finnish economy as indicated by domestic trade flows. We performed the analysis at the sectoral level, because that is the lowest level of aggregation at which networking information is available. Moreover, we completed the sectoral data with new information on the cluster structure of the Finnish economy (Luukkainen, 2001).

Several interesting patterns were detected in relation to the networking of the Finnish economy. There was a significant correlation between sectoral dependence and concentration of domestic trade. This reflected the fact that those sectors that tended to be highly dependent on their domestic trade partners also tended to have highly concentrated trade with them. The same pattern was found for both demand and supply linkages.

From the point of view of cluster policy, this finding is interesting. Successful cluster policy would obviously aim at increasing inter-sectoral collaboration and hence would likely lead to increased dependence. However, our results warn that increased concentration might be a possible by-product of increased dependence. The question is whether initiatives focused on increasing collaboration carry the risk of distorting the market by increasing concentration. We analysed this issue by performing a dynamic correlation analysis, where we tested whether changes in dependence were correlated with changes in concentration. Indeed, there were weak indications that such might be the case in demand networks. However, this issue should be studied and verified using longer time-series and firm-level data.

The second interesting finding concerns inter-temporal changes in networking. As expected, network structures tended to be fairly stable. Between the two periods (1992 and 1995), dependence and concentration were highly and positively correlated. However, we detected interesting patterns of dynamic change in networking. In particular, there was a surprisingly strong and significant stabilisation effect in dependence. Sectors with high initial dependence tended to decrease their dependence, while sectors with low initial values tended to increase their dependence. This effect was detected for both supply and demand linkages. This issue is interesting, but it should be borne in mind that our current understanding of the dynamics of network or cluster formation is rather limited. Our finding would suggest the need for further studies with longer time-series and micro data.

There were clear differences in the networking practices of the industrial clusters. The rapidly growing Finnish ICT cluster was strikingly different from other Finnish clusters. Dependence measures both backward and forward in the supply chains were very much lower than in other clusters. Moreover, in the ICT cluster, concentration of trade was significantly lower than in other clusters. The most obvious hypothesis would be that these issues must be connected with the highly international production and trade patterns of the Finnish ICT cluster. However, internationalisation is not a trivial explanation for the observed phenomena. In principle, it would be possible to scale up the cluster's activities internationally and still retain an arbitrary level of domestic concentration and dependence. Investigating why these measures are low in the ICT cluster strongly points to the need for further studies with sufficiently long time-series and firm-level data.

In clusters with lowest export intensity, *i.e.* construction and food, the results were quite the opposite to those found in the ICT cluster. These two sectors were highly dependent on domestic trade and they had concentrated trade patterns. The traditional Finnish export clusters, forestry and metals, demonstrated interesting networking patterns. One would have expected high backward dependence due to highly domestic supplier network and raw material production, and low forward dependence due to high export intensity. However, compared to other clusters, backward dependence was quite close to the average for all Finnish sectors and was lower, for instance, than that for the food or construction clusters. Furthermore, in the case of the forest cluster, forward dependence was surprisingly high (well above the Finnish average). These findings emphasise the importance of the Finnish forest cluster as a domestic supplier and not merely as an exporter, and they undermine the simplistic explanations of linear raw-material-based production processes of traditional Finnish "low-tech" clusters.

For cluster policy makers, our results highlight the fact that different sectors in the economy have radically different networking practices. These characteristics need to be clearly understood before specific cluster initiatives are implemented. Second, our results emphasise that clusters need to be considered as inter-temporal or dynamic phenomena. Cluster policy initiatives are uncertain investments that aim to achieve desired dynamic changes in networking practices. Successful cluster policy requires a full understanding of network dynamics, which must be complemented by well-planned monitoring and evaluation. Third, changes in collaboration and in market structure are parallel issues in networking. The downsides of cluster policies, especially if the initiatives are highly focused, are that: *i)* they may create collaboration with only minor, if any, additionality (Luukkonen, 2000); or *ii)* they may create market distortions. In a recent evaluation of Finnish cluster programmes (Pentikäinen, 2000), as many as one-third of respondents reported that new collaboration was created only to obtain public cluster-specific funding and that unwanted market distortions were created. Of course, this should not be interpreted as an indication of policy failure. However, it is an indication of the true cost of programmes that needs to be considered *vis-à-vis* the expected benefits of cluster initiatives.

A major issue for policy makers concerns the international dimension of clusters. This chapter was based on domestic statistics. However, our hypothesis is that the general findings that sectors are highly heterogeneous, that dependence and concentration are interconnected and that there are clear and predictable dynamic patterns in network formation, are likely to hold in international considerations and in other countries. Of course, this issue requires empirical validation and testing. However, especially when truly international sectors and clusters are considered, like the ICT or forest clusters in the Finnish case, the issue becomes more complicated. In these cases, neither policy nor analysis can be performed in distinct domestic settings. On the contrary, companies and other agents operate in international – if not global – settings, and both policies and analyses need to be performed at the same dimension.

For cluster researchers, three major issues require thorough analysis. First, the dynamics of network formation and networks' dynamic impacts require more theoretical and empirical work. Second, currently available network data are too aggregated; many of the most interesting issues cannot be thoroughly handled using current sectoral data. Third, international R&D, production and trade networks are becoming increasingly important, especially in small open economies and emerging high-technology fields. For empirically oriented research, this offers a huge challenge due to the problems involved in measuring international networks and trade links.

Annex Table A1. **Sectors and clusters of the Finnish economy, sector-specific dependence and concentration measures and nature of sectors' links to their primary clusters (1995 data)**

Description	NACE classification	cluster	nature of link	forward dependence	backward dependence	forward concentration	backward concentration
Manufacture of other products of wood	203, 204, 205	constr.	supply	0.62	0.76	0.55	0.19
Manufacture of plastic products	252	constr.	supply	0.62	0.50	0.08	0.12
Manufacture of glass and glass prods, non-refractory ceramic goods	261, 262	constr.	supply	0.39	0.51	0.08	0.06
Manufacture of ceramic tiles and bricks, cement, articles of concrete	263-268	constr.	supply	0.74	0.63	0.67	0.08
Collection, purification and distribution of water	41	constr.	supply	0.95	0.78	0.27	0.10
Building of complete constructions or parts thereof, service activ.	4501, 4509	constr.	dem+sup	0.27	0.77	0.23	0.13
Letting and operation of dwellings	7021	constr.	demand	0.00	0.94	.	0.39
Management of real estate on a fee or contract basis	7032	constr.	dem+sup	0.89	0.83	0.72	0.14
Other real estate activities	701, 7022, 7031	constr.	demand	0.64	0.77	0.18	0.11
Agriculture and related service activities	011-014	food	dem+sup	0.69	0.69	0.33	0.13
Hunting and fishing	015, 05	food	dem+sup	0.34	0.60	0.22	0.17
Produc., processing, preserving of meat, fish, prods thereof	151, 152	food	dem+sup	0.21	0.71	0.26	0.47
Processing and preserving of fruit and vegetables	153, 154	food	dem+sup	0.37	0.64	0.11	0.09
Manufacture of dairy products	155	food	dem+sup	0.24	0.95	0.20	0.52
Manufacture of grain mill products, starches and starch products	156	food	dem+sup	0.70	0.89	0.24	0.38
Manufacture of prepared animal feeds	157	food	dem+sup	0.71	0.79	0.86	0.17
Manufacture of other food products	158	food	dem+sup	0.30	0.58	0.19	0.07
Manufacture of beverages and tobacco products	159, 16	food	dem+sup	0.25	0.72	0.20	0.07
Hotels and restaurants	55	food	dem+sup	0.21	0.87	0.05	0.08
Air transport	62	food	supply	0.48	0.65	0.30	0.12
Activ. of travel agencies and tour operators; tourist assist. act.	633	food	demand	0.36	0.65	0.04	0.38
Forestry, logging and related service activities	02	forest	supply	0.88	0.58	0.43	0.27
Mining and quarrying of energy producing materials	10	forest	dem+sup	0.87	0.89	0.50	0.24
Sawmilling and planing of wood, impregnation of wood	201	forest	dem+sup	0.39	0.92	0.23	0.54
Manufacturing of veneer sheets, plywood, laminboard, particle board	202	forest	dem+sup	0.25	0.79	0.24	0.16
Manufacture of pulp, paper and paperboard	211	forest	dem+sup	0.16	0.54	0.10	0.11
Manufacture of articles of paper and paperboard	212	forest	dem+sup	0.53	0.73	0.10	0.34
Publishing and printing	22	forest	dem+sup	0.63	0.83	0.09	0.19
Manufacture of basic chemicals	241	forest	dem+sup	0.37	0.41	0.16	0.08
Manufacture of other chemical products and man made fibres	241-247	forest	dem+sup	0.47	0.45	0.08	0.07
Manufacture of furniture	361	forest	demand	0.25	0.72	0.21	0.09

Description	NACE classification	cluster	nature of link	forward dependence	backward dependence	forward concentration	backward concentration
Electricity, gas, steam and water supply	40	forest	dem+sup	0.59	0.33	0.12	0.10
Land transport; transport via pipelines	60	forest	supply	0.68	0.59	0.07	0.12
Water transport	61	forest	supply	0.09	0.64	0.06	0.19
Recreational, cultural and sporting activities	92	forest	dem+sup	0.32	0.73	0.31	0.24
Manufacture of office machinery and computers	30	ICT	demand	0.15	0.17	0.12	0.13
Manufacture of electrical machinery and apparatus n.e.c.	31	ICT	dem+sup	0.23	0.45	0.13	0.07
Man. of radio, television and communication equipment and apparat	32	ICT	dem+sup	0.15	0.23	0.07	0.11
Manufacture of medical and precision products	33	ICT	dem+sup	0.16	0.42	0.32	0.08
Post and telecommunications	641, 642	ICT	demand	0.68	0.72	0.14	0.10
Mining of metal ores	13	metal	supply	0.64	0.89	0.69	0.19
Manufacture of basic iron and steel and ferro-alloys	271	metal	dem+sup	0.36	0.31	0.18	0.09
Manufacture of basic metals n.e.c.	272-275	metal	dem+sup	0.20	0.34	0.13	0.12
Manufacture of fabricated metal products	28	metal	dem+sup	0.43	0.57	0.16	0.16
Manufacture of machinery and equipment n.e.c.	29	metal	dem+sup	0.14	0.43	0.06	0.08
Manufacture of motor vehicles, trailers and semi-trailers	34	metal	dem+sup	0.19	0.41	0.23	0.07
Building and repairing of ships and boats	351	metal	dem+sup	0.10	0.60	0.42	0.13
Manufacture of other transport equipment n.e.c.	352-355	metal		0.41	0.54	0.42	0.10
Recycling	37	metal	supply	0.65	0.58	0.45	0.14
Civil engineering	4502	metal	dem+sup	0.43	0.81	0.49	0.15
Road and railway maintenance	630	metal	demand	0.01	0.95	0.58	1.00
Other mining and quarrying	14			0.87	0.81	0.22	0.13
Manufacture of textiles	17			0.19	0.41	0.15	0.07
Manufacture of wearing apparel; dressing and dyeing of fur	18			0.21	0.53	0.07	0.15
Manufacture of leather and leather products	19			0.13	0.34	0.08	0.06
Manufacture of refined petroleum products, coke, nuclear fuel	23			0.46	0.18	0.07	0.09
Manufacture of rubber products	251			0.33	0.45	0.11	0.13
Manufacturing n.e.c.	362-366			0.22	0.49	0.12	0.05
Wholesale and retail trade	50, 51, 52			0.32	0.71	0.09	0.07
Other supporting transport and auxiliary activities	631, 632, 634			0.63	0.79	0.10	0.08
Financial intermediation and insurance	65, 66, 67			0.20	0.61	0.11	0.11
Business activities	71-74			0.61	0.73	0.04	0.14
Public administration and defence; compulsory social security	75			0.12	0.53	0.13	0.08
Education	80			0.06	0.70	0.11	0.13
Health and social work	85			0.05	0.51	0.09	0.07
Sewage and refuse disposal, sanitation	90			0.98	0.78	0.14	0.15
Activities of membership organizations n.e.c.	91			0.25	0.76	0.20	0.10
Other service activities	93			0.14	0.79	0.35	0.14

NOTES

1. For empirical findings, see Pentikäinen (2000).

2. In this chapter, the concept of networks should be understood as trade linkages that arise between component sectors of value chains and their propensity to facilitate the information and technology flows known to enhance innovation. Hence, this version of the network concept refers to cluster structures and innovation potentials, and not uniquely to specific patterns of sustained collaboration and co-operation among particular firms or other organisations.

3. The Herfindahl-Hirschman index is based on concentration of trade. For example, if a sector has domestic trade only with one sector, then the index value is one. The wider the distribution of trade, the lower the index value. The theoretical minimum of the index is zero (Tirole, 1998).

REFERENCES

Arrow, K.J. (1994), "The Production and Distribution of Knowledge", in G. Silverberg and L. Soete (eds.), *The Economics of Growth and Technical Change*, Edward Elgar, Aldershot.

DeBresson, C. (ed.) (1996), *Economic Interdependence and Innovative Activity. An Input-Output Analysis*, Edward Elgar, Cheltenham.

Economides, N. (1996), "The Economics of Networks", *International Journal of Industrial Organisation* 14, pp. 673-699.

Hernesniemi, H., M. Lammi and P. Ylä-Anttila (1996), *Advantage Finland – The Future of Finnish Industries*, Etla Series B 113.

Jacobs, D. and A.P. de Man (1996), "Clusters, Industrial Policy and Firm Strategy: A Menu Approach", *Technology Analysis & Strategic Management* 8, pp. 425-437.

Leontief, W. (1986), *Input-Output Economics*, 2nd edition, Oxford University Press, New York.

Luukkonen, T. (2000), "Additionality of EU Framework Programmes", *Research Policy* 29, pp. 711-724.

Luukkainen, S. (forthcoming, 2001), *Industrial Clusters in the Finnish Economy*.

Hauknes, J. (1999), "Norwegian Input-Output Clusters and Innovation Patterns", in OECD (1999), *Boosting Innovation: The Cluster Approach*, OECD, Paris.

Maddigan, R. (1981), "The Measurement of Vertical Integration", *The Review of Economics and Statistics* 63, pp. 328-335.

Mohnen, P. (1996), "R&D Externalities and Productivity Growth", *STI Review*, No. 18, pp. 39-66.

OECD (1999), *Boosting Innovation: The Cluster Approach*, OECD, Paris.

Pentikäinen, T. (2000), "Economic Evaluation of the Finnish Cluster Programmes, *VTT Group for Technology Studies Working Papers* 50.

Porter, M.E. (1990), *The Competitive Advantage of Nations*, The Free Press, New York.

Porter, M.E. (1998), "Clusters and the New Economics of Competition", *Harvard Business Review*, November-December.

Roelandt, T., P. Hertog, J. Sinderen and N. Hove (1999), "Cluster Analysis and Cluster Policy in the Netherlands", in OECD (1999), *Boosting Innovation: The Cluster Approach,* OECD, Paris.

Rosenberg, N. (1982), *Inside the Black Box: Technology and Economics*, Cambridge University Press, Cambridge.

Statistics Finland (1999a), *Input-Output 1995*, Tilastokeskus, Helsinki.

Statistics Finland (1999b), *Input-Output* (electronically ordered version), Tilastokeskus, Helsinki.

Tirole, J. (1998), *The Theory of Industrial Organisation* (10th printing), The MIT Press, Cambridge.

Williamson, O.E. (1987), *The Economic Institutions of Capitalism. Firms, Markets, Relational Contracting*, The Free Press, New York.

Wolff, E.N. (1991), "Dynamics of Growth in Input-Output Analysis", in *Technology and Productivity. The Challenge for Economic Policy*, OECD, Paris.

PART IV

POLICIES FOR CLUSTER-BASED INNOVATION

Chapter 17

CLUSTERING AND ECONOMIC COMPLEXITY: REGIONAL ICT CLUSTERS IN THE UNITED KINGDOM

by

David Charles and Paul Benneworth
Centre for Urban and Regional Development Studies, University of Newcastle

Introduction

Elsewhere in this volume, national ICT clusters have been examined, although typically these are relatively specialised within the broad parameters of ICT. In the case of the United Kingdom, a relatively large economy with considerable internal heterogeneity, the ICT production sector is complex. Different elements of ICT production have differing degrees of integration with each other, some linking more strongly into other sectors. Notably, there is a disjuncture between defence electronics and communications and information services. Furthermore, there is a distinct geography with very strong regional specialisations such that rather than a national cluster, the United Kingdom exhibits a range of more regionalised clusters that can be linked together to form a broader ICT industry. The choice of wording is deliberate, and the existence of a single, national ICT cluster is questionable, and hence the terms industry and cluster will be used cautiously in this chapter. Third, the United Kingdom is a very open economy, with a high proportion of foreign direct investment, but this has its own spatial division of labour following the geography of the indigenous industry.

This chapter also sets out to illustrate the reflexive interactions between clusters and policy in the United Kingdom, especially in the way that national policies of various forms have had regionally differentiated effects and how the development of the ICT clusters has affected national policies. Central to this has been the particular UK national system of innovation (NSI), taking the form of a distinctive culture[1] towards innovation and finance which has had a significant influence on the industry as a whole and on the particular regional clusters (Walker, 1993). The chapter begins by examining the policy environment and then explaining a little of the history of ICT in the United Kingdom, before examining examples of geographical clusters *within* the United Kingdom.

The key characteristics of the UK national system of innovation

The ICT industries in the United Kingdom have been profoundly affected by their relation to the state, especially by procurement and regulation decisions emanating from the government as well as public corporations; the British Broadcasting Corporation and the General Post Office were central to the creation of a technologically sophisticated market for electronics and IT (Baker, 1969). The industry has also been affected by the framework conditions through which government seeks to influence economic activity.

UK *company law* has for a long time stressed the importance of industry as a primarily financial concern (Dintenfass, 1992); its purpose is to generate wealth for the benefit of its owners, rather than providing a national strategic asset or being of value because of the employment created (Hirst and Zeitlin, 1988). This has a number of important consequences, the most regularly cited being the negative comment that British firms systematically under-invest in new products, attempting to maximise current profit, and so tend to fail to maximise the potential of their innovations (Costello *et al.*, 1989: Coates, 1994). However, there is a positive side to this particular system and that is a great flexibility in the economic base; the UK electronics industry has been characterised by rapid redeployment of resources in response to perceived profitable opportunities (Brummer and Cave, 1998). The aggregate effect of this is to encourage the ICT industry to be incredibly dynamic but volatile and subject to extensive closures during enduring periods of industrial downturn (Williams and Charles, 1986).

Government *procurement policy* has had a different set of impacts on the industry; until recently, defence contracts were underpinned by the principle that a defence contract had two purposes, to obtain a product, but also to provide a British capacity, and provide resources to build and maintain that product line. This stands in stark contrast to the financial principles stated above, in which industry was not considered to have inherent strategic value (Kennedy, 1987). It comes as no surprise that defence electronics was one of the more durable elements of the sector, and a corollary of this was the close relations with those ministries in and around London making those procurement decisions (Young, 1985). Indeed, the explicit spatial bias of London-based civil servants against the more remote locations of the United Kingdom (those more than 50 miles from London),[2] as well as the concentration of government defence research in the region led to much of the R&D capacity of defence firms being located in the South East (Buswell and Lewis, 1970; Keeble, 1992).

Another element of government policy which historically influenced the location of electronics and ICT industry was *spatial planning policy*, which in the post-war period was based around tight planning controls in the south eastern core, and an attempt to disperse industry to the periphery (Heim, 1988). However, the means employed were the use of industrial development certificates, through which companies in the South East could gain permission to develop in the South East but usually only for office and R&D activities, with manufacturing being pushed out to the development areas in the north or west of the country (McKay and Cox, 1979). This had the effect of reinforcing a spatial division of labour in the ICT industry and, although finally abandoned in the 1980s, this planning regime concentrated surviving elements of the UK-owned electronics industry in the south of England, as the UK-owned companies exited from large-scale production and closed manufacturing capacity in the periphery (Charles, 1992). This, in turn, created very strong positive externalities for the later growth of IT, software and computer consultancy activities.

Government *industrial policy* did little to help the industry overcome the problems of lagging UK industrial markets for ICT products raised by other countervailing policy measures and cultural attitudes. A particular confrontational attitude to labour relations actually hindered demand for new electronic technologies in manufacturing; managers were unwilling to invest in numerical control (NC) systems because of their financial implications, while organised labour saw those machines as a threat to the employment of their members (Newton and Porter, 1988). Hence, there was effectively a pact between managers and unions across manufacturing which militated against the introduction of new technologies. This weakened the domestic market for electronics firms involved in the development of industrial automation and control systems, increasing their volatility (Brummer and Cave, 1998). Although several of the largest UK electronics firms were involved in industrial controls, their interest withered as it became clear in the 1960s and 1970s that the domestic market for those products was very lethargic. When the government did intervene, it was only at the margin of this problem, addressing peripheral issues.

The final area where government influence is less perceptible is in the culture surrounding the *science, education and technology system* (Wilkie, 1991). A frequent criticism of the UK science base is its weakness in commercialising research, which has at its root a gulf between scientific and technical communities. It has been observed that British entrepreneurs at heart aspired to join the aristocracy and disdained the businesses which had generated their wealth (Wiener, 1981); those that persisted tended to be practically minded with no orientation towards embracing and transforming emerging scientific principles.

ICT in the United Kingdom: a historical perspective

The public sector and electronics production

During the 20th century, the experiences of war profoundly affected UK policy towards the electronics industry (Baker, 1969). Concerns to maintain an independent capability to supply the British armed forces, and the legacy of intense government intervention during wartime fostered a strong defence sector and a close relationship between firms and government. A consequence of this was that military R&D remains the greatest category of expenditure in UK R&D (40%), a percentage second only to that of the United States in the G7 countries.

Public sector procurement was also vital in providing profitable markets, with large, sophisticated monopsonistic purchasers of electronic products who could specify and fund development activity and innovation in a select group of companies. Naturally, as these new telecommunication technology companies grew, the search for new markets frequently took them back to the defence industry.

The certainty permitted by government and public corporations' orders in defence and telecommunications allowed growth by internal expansion as well as through acquisition of rivals (Brummer and Cave, 1998). In turn, this increased the financial pressures on those firms; strategic technological positions were less important than supporting the share price and the dividend. The 1970s and 1980s saw many smaller independent electronics firms bought up by larger firms and then rationalised (Young, 1985).

The orientation of the UK electronics industry towards fulfilling government contracts came at the expense of other activities, notably consumer electronics and when in the 1960s and 1970s this sector came under strong overseas competitive pressure, the UK companies divested (Jones and Marriott, 1970). In response to perceptions of rising overseas competition in the 1960s, government industrial policy focused on encouraging those industries and firms under the most pressure to improve their competitiveness through merger and reorganising to yield benefits of scale. The Industrial Reorganisation Corporation (IRC) was instrumental in the formation of a few dominant electronics companies in the United Kingdom, most notably the General Electric Company and International Computers Limited (ICL) (Hague and Wilkinson, 1983; Campbell-Kelly, 1989).

In the late 1970s, it became clear that there were whole sectors of the electronics industry that were being overlooked by the large firms. Although the absence of consumer electronics appeared of little consequence, as the United Kingdom attracted considerable Japanese inward investment, more worrying was a failure to develop a strong semiconductor industry. The UK Government feared that a failure by UK companies to invest in new enabling technologies would lead to a continued decline of the United Kingdom's competitive position. Therefore, the UK Government took the decision to invest in thc crcation of a microelectronics business, Inmos (McLean and Rowland, 1985). However, Inmos failed to generate a lasting competitive advantage for the United Kingdom despite early technical success because of an absence of government commitment to a long-term programme of investment.

Privatisation and deregulation: services rather than hardware

In 1979, the Conservative Government sought to increase the efficiency of the economy through exposure to free markets. Monetarist policies and recession combined to reduce manufacturing employment by one-third. In 1983, the government also embarked on a programme of privatisation. These policies were extremely disruptive to the established UK electronics industry; at the same time as the recession and its aftermath placed them under acute shareholder pressure, a significant share of their markets was opened to competition, and the government signalled its willingness for any uncompetitive businesses to either face closure or acquisition (Redwood, 1984; Kelly, 1987).

The European Commission adopted a stance that competitiveness required European-scale businesses to compete against Japanese and American firms. Deregulation of industrial support in the United Kingdom, in tandem with changes in European competition policy, led ownership of many flagship British companies to change hands. However, it did give UK telecommunications providers a head-start in developing competitive advantage as the single European telecommunications market gradually emerged (Charles, 1991).

However, although government policy abandoned direct support for individual firms and featherbedding through favourable procurement contracts, support for collaborative R&D continued. The largest programme was the Alvey Programme in the mid-1980s, which signalled both the move to collaborative pre-competitive research and a shift from hardware to software. Although a major element of the programme supported semiconductor design, much support also went to artificial intelligence and software applications (Oakley and Owen, 1990; Guy *et al.*, 1991; Georgiou, 1991).

Later, the establishment of the Technology Foresight Programme, with a sector panel for IT, electronics and communications, saw the continuation of a software focus, with an increasing priority given to the support for new markets for multimedia and information services.

The current state of the UK ICT industry

The UK ICT industry is conventionally defined as being the IT hardware, electronics components and systems, telecommunications and IT services sectors. No national ICT cluster definition has been produced, although a number of more narrowly defined clusters, such as defence electronics, have been identified. Consequently, the statistical evidence here uses a standard ICT sectoral definition.

In recent years, there has been a marked decline in the manufacturing of capital goods, although this has been greatest in the assembly of computers, given the relatively small size of the UK process control market (Table 1). The greatest increases have been seen in the growth of computer services, which have doubled since 1991 in employment terms, both in the sale of consultancy services, but to a greater degree in the rise of subcontract computing in which firms manage computer systems on behalf of their clients (Table 2).

Table 1. **Employment in the ICT industry in Great Britain in the 1990s, by sector**

ICT sector[1]	1991	1997	Change
Domestic electricals (TV, white goods)	61 900	63 800	3.1%
Capital goods (including computers)	110 100	96 800	-12.1%
Electrical apparatus and components	203 700	230 700	13.3%
Computer consultancy	118 400	234 300	97.9%
Subcontract and other computer services	51 000	117 400	130.2%
Total	545 000	743 000	36.3%

Note: Northern Ireland is not included in this table as data are collected differently.
1. These tables reproduce statistics provided from the *Annual Employment Survey*, which uses the 1992 Standard Industrial Classification. This classification was designed before the phenomenal rise of the Internet. The statistics do not capture the new media and electronic commerce business sectors which, although employing significant numbers in computer occupations, are currently classified with their legacy industries, and are not included in the totals presented.
Source: Census of Employment, 1991; *Annual Employment Survey*, 1997. © Crown Copyright Reserved ONS Statistics (NOMIS), 2000.

Table 2. **Growth of the computer services industry, 1981-97**

Year	Employment
1981	54 800
1984	78 700
1987	108 400
1991	169 400[1]
1997	351 700

1. There is a break in the series between 1987 and 1991 arising from the redefinition of SIC codes. The value of 169 400 is taken from the new series definitions (1992 SIC); under the 1980 SIC definitions, employment was calculated at 147 500.
Source: Census of Employment, 1981, 1984, 1987, 1991; *Annual Employment Survey*, 1997. © Crown Copyright Reserved ONS Statistics (NOMIS), 2000.

The ICT industry does not appear to be a particularly important sector of the UK economy, comprising only 3.3% of all employment. However, a significant element of that activity is in business areas in which the United Kingdom has world-class competitive advantages and globally competitive firms, such as communications infrastructure management. A critical element of this competitiveness is the expenditure on R&D, particularly made by businesses, but also government and university support for electronics and ICT firms. Businesses spent GBP 3.6 billion on ICT R&D in 1997, which comprises 38% of all business R&D; in comparison, the chemicals and pharmaceuticals industry, in which the UK has a definite R&D-based competitive advantage, has a research and development expenditure of GBP 2.8 billion. Proportionately, ICT R&D expenditure is split in favour of the service sector, arising from the United Kingdom's strengths in computer and telecommunications services (Table 3).

Table 3. **Composition and sectoral distribution of ICT business R&D, 1997**

In GBP millions

	All BERD	Capital	Salaries
Office machinery and computers	102	16	31
Electrical machines	424	25	167
Radio, TV, communications equipment	655	54	277
Precision instruments	336	19	162
Aerospace	893	28	308
Post and telecommunications	496	29	192
Computer services and related activities	703	84	301

Source: Economic Trends, 1999.

A further contributory factor to the competitive advantage of the UK cluster is derived from the concentration of particular specialisations in certain regions of the United Kingdom; in the South East, the industry contributes greatly to the economic success of its host region, while in Wales and Scotland, large branch plants are important in palliating the persistent unemployment in those regions. Table 4 shows the distribution of manufacturing and service ICT sub-sectors between the British regions. There is quite a strong split between manufacturing and services in the United Kingdom as a whole, with services concentrated in London and the wider South East, and manufacturing in Scotland, the North East and Wales.

Table 4. **Employment in the electronics and ICT industry in Great Britain by region, 1997**

	ICT manu-facturing	Manufacturing LQ	ICT Services	Service LQ	All ICTs	ICTs/ all industry	Service Index
South East	60 300	1.09	94 100	1.90	154 300	4.9%	174
Eastern	43 600	1.25	39 200	1.25	82 800	4.1%	100
Wales	33 000	1.97	5 300	0.35	38 200	4.0%	18
West Midlands	45 500	1.24	25 700	0.78	71 300	3.4%	63
South West	36 200	1.13	23 900	0.83	60 200	3.3%	73
Scotland	47 200	1.38	15 700	0.51	62 900	3.2%	37
North East	21 100	1.35	6 600	0.47	27 700	3.1%	35
London	22 100	0.64	79 700	1.48	101 800	2.9%	401
East Midlands	24 100	0.85	17 800	0.70	41 900	2.6%	82
North West	36 900	0.83	28 000	0.70	64 900	2.5%	84
Yorkshire-Humber	21 300	0.64	15 500	0.52	36 800	1.9%	73
Great Britain	391 300	1.00	351 700	1.00	743 000	3.3%	100

Note: Northern Ireland data is not included here as it is not collected in the same way. Northern Ireland comprises approximately 2.5% of the UK population and accounts for relatively little ICT employment.

1. This is the proportional importance of the service industry relative to its importance to the national ICT industry, expressed as an index. Thus, the Eastern region, with an index of 100, has the same proportional split between manufacturing and services in the ICT industry as does the United Kingdom as a whole, while in London, there is disproportionately high employment in services.

Source: Annual Employment Survey, 1997. © Crown Copyright Reserved ONS Statistics (NOMIS), 2000.

The greatest strengths in the UK electronics industry are where particular elements have achieved a coherence that has given them durability. Sectorally, telecommunications and defence electronics and ICT have been afforded a degree of durability by government funding regimes. Spatially, stability has also been provided through the labour market, because of the relative inertia of highly skilled employees, which mean that clusters experience very strong positive externalities from the availability of skills. Conversely, those areas without those high-skill labour markets have experienced difficulty in developing them and have had difficulty in competing with those strong regions.

Splitting the industry into manufacturing and software, location quotients present two very distinct geographies, with a concentration of manufacturing in the peripheral regions, while the software elements of the industry are strongly concentrated around the South East. This raises the question of whether there are distinct geographical electronics and/or ICT clusters in the United Kingdom or whether there is a national cluster that is geographically differentiated internally.

The sub-national clusters of national competitiveness

Three areas of the United Kingdom are used below to illustrate durable clusters of the ICT industry which contribute to the competitiveness of the UK economy. These are not necessarily the only ICT clusters in the United Kingdom, and in describing their orientation we focus on the main characteristics only, omitting some elements, but they do illustrate broader tendencies. In each of the three regions, there are very different reasons for their rise, with Scotland's cluster, supported by government to alleviate the problems of the decline of heavy industry, much more defensive in its nature than either the Thames Valley or Cambridge clusters.[3] Each of these clusters has benefited from government policy support, and their state today remains heavily influenced by government policy. As successful clusters, each of the three has in turn influenced government thinking in a range of policy areas. In this section, we present an overview of each of the three ICT clusters, explaining:

- How their current state relates to the policy framework within which they developed.
- How they have developed competitive advantage, their characteristic forms of innovation, and their contribution to the UK economy.
- How their success has influenced *ex post* government thinking and policy making.

Scotland and the rise of Silicon Glen

Origins. Since the first regional aid map was drawn in the United Kingdom to identify regions with poor economic performance in 1934, Scotland has been the beneficiary of significant public investment to effect industrial modernisation (Hargreave, 1985). The fillip to this process came with the Second World War, where Scotland's distance from the main bombing targets made it a suitable location for a Ferranti defence electronics factory. This introduced a set of new technologies into Scotland and the pressures of warfare encouraged the development of a skilled workforce able to deal with the demands of electronics (Charles, 1992). In the post-war period, Scotland was the recipient of considerable inward investment from British and overseas-owned companies creating manufacturing facilities in Scotland to offset their IDC requirements in the South or to benefit from incentives, drawing on the emerging pool of trained labour.

Ferranti was an important player in this process because they realised that for their Scottish factory to succeed and be profitable, they had to develop indigenous talent. Ferranti also realised that diversification away from war-time products was critical, and hence they developed a radar systems business at their Edinburgh plant. They also managed a government-funded laboratory, whose purpose was to develop staff with a combination of engineering, entrepreneurial and academic skills, which created a pool of highly capable senior managers. The presence of highly skilled operators and engineers, as well as an extremely favourable grant regime attracted many overseas investors, and this mixture of high-value-adding manufacturing and design functions has been dubbed "Silicon Glen".

Cluster policy. More recently, Scottish Enterprise, the regional development agency for Scotland, identified electronics as a key industrial sector worthy of support, and designed a cluster strategy for electronics firms to capitalise on the advantages offered by Scotland. The competitive advantage of the Scottish electronics cluster is undoubtedly in high-skill, high-value-added and low-cost manufacturing. There is a significant presence of microelectronics firms whose competitive advantage in manufacturing is based around maximising the quality of the production process. There was significant emphasis in the Scottish Enterprise strategy on maintaining this skill base, and the ability of workers to deal with the most modern manufacturing technology available.

The Scottish cluster has a strong dynamism, although the demise of Ferranti[4] highlighted the fact that the sector has latterly come to be dominated by foreign-owned firms, whose manufacturing facilities were inherently unstable and prone to closure during downturns. The nub of the problem is that Scotland tends to specialise in manufacturing the most technologically mature products; thus when technology advances, plants are left obsolete and it is often not cost-effective to upgrade them. Hence, when electromechanical technologies for office machinery were replaced with electronics, around 15 000 jobs were lost in Scotland. Relying on cost competition of mass product markets (which is what many microelectronic products are) means that adverse market movements can make plants rapidly uncompetitive, and in no position to adapt to price pressures. The collapse of computer memory prices in 1997-98 and the contemporaneous Asian monetary crisis led to the mothballing of a new semiconductor facility before completion. However, the continued availability of grant aid and the undoubted quality of the workforce have meant that after each collapse, Scotland has recovered and inflows of investment have resumed.

Innovation style. Innovation in the Scottish electronics cluster is profoundly affected by the dominance of foreign-owned businesses, the research strengths of Scottish universities, and the mismatch between the two. There are relatively few strong and extensive science-technology-market networks in the Scottish cluster because of the disjunctures associated with the techno-economic structure. On the one hand, the Scottish universities pursue the commercialisation of their knowledge base with partnerships on a global scale, limiting their local involvement to technology transfer and technical assistance. On the other hand, there are large multinational firms present in Scotland, whose Scottish locations are involved in cross-border R&D projects, particularly the *in situ* downstream elements of development and implementation.

Policy effects. The case of the Scottish ICT cluster has been a great influence on British industrial policy for peripheral regions. Although a policy of modernisation through inward investment was attempted in Scotland, Wales and the northern areas of England, Scotland has been used as a best practice case study for those other regions. The key essence of the policy was the intelligent use of support to generate capacity of generic use to the industry, investing in schemes which employed people in high-value occupations, and providing tailored support for individuals to fill those positions. Furthermore, support has been provided to encourage firms to embed themselves in the region, and supporting the upgrade of the elements of the manufacturing process present in the region. The attraction of Cadence to Scotland in 1998 to establish a design centre was seen as proof that this policy was producing tangible results. Scottish universities were awarded GBP 4 million under the Science Enterprise Challenge Fund to co-operate in microelectronics to provide opportunities to work closely with inward investors.

Although inward investment in mature manufacturing has a degree of inherent volatility, it is clear that Scotland has a competitive position in a niche for high-value manufacturing, and policy support is directed at ensuring that this is sustained in the longer term and attempting to use this activity to generate endogenous potential. Thus, the attraction and embedding of high-value-added and high-technology inward investment has formed a key element of the modernisation of Wales and the North East. In Wales, this policy has been slightly more successful than in the North East. In Wales, a great deal of effort was expended to bring LG to Newport as a flagship investor; the same factors which led to the mothballing of Hyundai's plant in Livingston led to plans for a semiconductor fab at the Newport site being shelved (Phelps and Tewdwr-Jones, 1998). In the North East, two lead semiconductor firms had been attracted, Siemens and Fujitsu, only for both of those firms to close their factories during the market downturn in 1998 (Charles and Benneworth, 1999).[5]

Defence electronics and IT services in the Thames Valley

Origins – intersecting policies. The growth of the IT service cluster in the corridor to the west of London was much less a product of direct state intervention. However, it is clear that a number of government policies were instrumental in leading to the development of the cluster in the Thames Valley, initially in electronics, which subsequently supported the emergence of ICT services. The framework within which this cluster grew was post-war planning policy, in which a strategic decision was taken to develop the area west of Heathrow airport beyond the London green belt (ECOTEC, 1999). Proximity to London and the airport would make the location attractive to industry while taking some of the pressure off the capital's land market.

There were a number of other policies which supported the rise of this cluster. Military procurement policy was very important in funding the growth of the industry. Military contracts were placed with extant companies and when seeking expansion in the south of England, the area west of Heathrow was a natural location for development establishments, a position reinforced by the locations chosen for the military's own defence research establishments (Breheny *et al.*, 1985). Industrial policy in the 1960s was based around the promotion of competitiveness through merger and creation of giant companies, which had the effect of concentrating research capacity around strong locations. The aggregate effect of this was to create the conditions in the Thames Valley as a site of a highly competitive although defence-oriented electronics sector.

Science policy has also benefited the Thames Valley, which is home to a number of flagship government research centres (Heim, 1988; Charles and Benneworth, 2001). The strength of government-funded civil R&D was critical in supporting the development of a workforce with particular expertise in the development of new technologies.

Deregulation of the telecommunications market in the United Kingdom also made a significant contribution to the competitiveness of the industry. It removed the British Telecom monopoly on land line services, and allocated frequencies for the development of mobile telephony. Contracts to develop the infrastructure were allocated on the basis of a proven capacity to manage the network rather than to develop the infrastructure. This provided the natural spur for defence electronics firms to diversify into commercial technologies, and indeed one defence company, Racal, set up Vodafone as a subsidiary business. As the Thames Valley was already a home to many of those highly skilled professional staff with experience in the development of telecommunications systems, the deregulation stimulated the growth of new firms.

In the last decade, government policy has also encouraged a shift into services, while the adoption of commercial procurement conditions in the defence sector has reduced the attractiveness of defence contracts. Government has involved the private sector in the development and management of new IT services for ministries as well as new executive agencies. This both stimulated new markets for existing firms, as well as encouraging new investment by mainly US-based IT firms in the Thames Valley. In parallel with this, the explosion in the Internet has benefited telecommunications firms.

Although part of this benefit comes from their role in selling access to the Internet, their expertise in network management has obvious commercial spin-offs.

Innovation styles. The dominance of large and formal research activities in the region has had an impact on the pursuit of innovation in the Thames Valley. There are many large business, higher education and government research centres located in and around the Thames Valley area, although there is little evidence of strong interdependence other than through labour markets and common infrastructure requirements. Furthermore, it is important to stress the effect of Racal on the cluster, whose own attitude to innovation remains important on its operation; Racal expanded successfully into new and innovative fields by establishing new operations as independent businesses, creating internal capacity for innovation. These firms pursued their own innovative activities independently from other business units in the group, thus enabling each site to remain at the forefront of technological development, and removing the possibility of the emergence of an inter-site division of labour. Given the need to populate the new subsidiary businesses with staff from existing businesses, the new operations were always established close to the old (Charles, 1992). The success of this approach is demonstrated by the success of the Racal offshoot company, Vodafone.

Cluster coherence? In the case of the Thames Valley, there are questions which must be raised as to the degree of geographical coherence of the cluster (Coe and Townsend, 1998). Certainly, it is very hard to decide exact definitions, given that the "M4 corridor" runs 100 km further west to Bristol, and there is a crescent from the Thames Valley which runs north-east to Hertfordshire, itself an important site for electronics and IT services. Despite its geographical vagueness, the Thames Valley is the closest of the three cluster cases here to a national-scale cluster; there are many R&D locations for UK multinationals and a significant element of civil and defence public R&D is performed within the cluster. There is a great deal of co-ordination work for R&D and innovation networking which is located in and around this cluster and, so far, more than "Silicon Glen" or "Silicon Fen", the Thames Valley is an exemplar of a national cluster of ICT competitiveness.

Effects of the cluster on government policy. This fact is reflected in the two main policy areas where the success of the cluster has significantly influenced central government thinking, although there is a third change which has particular spatial policy implications. The first is in what remains of industrial policy, which is heavily influenced by conceptions of national competitiveness. It is clear that this cluster is globally competitive; the take-over of Mannesmann (a diversified German engineering concern and one-time owner of Orange, a second UK mobile telephony provider) by Vodafone, itself a spin-off from Racal, indicates the strength of the cluster. The 1998 Competitiveness White Paper (DTI, 1998), *Our Competitive Future: Building the Knowledge-driven Economy*, stresses the importance of ICTs to the development of a new knowledge economy and, although the concept is generalised to cover industries across the United Kingdom, it is clear that at least part of the model is derived from the strength in the Thames Valley communications and IT services cluster.

The second area where the cluster has influenced policy is in the area of science. It is clear that one of the competitive advantages of the cluster is derived from the proximity of scientific research establishments to high-value-adding manufacturing and service firms, creating a common labour market and supporting the innovative culture. The government appears to have accepted the need to concentrate scarce scientific resources to the benefit of the United Kingdom as a whole, something borne out by the recent decision to locate a new synchrotron in Oxfordshire rather than in Cheshire in the North West. Although the predisposition for the location of government R&D in the South East has endured since the 1950s, the success of the Thames Valley cluster in contributing to national competitiveness has vindicated and helped to sustain the orientation of national SET expenditure.

The third policy area to which the Thames Valley cluster has made a significant contribution is in the area of land use planning, something with ramifications beyond central government because of the organisation of planning in England. The central government and regional groupings of local authorities, the Regional Planning Conferences, agree on a decennial target for house building in each of the regions, balancing the need for nature conservation, sustainable communities and the housing demands of economic development. The success of the Thames Valley cluster has led to rapidly rising demand for housing stocks and the argument has been advanced, very successfully, that more land must be released for house building to accommodate rising demand from the industries. The economic argument is reinforced with the argument that if more housing is not forthcoming, then the South East, and the United Kingdom in general, will lose out to other countries with more flexible spatial planning regimes. The latest version of *Planning Policy Guidance Note No. 3: Housing*, emphasises that it is important to provide houses to sustain economic dynamism, and "economic growth should not be frustrated by a lack of homes for those wishing to take up new employment opportunities" (DETR, 2000).

Cambridge – embodying professional expertise in new media

Origins. The 1985 report by consultants Segal, Quince and Partners signalled the emergence into the popular and public consciousness of what they termed *The Cambridge Phenomenon*, the existence in and around Cambridge of a highly dynamic and innovative cluster of high-technology industries (Segal, 1985). The city of Cambridge was host to one of the United Kingdom's most prestigious and research-active universities, but it was clear from the report that the relationship between this cluster and the university was more complex than merely as a source of highly skilled employees and new high-technology spin-off firms.

There were two main policy measures which had historically contributed to the growth of the cluster (ECOTEC, 1999). The first element was a very stringent planning regime both within the city area of Cambridge as well as in the wider county of Cambridgeshire in which new industrial premises are only made available if there are compelling reasons for location in Cambridge. The selection and concentration of high-technology firms has the potential to create clustering benefits from proximity to competitors as well as the Cambridge location boosting the prestige of small companies. The contribution the university made to this situation was the early creation of a science park by one of its colleges, offering serviced areas where small firms can grow and thrive; although science parks are now common across the United Kingdom, the age of the park, which opened in 1970, means that a number of firms born on the science park have since matured into dynamic and innovative companies.

The second policy element was government science policy (Wilkie, 1991); Government policy favoured the development of high-technology bioscience laboratories in the Cambridgeshire area, drawing on university research expertise as well as the natural evolution of agriculture research into biotechnology. Research in this area was revolutionised by the potential for analysis offered by computer-based technologies. Electronics and instruments firms located in Cambridge therefore benefited from a highly demanding and inquisitive market with whom cutting-edge products could be developed and sold globally. The Babraham Institute public research centre has also had a number of spin-off firms who have drawn extensively on the bioscience base of the Cambridge area to develop new ICT-derived products made commercially feasible by advances in ICT processing power and software programming languages.

Cluster characteristics. The essence of the competitive advantage of these firms and of the cluster is their dynamism and innovation in emerging activities. Although there is a distinctive Cambridge ICT cluster, success is spread across a number of sectors. This effect has been wider than purely the biotechnology sector; Cambridge is host to a large number of technical and technological service firms, and each has had their market dramatically changed by new information and communications technologies. Generally speaking, much of the ICT growth has come from firms whose primary expertise is not necessarily in computers or IT, but in some other area such as geophysics, medical diagnostics or machine vision. These firms have developed electronic and software products which embody their expertise.

The other element of the competitive advantage of the cluster is that it has been able to take advantage of the financial flexibility made possible by venture capitalists. Although venture capital remains extremely volatile as a form of funding, there has been an expectation from financiers that because ICTs are a transforming technology, with long-term potential, developmental firms should be allowed to run medium-term deficits. Venture capitalists have plugged these gaps in anticipation of future rewards, and in recognition of the strengths which the Cambridge environment offers. The Cambridge economy is supportive of high-technology SMEs in a range of ways. A Cambridge location has a degree of cachet, which can ease accessing global markets. Close proximity to competitors allows new waves of products to be launched against each other, which raises market impressions that they represent a new generation, increasing the imperative for their purchase, and increasing the sales of the firms.

Innovation style. The style of innovation in the Cambridge cluster bears much resemblance to the behaviour of two of the groups from whom entrepreneurs and innovators are drawn. The small high-technology firm approach to innovation is culturally very similar to the team-based approach to research in universities, and the semi-autonomy of fee-earners in professional partnerships. Both those occupations frequently blur the distinction between work and leisure (Massey *et al.*, 1992) because of the intensity of the employment, and it is this approach to innovation which encourages small-firm-based success. Innovation is also highly reflective in the cluster, with similar firms tending to be aware of how rivals are performing because of the advantages of roughly synchronous product launch. Indeed, this culture, characterised by autonomy and dedication, is important to the success of larger R&D-based organisations, such as the Microsoft Research Laboratories.

In policy terms, Cambridge is unequivocally perceived as a success and hence an exemplar (DTI, 1998; DTI/DfEE, 2001). This is because it has overcome all the weaknesses inherent in the UK system of innovation, maximised the potential of opportunities and capitalised on its strengths, with a very limited amount of government industrial intervention. The Cambridge system of innovation is underpinned by a culture in which traditional divides between academic work and entrepreneurship have been broken down (Massey, 1995).[6] Traditional financial short-termism has been overcome because the huge potential of many of the firms has attracted novel forms of finance, particularly venture capital and listing on the Alternative Investment Market. Cambridge has managed to mix British and foreign ownership with none of the traditional drawbacks of inward investment, particularly in terms of dealing with rationalisation – the business competitiveness strengths of the Cambridge-located firms are embodied in individuals, making their development contingent upon maintaining location in Cambridge.

Effects on policy. The identification of the Cambridge phenomenon has been followed by its widespread incorporation into UK industrial policy, because it demonstrates that industrial development and modernisation is possible within the UK innovation system at a relatively low cost to the public purse. Indeed, because it is argued that most of the investment necessary to secure the benefits of an ICT cluster is already present across the United Kingdom in the form of universities, innovative businesses and government R&D, the level of government expenditure required to increase regional competitiveness is relatively low. Much of the tone of the 1998 Competitiveness White Paper has been influenced by analyses of the situation in Cambridge, particularly in terms of government attempts to reproduce the entrepreneurial culture of Cambridge. While the Thames Valley is an example of a national cluster of sectoral competitiveness (albeit concentrated within a small geographical region), DTI policy on sub-national territorial competitiveness is derived from an analysis of the highly place-specific success of the Cambridge region.

The other important element of policy which has been informed by the case of Cambridge is science and technology policy and particularly the absence of any kind of spatial consideration in the allocation of funding (Charles and Benneworth, 2001). Table 5 presents the R&D index for the British regions, which shows that the Eastern and South East Government Office regions have a higher share of R&D than would be expected by their contribution to regional wealth. The success of both Cambridge and the Thames Valley have justified the argument that national competitiveness is well-served by the existing allocation of funding. The argument is that the pattern of funding follows the efficiency of the users; therefore, raising the technological capacity of the poorer regions will improve their performance without undermining the competitive strengths of the United Kingdom as a whole.

Table 5. **The R&D concentration index for the regions, 1997**

	GERD (in GBP millions)	R&D index
United Kingdom	14 462	100
England	13 185	107
Eastern	2 767	208
South East	3 415	149
South West	1 146	98
North West	1 503	98
East Midlands	899	93
London	1 593	73
West Midlands	988	81
Scotland	876	73
North East	334	64
Yorkshire and the Humber	540	49
Northern Ireland	145	44
Wales	257	43

Note: This index indicates the relative level of gross expenditure on R&D (GERD) as a proportion of all UK expenditure relative to the proportion of GDP produced by that region. A value of 100 indicates that the proportion of R&D that takes place in that region with respect to the United Kingdom is equal to its share of R&D; a higher value indicates a concentration of R&D, while less than 100 indicates a relative absence. Regions used are government office/regional development agency regions.
Source: Economic Trends, 1999.

The ICT cluster and cluster policy

The lead responsibility for cluster policy has lain with the DTI. The key concern of the department is with competitiveness, and its periodic White Papers have set the framework for the bulk of routine intervention in the economy. Prior to 1997, there was predominantly a sectoral concern, and electronics was dealt with in the Foresight exercise as a national-scale cluster. The 1998 Competitiveness White Paper, however, moved from a purely sectoral shift to thinking about clusters, particularly in terms of the non-material linkages between the various firms in those clusters.

Much of the cluster policy put forward and developed after the 1998 White Paper relates to the biotechnology cluster, establishing a Minister-led team, developing a "Genome Valley" in UK regions and the "Biotechnology Means Business" initiative. However, there are specific policy elements in current government thinking which have their origins in the three UK ICT clusters. Clustering is an important element of DTI policy, particularly in terms of the delivery of their regional competitiveness agenda. Thus, when the newly formed regional development agencies (RDAs) submitted their regional economic strategies to central government, all were urged in the strongest possible terms to "to facilitate cluster development by seeking to promote informal networks rather than create new clusters" (DETR, 2000).

First, the Scottish cluster is exemplified as an example of best practice in the collaborative exploitation of research, in assisting a peripheral region to overcome the problems caused by the absence of network linkages between universities and sophisticated multinational users located in Scotland. "Project Alba is a unique collaboration between public and private sector, industry and academic partners which will propel Scotland to the forefront of world-class research and design of next generation semiconductors" (DTI, 1998, p. 44). Subsequent programmes and funding stressed the importance of building linkages between centres of excellence in universities and local businesses rather than creating research excellence. In particular, this appeared to indicate an appreciation within government of the spatial element of clustering, and the importance of proximity in allowing access to technical knowledge.

Second, the success of the Thames Valley cluster, and in particular the degree to which the regulation framework has contributed to its competitiveness has influenced government policy towards world-class firms. In the Thames Valley, large defence contractors were critical to the development of businesses which could compete globally. Government policy towards defence therefore hinges around encouragement of large defence speciality firms, such as BAE Systems (formed from the merger of BAe and Marconi Electronic Systems) and the Rolls Royce take-over of Vickers. More critical to the industry, however, may be the GBP 530 million provided in launch assistance to BAE Systems to develop, within Airbus Industrie, the A3XX very large aircraft, which reprises the policy of providing stability to the industrial base by removing a degree of the market pressure from the developers.

Third, the success of Cambridge has led DTI, which has since 1997 had responsibility for Science, Engineering and Technology (SET), to reaffirm its commitment to the funding for basic science. Although applied research and technology transfer are important elements of the national science base and contribute to competitiveness, the attraction of the Microsoft Research Centre to Cambridge demonstrated that multinationals have a need for basic research. New funding for basic research in the universities, including a series of new university innovation centres around the United Kingdom seems to suggest that the importance of the knowledge base to the emergence of successful clusters in all the UK regions is being recognised as a priority.

What are the wider lessons of this for cluster-based innovation policies? First, it is clear that in relatively large economies such as the United Kingdom, cluster-based innovation systems are likely to emerge at a regional scale and, indeed, the concentration of particular clusters in particular regions may be an unforeseen consequence of public policy. These regional innovation systems have a particular set of characteristics that differentiate them from scaled-down national innovation systems. They are more specialised around focused clusters, they draw heavily on national inputs that may lie outside of the region, and their boundaries may be less determined by administrative territories and more by interactions and flows between firms. Thus, the mapping of such regional clusters cannot be restricted to a specific region, but are often trans-regional (see Pinch and Henry 1999 on motor sport as another example).

The other main conclusion is that clusters are affected by a wide variety of policies, many of which lie outside of formal cluster policy, or even industrial and technology policies. Each cluster, and even different ICT clusters here, is affected by a complex interplay of policies influencing the trading environment, sources of innovation, the nature of places where cluster resources come together and the regulation of the cluster. A focus purely on innovation or industrial policy, without such a broad and historically contextualised perspective will yield a narrow and myopic view. Having identified the existence of a cluster and made a decision to support it, governments must assess how different policies have affected the cluster and might affect its future trajectory. In this way, the cluster provides an integrating perspective within national innovation systems to turn the theory into a more pragmatic and practical application. The heterogeneity of clusters, and their complexity, provides a challenge for policy, but also a focus for integration and greater effectiveness.

NOTES

1. Although culture may not always be thought of as an element in a national innovation system, it is clear that different national cultures affect styles and forms of innovation. Hence, culture is both an element in a national innovation system framework, as well as affecting the nature and operation of other elements within that system, such as finance or particular institutions.

2. Apocryphally, this limit relates to the distance for which civil servants could claim a (very generous) mileage allowance; beyond this distance they were required to take the train, although the contracts they awarded did have allowances for their entertainment and accommodation while away from home.

3. Although Wales has a strong electronics manufacturing sector, it is necessary to discount this as a geographically bounded regional cluster. The Welsh electronics industry was based on the attraction of inward investment to replace the rapidly declining heavy engineering and extractive sectors. Inward investors were attracted into what was an entirely greenfield location (fresh land and employees), with no real history of electronics, which drove the firms there to compete effectively on the basis of their cost competitiveness (Phelps and Tewdwr-Jones, 1998).

4. Ferranti was an successful UK-owned firm with strengths in defence systems, components and industrial electronics. It sought to achieve greater market share in the defence sector in the late 1980s by buying a US-based defence contractor, but this turned out to have been trading fraudulently and the resultant losses forced Ferranti into receivership and break-up. At the time of the collapse, Ferranti was the biggest UK-owned firm in Scotland and accounted for 20% of all electronics jobs in Scotland.

5. Both factories have since been re-opened by smaller and more niche-oriented firms, Atmel and Filtronic.

6. Although the degree to which this is true is debatable. Certainly, much of the new firm formation has not come from direct academic spin-off, and only the largest of the multinationals located in and around Cambridge have the capacity to undertake much collaborative world-class research with the university. Many of the "academic" spin-offs come from either entrepreneurial graduates of the university or from government-funded research institutes which have been subject to market imperatives since the 1970s – much longer than is the case for university-sector institutions.

REFERENCES

Baker, W.J. (1969), *A History of the Marconi Company*, Methuen & Company, London.

Breheny, M., P. Cheshire and R. Longridge (1985), "The Anatomy of Job Creation? Industrial Change in Britain's M4 Corridor", in P. Hall and A. Markusen (eds.), *Silicon Landscapes*, Unwin Hyman, London.

Brummer, A. and R. Cave (1998), *Weinstock — The Life and Times of Britain's Premier Industrialist*, HarperCollinsBusiness, London.

Buswell, R.J. and E.W. Lewis (1970), "The Geographical Location of Industrial Research Activities in the UK", *Regional Studies* 4, pp. 296-307.

Campbell-Kelly, Martin (1989), *ICL: A Business and Technical History*, Clarendon Press, Oxford.

Charles, D.R. (1991), "Procurement in European Telecommunications", *Utilities Policy* 1, pp. 134-143.

Charles, D.R. (1992), "The Corporate Dimension of Research and Development Location", unpublished PhD thesis, University of Newcastle upon Tyne, Newcastle.

Charles, D.R. and P.S. Benneworth (1999), "Plant Closure & Institutional Modernisation: Siemens Microelectronics in the North East", *Local Economy* 14, pp. 200-213.

Charles, D.R. and P.S. Benneworth (2001), "Are We Realising our Potential – Joining up Science and Technology in the English Regions", *Regional Studies* 35, pp. 73-79.

Coates, D. (1994), *The Question of UK Decline: The Economy, State and Society*, Harvester Wheatsheaf, Hemel Hempstead.

Coe, N.M. and A.R. Townsend (1998), "Debunking the Myth of Localised Agglomerations – The Development of a Regionalised Service Economy in South East England", *Transactions of the Institute of British Geographers* 23, pp. 385-404.

Costello, N., J. Michie and S. Milne (1989), *Beyond the Casino Economy*, Verso, London.

Department for Environment, Transport and the Regions (2000), *Planning Policy Guidance No. 3: Housing*, DETR, London.

Department of Trade and Industry (1994), *Competitiveness:- Helping Business to Win,* Her Majesty's Stationary Office, London.

Department of Trade and Industry (1996), *Competitiveness – Creating the Enterprise Centre of Europe*, Her Majesty's Stationary Office, London.

Department of Trade and Industry (1998), *Our Competitive Future: Building the Knowledge-driven Economy*, Her Majesty's Stationary Office, London.

Department for Trade and Industry/Department for Education and Employment (2001), *Opportunity for All in a World of Change: White Paper on Enterprise, Skills and Innovation*, The Stationary Office, London.

Dintenfass, M. (1992), *The Decline of Industrial Britain 1870–1980*, Routledge, London.

ECOTEC (1999), *Planning for Clusters*, Department of Environment, Transport and the Regions, London.

Georgiou, L (1991), *Evaluating Alvey, Britain's National IT programme*, Manchester Statistical Society, Manchester.

Guy, K, P. Quintas, M. Hobday, L. Georgiou, C. Cameron and T. Ray (1991), *Evaluation of the Alvey Programme for Advanced IT*, Her Majesty's Stationary Office, London.

Hague, D. and G. Wilkinson (1983), *The IRC – An Experiment in Industrial Intervention*, George Allen & Unwin, London.

Hargreave, A. (1985), *Silicon Glen — Reality or Illusion: A Global View of High Technology in Scotland?*, Mainstream, Edinburgh.

Heim, C.E. (1988), "Government Research Establishments, State Capacity and Distribution of Industry Policy in Britain", *Regional Studies* 22, pp. 375-386.

Hirst, P.Q. and J. Zeitlin (1988) (eds.), *Reversing Industrial Decline? Industrial Structure and Policy in Britain and Her Competitors*, Berg, Oxford.

Jones, R. and O. Marriott (1970), *Anatomy of a Merger – A History of GEC, AEI & English Electric*, Jonathan Cope, London.

Keeble, D. (1992) , "High-technology Industry and the Restructuring of the UK Space Economy", in P. Townroe and R. Martin (eds.), *Regional Development in the 1990s: The UK in Transition*, Jessica Kingsley, London.

Kelly, T (1987), *The British Computer Industry – Crisis and Development*, Croom Helm, London.

Kennedy, W.P. (1987), *Industrial Structure, Capital Markets and the Origins of British Economic Decline*, Cambridge University Press, Cambridge.

Massey, D. (1995), "Masculinity, Dualisms & High Technology", *Transactions of the Institute of British Geographers* 20, pp. 487-499.

Massey, D., P. Quintas and D. Wield (1992), *High Tech Fantasies: Science Parks in Society, Science and Space*, Routledge, London.

McKay, D.H. and A.W. Cox (1979), *The Politics of Urban Change*, Croom Helm, London.

McLean, M.and T. Rowland (1985), *The Inmos Saga: A Triumph of National Enterprise*, Pinter, London.

Newton, S. and D. Porter (1988), *Modernisation Frustrated — The Politics of Industrial Decline in Britain Since 1900*, Unwin Hyman, London.

Oakley, B.W. and K. Owen (1990), *Alvey*, HIT Press, London.

Phelps, NA. and M. Tewdwr-Jones (1998), "Institutional Capacity Building in a Strategic Policy Vacuum: The Case of the Korean Company LG in South Wales", *Environment and Planning C: Government and Policy* 16, pp. 735-755.

Pinch, S. and N. Henry (1999), "Paul Krugman's Geographical Economics, Industrial Clustering and the British Motor Sport Industry", *Regional Studies* 33, pp. 815-827.

Redwood, J. (1984), *Going for Broke: Gambling with Tax Payers Money*, Blackwells, Oxford.

Segal, N. (1985), *The Cambridge Phenomenon — The Growth of High Technology Industry in a University Town*, Segal, Quince and Partners, Cambridge.

Walker, W. (1993), "National Innovation Systems: The United Kingdom", in R.R. Nelson (ed.), *National Innovation Systems: A Comparative Analysis*, Oxford University Press, Oxford.

Weiner, M.J. (1981), *English Culture and the Decline of the Industrial Spirit*, Cambridge University Press, Cambridge.

Wilkie, T. (1991), *British Science and Politics since 1945*, Blackwells, Oxford.

Williams, H. and D.R. Charles (1986), "Electronics in the North East of England: Growth or Decline", *Northern Economic Review* 13, pp. 29-38.

Young, P. (1985), *Power of Speech: A History of Standard Telephones and Cables 1883–1983*, George Allen & Unwin, London.

Chapter 18

DANISH CLUSTER POLICY: IMPROVING SPECIFIC FRAMEWORK CONDITIONS

by

Mette Holm Dalsgaard
Danish Agency for Trade and Industry

Introduction

Theory and research has shown the importance of clusters to a successful economy. Clustering stimulates the start-up of companies, as well as their competitiveness through stronger innovativeness and productivity (see, for example, Porter, 2000).

The challenge for policy makers such as the Danish Ministry for Trade and Industry is to transform cluster theory and analysis into visible policy initiatives that can strengthen firms. A related aspect is for industrial policy to respond to the way businesses are organised today. This calls for broad based co-operation based on a practical version of the cluster concept. Denmark has pursued this strategy for almost a decade now and is still on a learning curve.

This chapter will elaborate on the Danish experiences with cluster policies. It will highlight the importance of dialogue throughout the entire process – from designing clusters, analysing their performance and specific needs, discussions with companies and other relevant actors, to creating and implementing new policies. As will be explained, developing cluster policies in Denmark is related to the need for specific framework conditions.

Although the focus is mainly on the political processes, consistent analysis is an integral part of the cluster project; this chapter will briefly touch upon how this analysis is carried out in Denmark. For a more thorough presentation of the theoretical background and characteristics of Danish cluster analyses, see Drejer *et al.* (1999).

Specific framework conditions as a competitive advantage

The main aim of the cluster approach is to support Denmark's position as one of the most prosperous regions of Europe, both today and for the future. Firms locate their activities in those regions where the conditions are most appealing, such as the strength of the competences on which they depend. It is important to provide the enterprises that are to sustain Denmark in the years to come with the necessary environment. Trade and industry policies must take into account the increasing competition among "centres of excellence" around the world.

Such is the aim of the new strategy for trade and industry – *.dk21* – set out by the Danish Government (Danish Ministry of Trade and Industry 2000). It involves a joint effort between nine Danish ministries, led by the Ministry of Trade and Industry. The key elements of .dk21 are specific framework conditions for Danish trade and industry: Knowledge and competences, Global outlook, Committed venture capital, Well-functioning markets, Quality of regulation and Social responsibility of companies.

With .dk21, the political focus has shifted away from general initiatives at macroeconomic level and direct subsidies to certain industries towards tailored policy initiatives providing indirect support to industries through networks.

Macroeconomic conditions, such as interest rates, exchange rates, taxation and inflation, remain vital for competitiveness. However, policy framework conditions that apply to particular areas of the economy – specific framework conditions – are gaining in importance. These include all those circumstances that directly or indirectly influence the production situation of trade and industry, and frequently originate in the interaction between business and the public sector. For example, research in the biotechnology field has a significant impact on the Healthcare cluster, while it does not apply to the Construction/housing cluster. Similarly, construction regulations have little bearing on the Healthcare cluster.

This reflects the fact that a country or region can to a large extent create its own specific framework conditions, whereas macroeconomic conditions are increasingly determined by external factors.[1] If a country or region provides better specific framework conditions than are found abroad, the competitiveness of business in that country or region increases. The specific framework conditions become *the* competitive advantage.

.dk21 is unique in that it represents a joint strategy of the Danish Government. It involves co-operation with other ministries in many new projects linking industrial policies to investments and strengths in other political jurisdictions. The goal is to combine competitiveness of the Danish trade and industry with the general welfare of society, *i.e.* through projects involving the Danish Ministry of Education and the Ministry of Trade and Industry aimed at stimulating regional knowledge centres.

Co-operating with many partners

The cluster project based in the Danish Ministry for Trade and Industry operates with many partners, as shown in Figure 1. Co-ordinating these inputs demands considerable resources, but such dialogue is necessary for informing the decisions of policy makers and disseminating the results.

The *Board of Referees* consists of 20 top management executives, appointed for their prominent positions and valuable experience in Danish business and industry. The Board of Referees is responsible for highlighting the conclusions that can be drawn from the analysis. The objective is to specify how appropriate policies might be initiated based on the analyses.

The Danish cluster project is supported by a standing committee, the *expert panel*, which serves to ensure the coherence and depth of the work in progress, and contributes analytical insight and professional experience. The composition of the expert panel represents the various domains of professional expertise.

Figure 1. **Partners in the Danish cluster project**

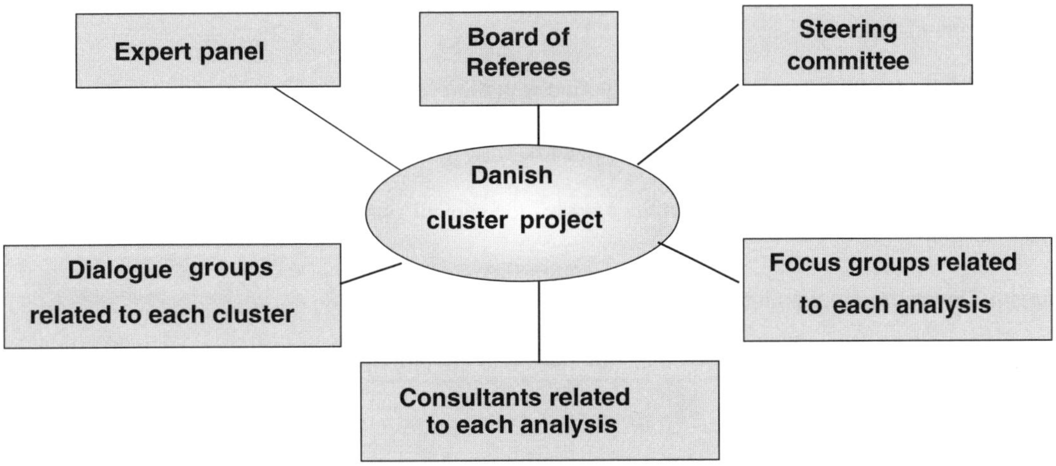

Furthermore, there is ongoing co-operation with the project's *steering committee,* which is headed by the Permanent Secretary of the Ministry of Trade and Industry. Policy initiatives are formulated independently by the Agency for Trade and Industry, in close co-operation with the Department of the Ministry.

The analyses and reports are performed by private *consulting firms* working on a contract; they are assisted by a *focus group* of key persons having in-depth knowledge of the cluster under study.[2]

Once the analysis has been carried out, dialogue is launched with companies representing each of the clusters with regard to the appropriate policy recommendations. These companies participate in *dialogue groups*, together with representatives from knowledge centres and relevant public authorities. Participants are invited to attend meetings and seminars, and are further qualified for participation in various working groups with specialised agendas. The Ministry of Trade and Industry uses its knowledge and networks in the business environment to seek out relevant participators with interest and influence.

A policy-making process based on analysis and dialogue, with equal importance being given to each, helps to develop "best practice policies" in a very dynamic sense. Analysis and dialogue are an efficient way of identifying barriers to and opportunities for innovation: a unique and constantly changing set of factors that "triggers" each cluster. Examples include the tailoring of research centres, vocational training and other educational institutions, patent systems, development contracts, competition laws, and technology incubators to the specific needs of the clusters.

Designing the clusters and the policies

Before going into greater detail about the current cluster analyses and dialogues, this section sketches out the structure of the Danish cluster project. The cluster project was initiated in the Danish Ministry for Trade and Industry, which published eight mega-cluster reports in the early 1990s, followed by dialogue with companies and other relevant stakeholders. The result was almost 170 policy initiatives, of which Box 1 lists a few examples which exemplify specific framework conditions.

Box 1. **Examples of first-generation cluster policies**

- *IT/Communications:* IT growth centres in Copenhagen/Malmoe and flexible tender procedures for IT projects in the European Union.
- *Agro-food:* International ethical rules and dialogue with the population concerning gene technology.
- *Transport:* New education for transport and logistics.
- *Healthcare:* Four "research centres without walls" for clinical pharmacology and appliances, neurology and growth/regeneration.
- *Tourism:* Development of an Internet portal.
- *Operational services:* Integration of ethnic minorities into the labour market.

The second generation of cluster activities in the Ministry of Trade and Industry in 1999-2002 involves updating these reports as well as carrying out new analyses, such as cross-sectional studies and clusters of competence. Again, this will be accompanied by dialogue and formulation of proposals for policy, as shown in Figure 2.

Figure 2. **Activities and structures of the Danish cluster project**

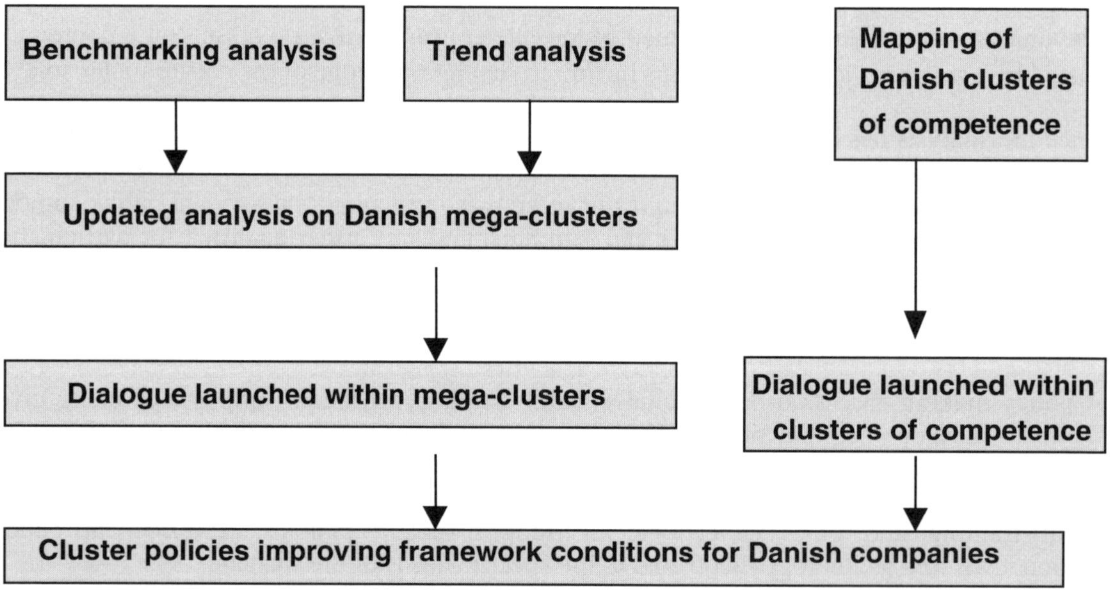

As Figure 2 illustrates, the process from analysis to policy is now pursued in two parallel trajectories – *mega-clusters* and *clusters of competence.* Box 2 provides a definition of the Danish clusters that will be dealt with in greater detail in the following sections.

Box 2. **Definition of Danish clusters**

Networks of firms and knowledge institutions with common framework conditions. Their interrelations create shared competencies, enabling production with a relatively high value added.

Approached at two levels:

- 12 Danish mega-clusters.
- Large number of Danish clusters of competence.

The Danish clusters are not all defined by vertical linkages in a value chain.[3] Other important elements can link the actors, some of which are not (yet) measurable in a quantitative manner. These include horizontal spillovers, mobility and common framework conditions.

A pragmatic approach has been used to choose the best methods for identifying the Danish clusters. As cluster policy making in Denmark is primarily focused on co-operation, it is crucial to achieve acceptance and participation of all actors in the design of clusters and cluster policies. The clusters are defined in co-operation with the expert panel and other partners with an in-depth knowledge of Danish trade and industry.

It is important that participants are familiar with the constantly changing circumstances of Danish businesses in order to be able to identify promising business areas as well as promising policy initiatives that can have a significant impact with only limited resources. The Danish Ministry of Trade and Industry tries to retain a general approach when choosing between different angles. An important consideration relates to what can be changed in practice. In particular, this involves the framework conditions set up in the government strategy for Danish trade and industry – .dk21.

The design of the clusters and policies is linked wherever possible to the broader aims of government, such as healthcare, social inclusion and improved educational performance. As mentioned above, it is of increasing interest to link trade and industrial policies to strengths and investments in other areas, in order to benefit from these. It should be borne in mind that the aim of the Danish cluster approach is to improve the business environment and performance and not, for example, to improve the national health service, which should have its own "sponsors". However, there can be mutual benefits which can help to promote policy. This reflects the unique concept of .dk21, which replaces the narrow approach of industrial policy with one based on co-operation across ministerial jurisdictions.

The Danish mega-clusters

Examples of policy results from the first generation of mega-clusters were given above. The 12 mega-clusters which will be investigated during the next couple of years are shown in Box 3.

These mega-clusters cover 90% of the Danish economy; they are, to a certain extent, based on the concept proposed by Porter.[4] Companies are classified into a series of broad industry domains interrelated across branches and sectors. Compared to traditional analyses of isolated industries, such groupings better reflect the ties spun between related industries. They take into consideration the mutual linkages between, for example, producers of construction materials and the construction industry.

Box 3. **Danish mega-clusters**

- Agro-food
- Tourism
- IT/communications
- Energy/environment
- Trade and retail
- Logistics and support services
- Furniture/clothing
- Construction/housing
- Transport
- Healthcare
- Knowledge-based services
- Metal industry/Processing technology

The Danish Ministry of Trade and Industry is now working to sharpen the mega-cluster studies. An important lesson to be drawn from the first generation of Danish cluster policy was the need to become more familiar with cluster dynamics and critical framework conditions, and to avoid "holding the microphone", in the sense of providing solutions to opportunistic suggestions by firms, etc.

The benchmarking and trend analyses shown in Figure 2 concern the current performance and future challenges of the Danish clusters related to increasing globalisation, and provide a stronger point of reference for decision makers wishing to design policies for the mega-clusters (Danish Agency for Trade and Industry, 2000b, 2000c).

The Construction/housing mega-cluster was the first of the second-generation Danish cluster studies (Danish Agency for Trade and Industry, 2000d). It concluded that the Construction/housing cluster is facing "lock-in": factors such as culture, organisational set-up, regulations, low market transparency and lack of competition seem to present serious barriers to innovation. Many companies operate with a high degree of professionalism and face significant demand on international markets. However, large parts of the cluster have problems in terms of productivity, price level and quality.

Following the analysis, the Danish Ministry for Trade and Industry and the Ministry of Housing and Urban Affairs established a taskforce. The results were published in Danish Ministry of Trade and Industry and Danish Ministry of Housing and Urban Affairs (2000). The Danish Government presented 28 policy initiatives aimed at improving the framework conditions for innovation and productivity as well as an output of a higher quality and lower prices in the Construction/ Housing cluster. Box 4 lists the main proposals for new policy initiatives.

The policy approach for the Danish Construction/housing cluster has stimulated intense debate about the prospects for this area of the Danish economy. The final outcome is yet to be seen. So far, the activities relating to the Construction/housing cluster have been successful and the Danish Ministry of Trade and Industry has started implementing the proposals in co-operation with the Ministry of Housing and Urban Affairs.

From a strictly theoretical point of view, the Danish mega-clusters are not sufficiently focused for an ideal cluster approach. They serve other important purposes, although they do constitute the most comprehensive business analyses of competitiveness and growth conditions in Danish trade and industry. In addition, they cover most of the Danish economy, thus ensuring broad acceptance when they were introduced in the early 1990s. The Danish Ministry of Trade and Industry works continuously to improve the approach; business life and its conditions have undergone significant change since the 1990s, requiring revised and updated designs and analyses of the Danish mega-clusters.

Box 4. **Main proposals in the policies for the Danish Construction/housing cluster**

Strengthening the role of the client, through, *e.g.* development of tools for BOOT projects, and stronger incentives related to the Danish Building Defect Fund concerning quality aspects in public construction investments.

Increased competition and well-functioning markets for construction materials, through, *e.g.* improving market transparency by collecting and publishing information in a "Construction Consumers Information Spot" on price and quality for similar products in different countries, and through efforts to open the national market to construction materials from neighbouring countries.

Development of a co-operation culture in the Construction/housing cluster, through, *e.g.* changing tender procedures, developing guidelines for partnering, and supporting building projects that experiment with new forms of organisation.

An effort to improve innovation through better learning, through, *e.g.* a concerted strategy for the Danish research institutions, a Centre for Evaluation, and ICT standards for communication.

Opportunities for a more focused policy approach are now arising, making it legitimate to introduce the notion of "clusters of competence". This would not have been possible ten years ago. However, work continues on updating the mega-cluster studies and policies since industrial policy should always have a broader base to complement the more focused approach of the smaller clusters.[5]

Danish clusters of competence

The Danish clusters of competence are a new concept which was introduced with a view to stimulating a more focused policy formulation process. The clusters of competence are areas in Danish economy which show excellence in the international as well as the national context.[6] The results of a comprehensive analysis of Danish clusters of competence were published and discussed at a conference held in Copenhagen in February 2001 (Danish Agency for Trade and Industry, 2001. Figure 3 shows the 29 Danish clusters of competence identified to date.

The Ministry of Trade and Industry is now faced with further analysing and designing policies related to the clusters of competence. A new idea is for the Ministry to form a taskforce supporting the self-knowledge of clusters as well as stronger co-operation between national and regional policy makers.

Children's Play & Learning is the first cluster of competence with dialogue experience. It was a development project for this kind of policy making. The same approach will not necessarily be taken for the other clusters of competence as they are very different in nature. However, many of the elements of the process provide valuable experience for further work with the Danish clusters of competence.

Figure 3. **Danish clusters of competence: 29 examples**

	National	Regional
Existing	• Thermal technology • Technical appliances for the disabled • Pork • Dairy products • Water environment • Fur • Seed-growing • Power electronics • Hearing aids • Wind technology • Maritime industry	• Mobile/satellite communication in Northern Jutland • Business Tourism in the Capital region • Stainless steel in Eastern Jutland • Horticulture at Funen • Pharmaceuticals in the Oeresund region • Textiles/clothing in Herning-Ikast • Offshore industry in Esbjerg • Furniture in Salling • Transport in Eastern-Southern Jutland
Emerging	• Organic food • Children's play & learning • Waste management • Sensor technology • Bioinformatics	• Movies/broadcasting in the Copenhagen region • Oeresund Food Network • PR/Communication in the Copenhagen region • Pervasive Computing in Copenhagen and Aarhus

The notion behind the Children's Play & Learning cluster is an acknowledgement of the extensive expertise which exists in Denmark in terms of children's development and needs. There is a long pedagogical tradition in Danish day-care institutions, with extensive research being carried out in this area, as well as the competitiveness of Danish companies producing high-quality toys, etc., for children. However, until recently there had been limited co-operation among the various private-public actors in this field. Box 5 illustrates the general process of the Danish cluster project, from analysis to policies.

Box 5. **From cluster analysis to policy**

- Use analysis to identify possible actions.
- Set up focus groups to formulate policy proposals: generate ideas with firms/organisations/ researchers and clarify potential.
- Create policy initiatives and co-operate with other ministries.
- Plan the budget.
- Follow up on the initiatives.

Due to the way in which the cluster is defined, generating ideas for improving the critical framework conditions for Children's Play & Learning involves many different actors. The working group included firms ranging from, for example, minor IT producers, kindergarten services and the media corporation Egmont, to LEGO and the national broadcasting channel DR. It also included organisations such as the Danish National Federation of Early Childhood and Youth Education, and the Industrial association for Producers and Importers of Products for Children. Finally, researchers working on childhood education and on the linkages between children's learning and toys were also involved.

In the policy-making process which followed, various ministries were involved once the ideas were formulated; these included the Danish Ministry of Social Affairs and the Danish Ministry of Education. To implement the policies, the proposals have to be put to the various Danish ministers for policies involving their jurisdiction. The same holds for the budget.[7] To ensure that the initiatives achieve their full potential, the Ministry of Trade and Industry follows up on the process, an important and demanding task.

The Danish Ministry for Trade and Industry has released a report showing the results from the first year's activities with Children's Play & Learning; seven policy proposals are now being implemented (Danish Agency for Trade and Industry, 2000a). A general result of the work with Children's Play & Learning was establishing new mutually beneficial linkages inside the cluster. The policy initiatives are shown in Box 6.

Box 6. **Policies for Children's Play & Learning**

- Private-public research consortium in Learning Lab.
- IT master education for pre-school teachers.
- Guidelines for private-public co-operation.
- Kindergarten of the future.
- National IT strategy – focus on children.
- Business PhDs for small companies.
- Support for the development of Danish movies for children.

The learning lab is inspired by the MIT Media Lab; it represents the first arena for co-operation among the different parties in the cluster and is financed by the companies themselves. The IT training of pre-school teachers is a new educational initiative, combining different competences to develop new approaches. The guidelines for private/public co-operation are most important because of the ethical aspects of this area; they provide a good learning example for more private/public co-operation to come. The kindergarten of the future will focus on the development of state-of-art equipment and services.

The report on Children's Play & Learning immediately made the front page of one of the largest Danish newspapers, a clear illustration of the broad appeal of this approach. This highlights the importance for government to add some PR value to industrial policy and build on goodwill, both as a prerequisite and as part of a cluster approach, and it provides a good example of diffusing industrial policy in areas of relevance to a broader audience.

Another important policy-making experience from Children's Play & Learning was the benefits to be had from working with a much smaller group than was normally the case, thus implying closer relations and a better visibility. There were positive reactions from all the parties involved, including some rather unexpected praise.[8] The cluster actors were delighted to find new partners for their development projects. Demonstrating the common interest to the cluster actors is very important in any policy-making process.

However, creating new policies in this way is extremely resource-intensive. This is especially the case where there was no prior meeting point for the actors involved. The resources required of policy makers include a great deal of creativity, clarification of the different competences and interests, and the need to pass on some tasks to others at the right time. It is vital that the right people be involved in the project, and it is necessary to have a realistic time horizon and planning. The cluster policy makers play the most active role and must not expect too much from others; this involves setting the agenda to ensure that the process moves forward in a focused way.

Concluding remarks on policy learning

The experience of the Danish Ministry of Trade and Industry shows that the concept of clusters deals not only with innovation in industries, but also with innovation in policies. This implies a number of major challenges for policy makers.

First and foremost, the cluster approach involves transforming ideas from firms and other cluster actors to specific policy initiatives within existing – and sometimes quite restraining – structures such as the budget and the political system.

Second, policy making involves a number of dilemmas. One is to what extent the entire political system should be focused towards clusters. An educational or research policy can be too specialised, leading to less flexibility for the needs of the future, and resulting in a lack of the critical mass needed to become self-sustaining. This is especially true in relation to clusters of competence. Also, the extent to which the policies should reflect the needs of the clusters today, compared to future needs, is a risky business. These are difficult dilemmas to solve at a general level. A thorough understanding of the clusters and dialogue with the companies involved can help to find a balance between specific and general policies, as well as between present and future needs.

Third, a broad acceptance and participation in the cluster activities is crucial. The creditability and impact of the cluster analyses and of the policy makers involved are very important in this context. Promoting the cluster idea in a broader context for industrial policies, and finding the right partners and timing mean bringing different actors together by illustrating the potential benefits of closer co-operation. An important task for cluster policy makers is to change attitudes and improve the awareness of trade and industry and the traditional organisations. It is vital to create a joint agenda for industrial development, thus motivating companies to take their own actions.

Fourth, there is an enormous task related to cluster policy making within the political system itself. The Danish experience has shown the need for greater co-ordination on a horizontal level between different political jurisdictions, as well as vertically between state agencies and regional agencies, in policy making for the Danish clusters. Cluster policies should not be distinct from other policies, but rather should form the "backbone" of our understanding of trade and industry. Here, there is still room for improvement in Denmark.

As regards co-operation, utilising and adding to international cluster experience is required, since many clusters have influential international dimensions. This could imply, for example, cross-border policy co-operation for cross-border clusters, and learning about "good practice cluster environments" elsewhere.

Finally, it should be acknowledged that developing cluster approaches and policies is rarely a rational linear process. Policy making related to clusters involves a great deal of "trial and error". What is important, then, is to evaluate experiences as systematically as possible. This can also help in designing specific approaches for each cluster, depending on the geographical and industrial context.

The Danish approach is pragmatic and flexible, adjusting to the possibilities, demands and the constantly changing reality. Past experience of cluster policies is captured on a rather informal basis and there is still room for improvement, although the Danish cluster project will always focus more on setting things in motion than on describing each of the steps taken. However, there are increasing demands – and also very good reasons – for policies to be well managed and the results measurable.

There is no magic recipe for the success of cluster policies, it is a constant process of learning and improving. Most important for the cluster approach is to create awareness and a constructive debate about the situation in each cluster. The ideal outcomes are a common agenda and policies that strengthen the competences and competitiveness of the clusters.

NOTES

1. In the national economy, stable macroeconomic (*i.e.* inflation and equilibrium) and general business conditions (*i.e.* stock exchange regulations and corporate taxes) are of greater importance for firms than are specific framework conditions. In a global economy, on the other hand, the specific framework conditions (*i.e.* competent venture capital institutions, world-class education and research environments) of each region take on greater significance since other conditions tend to be harmonised across countries.

2. However, the Danish Ministry for Trade and Industry set the guidelines for the analyses since they are designed to serve as inputs to the policy dialogue. This implies that the reports must be reliable, recognisable, future-oriented and able to stimulate debate. The cluster analyses involve a SWOT analysis of economic performance, structural conditions, market demands, competitive advantages, specific framework conditions and international benchmarking. Identification of the key barriers faced by each of the clusters is at the centre of the cluster analyses and dialogue.

3. Production organisation in a "New Economy" could indicate that vertical clusters are split up by global sourcing, making the clusters more horizontal with other cohesive forces than pure trade relations.

4. A number of Danish researchers and consultants were involved in early studies of strongholds in Danish industry (Porter, 1990). The Danish knowledge base on clusters also embraces a range of Scandinavian research and theories. See also Drejer *et al.* (1999), for a thorough introduction to the background for Danish clusters.

5. In Spring 2001, the Danish Ministry for Trade and Industry published reports on the three mega-clusters, Healthcare, IT/Communications, and Metal industry/Processing technology, all of which will be followed by new policy initiatives.

6. As such, they are the true expression of the cluster definition given in Box 2.

7. Implementation of policies is generally financed through the national Budget. Some policies are adjustments of existing initiatives, a less costly option. Some are financed by the Ministry of Trade and Industry, others by other ministries. There is no automatic funding for policies and requests for funding have to be put forward on a case-by-case basis.

8. For example, the union involved had no prior experience of involvement in this kind of policy formulation.

REFERENCES

Danish Agency for Trade and Industry (2000a), "Børns Leg og Læring. Initiativer til et styrket erhvervsområde" [Children's Play and Learning. Initiatives for a Stronger Business Domain], Danish Agency for Trade and Industry, Copenhagen.

Danish Agency for Trade and Industry (2000b), "International benchmarking af dansk erhvervsliv. En statistisk analyse af de danske ressourceområder" [International Benchmarking of Danish Businesses. A Statistical Analysis of the Danish Mega-clusters], Danish Agency for Trade and Industry, Copenhagen.

Danish Agency for Trade and Industry (2000c), "Erhvervsudsigten: Trends der vil påvirke dansk erhvervsliv de kommende 10 år. Analyse af de danske ressourceområder" [The Business Forecast: Trends that will Influence Danish Business Life over the Next 10 Years. Analysis of the Danish Mega-clusters], Danish Agency for Trade and Industry, Copenhagen.

Danish Agency for Trade and Industry (2000d), "Bygge/Bolig – en erhvervsanalyse" [Construction/housing – A Business Analysis], Danish Agency for Trade and Industry, Copenhagen.

Danish Agency for Trade and Industry (2001) "Kompetenceklynger i dansk erhvervsliv – En ny brik i erhvervspolitikken", Danish Agency for Trade and Industry, Copenhagen. Translated edition: "Clusters of Competence in Denmark – A New Aspect of Industrial Policy", Danish Agency for Trade and Industry, Copenhagen.

Danish Ministry of Trade and Industry (2000), "Government Industrial Development Strategy .dk21", Danish Ministry of Trade and Industry, Copenhagen.

Danish Ministry of Trade and Industry and Danish Ministry of Housing and Urban Affairs (2000), "Byggeriets Fremtid" [The Future of Construction], Danish Agency for Trade and Industry, Copenhagen.

Drejer, I., F.S. Kristensen and K. Laursen (1999), "Studies of Clusters as the Basis for Industrial and Technology Policy in the Danish Economy", in OECD (1999), *Boosting Innovation: The Cluster Approach*, OECD, Paris.

Porter, M.E. (1990), *The Competitive Advantage of Nations*, Free Press, New York.

Porter, M.E. (2000), "Location, Competition and Economic Development: Local Clusters in a Global Economy", *Economic Development Quarterly*, Vol. 14, No. 1, pp. 15-34.

Chapter 19

TOWARDS SECOND-GENERATION CLUSTER POLICY: THE CASE OF THE NETHERLANDS

by

Victor A. Gilsing*
Dutch Ministry of Economic Affairs

Introduction

In the early 1990s, the Netherlands was one of the first countries to adopt a cluster perspective as part of its innovation policy. Over the years, the cluster concept has grown into a concrete range of policies covering a broad set of instruments. Due to the novelty of the concept, a great deal of experimentation and learning has taken place over the years, allowing a number of policy lessons to be drawn. These lessons mainly concern the Ministry of Economic Affairs as the involvement of other ministries in cluster policy has been limited to date.

In this chapter, we will first provide an overview of Dutch cluster policy, based on which we will discuss some of the most important lessons learned. From these lessons and new insights, we will discuss how Dutch cluster policy is being upgraded to become a second-generation cluster policy framework. We will then briefly highlight the challenges facing the new framework. The chapter ends with some conclusions.

Overview of Dutch cluster policy

The cluster concept was initiated in 1991. Inspired by Michael Porter's *Competitive Advantage of Nations* (1990), the Netherlands has been among the first countries to shift from an industrial, sectoral policy approach to a cluster-based approach. A number of studies followed with the aim of making the cluster concept more transparent to both firms and policy makers. In 1993, *Industrial Policies in the 1990s*, together with the policy paper, *Competing with Knowledge*, further elaborated on the newly developed notions by focusing on the importance of co-operation, both between firms as well as with the public knowledge infrastructure. In 1995, *Knowledge in Action* set out the points of innovation-oriented policy in greater detail and stressed the importance of building closer links between public and private research. Simultaneously, a number of R&D subsidy schemes were developed. While the

* The author wishes to thank Dr. Paul M. op den Brouw of the Ministry of Economic Affairs for useful comments. The text presented here is the sole responsibility of the author and does not necessarily reflect the view of the Dutch Ministry of Economic Affairs. The usual disclaimer applies. The author can be contacted at: v.a.gilsing@minez.nl.

original focus of these schemes was on stimulating co-operation between firms, they were later broadened to encompass technological co-operation within clusters. In 1997, a further refinement was introduced, with the definition of the three policy roles described below.

Framework policy

Framework policy is based on the notion that innovative clustering can only prosper in a sound macroeconomic environment. The objective of framework policy is to create a favourable and stable entrepreneurial climate. Only then can firms focus on improving their competitiveness and innovation potential. More specifically, the framework policy comprises four elements:

- Competition policy and deregulation.
- General technology policy.
- Macroeconomic policy.
- Solid and reliable physical infrastructure.

Organising stakeholders

This policy deals with the identification and stimulation of innovative clusters. It aims to create a meeting place for businesses, institutions and government agencies. However, it is the responsibility of cluster participants to decide on the contents of such meetings. More specifically, this role is fulfilled through various instruments:

- *Strategic information.* A cluster crosses sectoral boundaries; thus, it is not always easy to understand its (changing) structure. Government should try to enhance awareness by providing strategic information. Such information is often generated through close interaction and information exchange among policy makers, firms and knowledge institutes. It may cover a wide range of topics, ranging from new market developments, technological progress, other forms of co-operation, non-technological innovations, to the impact of the Internet, etc.

- *Organisational capacity.* In addition to information, government can also become involved through its organisational capacity. A cluster may have a need for a "process manager", capable of facilitating co-operation among firms. Due to the boundary-crossing nature of clusters, firms may not always know each other very well, in which case government may act as a neutral mediator. Examples of relevant instruments include brokerage events, conferences and platforms.

  The above instruments of strategic information and organisational capacity are aimed at specific clusters.

- *Subsidies.* In addition to the above-mentioned tools, subsidy schemes are available to provide incentives for R&D undertaken in co-operation with other firms or (public) research institutes.

  In this respect, subsidy schemes are generic in scope and target all clusters.

Innovative public procurement policy

Through its role as a demanding customer, government can stimulate – and sometimes even initiate – innovative clustering by matching specific societal needs. If they are to provide suppliers with incentives to innovate, procurement policies need to articulate specifications which are more functional than technical.

Following this brief illustration of how cluster policy has evolved to become an essential feature of policy making, we will now take a more reflective stance and turn to the "lessons learned".

Lessons learned

As stated above, clusters represented a novel concept in Dutch innovation policy. There was thus a strong need for learning from other countries' experiences. At the time, few countries had adopted the cluster approach, thus limiting the opportunities for benchmarking and mutual learning. Substantial policy experimentation and learning has taken place, which can be characterised as "learning-by-doing" rather than learning through analysis or planning. Over the years, this process of policy experimentation and learning has allowed a number of lessons to be drawn.

A first lesson is that clustering is a market-induced process which, in the end, should contribute to the "bottom line" of the participating firms. Depending on the cluster, there may be a role for government. Another important lesson regarding this public role is the fact that there is no single "best way" to cluster policy. A "one-size-fits-all" approach is usually counterproductive. Therefore, over the years, various policy roles have developed (Figure 1).

Figure 1. **Different roles in different clusters**

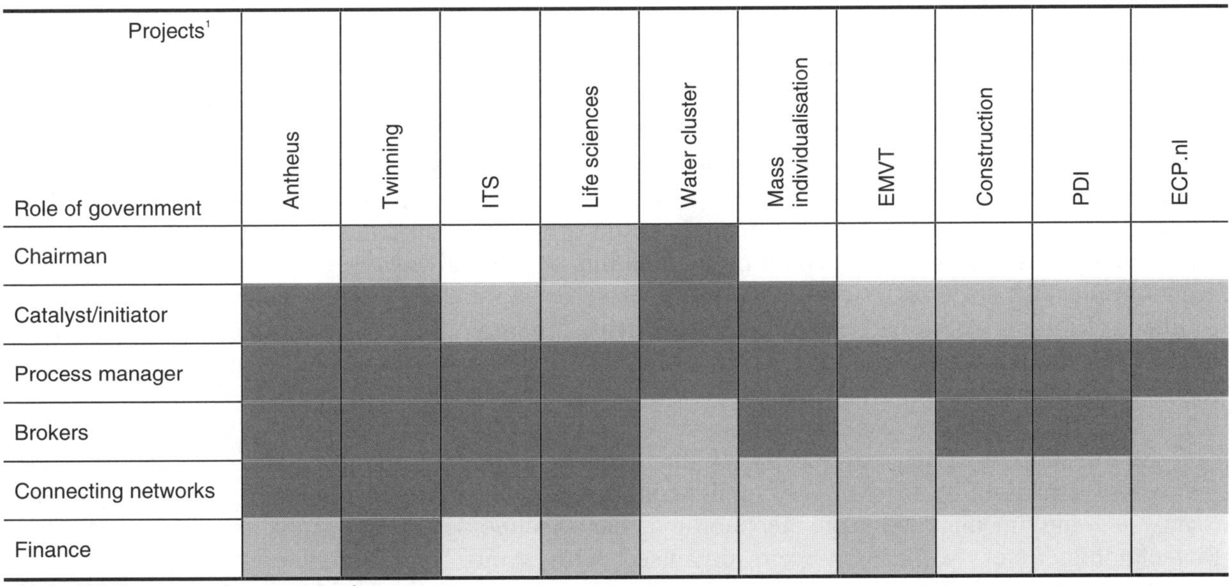

Note: White = no role; grey = role.

1. Antheus is a regional cluster project at the micro level, aimed at increasing co-operation between a large aluminium plant and the smaller (aluminium-using) firms surrounding it. ITS stands for Intelligent Transport Systems. EMVT is the Dutch abbreviation for Electro Magnetic Power Technology. PDI stands for Product Data Interchange, a project mainly aimed at supporting this technology in the chemicals cluster. The other projects are described in the main text.

As stated above, strategic information and organisational capacity are key elements in "organising stakeholders". The roles shown in Figure 1 provide a more detailed breakdown of these two elements.

Figure 1 shows that government fulfils different roles in different clusters. We will elaborate on these different roles using three examples (without providing a detailed description of each of the projects). In the case of ECP.nl, a group of Dutch banks and retailers set out to create a platform for electronic commerce, the goal of which was to build leverage for electronic commerce by addressing some technical issues (*e.g.* reliable payment systems) and to stimulate confidence among professional users and consumers. Despite all the efforts, this platform did not materialise. The initiators approached the government to facilitate this process which finally resulted in the creation of a platform, E-commerce Platform Nederland.[1] In this case, the role of government was mainly one of managing the process among stakeholders.

The opposite example of a top-down approach with substantial governmental involvement is illustrated by the case of Twinning. The policy objective was to increase the number of technological start-ups in the ICT field. The government created an incubator to support young ICT start-ups by providing funds and managerial support as well as offering various (technical) facilities.[2] It thus fulfilled a broad range of roles which ranged from being a prime initiator, acting as a broker, connecting different networks (banks, universities, entrepreneurs, venture capitalists) to providing finance.

In the case of the Dutch water cluster, the government acted more as a chairman and catalyst. Historically, the Netherlands has strong technological capabilities in the field of water treatment and management. However, very few of these skills and services were exported. To address this issue, government, in close co-operation with industry, founded a new firm which co-ordinates export activities from the Netherlands. Due to the small size of most of the firms in this cluster, the ministry acted as chairman of the process.

Each cluster is unique and the appropriate role for government depends on the specific context of the cluster. Regardless of the approach followed in a specific cluster, it is not always easy in practice to ensure that the activities aimed at strengthening a cluster are actually carried out. Even where cluster participants and government agree on a shared *vision* for strengthening their cluster (often after some lively discussion), *follow-up* is not always automatic. Follow-up is a shared responsibility, which explains why not everybody feels sufficiently responsible for such collective action. In some cases, intermediary parties can play a facilitating role but this may not always be a viable option. Often, such organisations take a sectoral view and do not necessarily represent a cluster. Therefore, in organising follow-up, a more viable path is to collaborate with "champions", *i.e.* firms or institutes willing to invest (time, activities and, sometimes, money) in their cluster. As the results of these efforts become more visible, the interest of other stakeholders will be awakened and their involvement may grow.

An illustration of the complexity of this process is the follow-up to the Technology Radar workshops organised by the Ministry of Economic Affairs in 1999. In several key technology areas (catalysis, multimedia, language technology, nano-technology, data and knowledge systems, biotechnology, etc), workshops were organised with senior representatives from business and knowledge institutes. The aim of these workshops was to improve the linkages between the two parties. The workshops were well prepared and most were highly successful. In the majority of cases, follow-up went smoothly, although in a few cases it did not materialise at all. In the latter case, an in-depth evaluation study revealed that perceptions and expectations with regard to follow-up diverged between workshop participants, on the one hand, and the ministry, on the other. The perception of the ministry was that things had been made sufficiently clear in the final workshop and that industry and the public knowledge infrastructure would pick up on the recommendations. The reason why this did not happen to the extent expected is related in part to our earlier observation about the difficulty of shared responsibility. Another reason is that some clusters were too broadly defined, making it difficult to agree on follow-up which be beneficial to all stakeholders (Willems and van den Wildenberg, 2000). This example provides a valuable lesson; namely, that it is important to understand which parties are core players in the cluster and which belong on the periphery. Depending on the type of cluster, as well as on the specific issues at hand, it may not be essential to involve all of the parties.

With the benefit of hindsight, we can observe a tendency to become more involved in "high-tech" clusters, such as multimedia, biotechnology, nano-technology, etc. Although the importance of such new clusters is beyond doubt, government does not always have a role to play. At the same time, more traditional clusters may sometimes receive less attention than is warranted. An example is the construction cluster in the Netherlands which comprises a value chain consisting of construction firms, specialised suppliers, architects and (professional) users. Total turnover is estimated at NLG 100 billion (about 10% of total Dutch gross national income). It is estimated that about 10% (NLG 10 billion) of this total turnover can be attributed to so-called "failure costs": substantial macroeconomic losses originating from the inability to meet technical and customer requirements.[3] The reasons for these failure costs do not pertain to the construction firms only. Rather, they stem from the way in which firms in all parts of the value chain co-operate as a cluster (den Hertog and Brouwer, Chapter 10 of this volume). Cluster policy does address these issues but, in general, the measures taken are less effective than those aimed at more high-tech clusters.

The example of construction shows that innovation is not always related solely to technology but may stem from non-technological innovations as well. Such innovations relate to innovations in, *inter alia*, marketing, strategy or the organisation of inter-firm co-operation (*e.g.* through alliances or joint ventures). An example of cluster policy which explicitly deals with non-technological innovations is in the machine-building cluster. This cluster is formed by large OEM firms such as Stork, ASML and Oce and various (main) suppliers. Traditionally, the relationships between subcontractors and suppliers were strongly focused on constantly lowering costs. The growing need for innovation requires a shift in direction away from a sole focus on efficiency towards innovation. To this end, some cluster champions, together with the ministry, have constructed a business roadmap: a joint vision of the future has been developed and translated into a novel model of co-operation which encourages cost efficiency as well as knowledge sharing, learning and innovation.

Although these are examples of non-technological innovations within a cluster, we believe that cluster policy in the Netherlands has a certain bias towards technological innovation in technology-based clusters. In this respect, non-technological innovations in clusters remain relatively unexplored territory in the Netherlands.

Another lesson which follows from the above observations is that it is useful to distinguish between newly emerging clusters *vs.* more mature clusters. In general, the challenge in dealing with newly emerging clusters is that the boundaries are not easy to determine. This sometimes makes it difficult to identify all the relevant players, especially as formal institutions such as industry associations may not yet be present. These formal structures often abound in more mature clusters, although this may pose another problem. Such formal institutions are often defined according to sectoral structures rather than cluster structures and may not necessarily represent a relevant cluster community. Due to the small scale of the Netherlands, the role of various regional intermediaries is often limited. The value chain of a cluster tends to cross regional boundaries – which can mean that a regional association does not naturally view itself as the "owner" of any follow-up.

Finally, a further lesson is that the period of extensive policy experimentation is now coming to an end. Increasingly, cluster policy should consistently deliver its promises, *i.e.* the results should justify the inputs. The results of cluster policy will differ depending on the type of cluster. To better understand how cluster policy can be further improved, it is useful to further elaborate on the differences among clusters and the subsequent varying governmental roles.

Different roles in different clusters

As already argued, an important lesson to be drawn from the exercise is that differences among clusters can be substantial. We have already differentiated between emerging, high-tech clusters *vs.* more mature, traditional clusters. These characterisations are useful but remain very general. At a deeper level, clusters can differ along various dimensions. One important dimension is demand, which can range from homogeneous to differentiated or from advanced to more standard (Porter, 1990). Another important dimension is the type of knowledge in a cluster. Relevant elements are, for example, whether knowledge is tacit *vs.* more codified, or embodied *vs.* disembodied (Malerba and Breschi, 1997). An interesting example is the Dutch multimedia cluster. The newness of the technology means that knowledge is often highly tacit in nature and hence resides in people's heads rather than in codified form (*e.g.* in documents). Not surprisingly, the multimedia cluster shows strong regional concentration. Geographical proximity allows for quick and easy interactions among people, thereby facilitating knowledge sharing, learning and innovation.[4]

In addition, differences can also relate to the ways in which clusters create new knowledge. Science-based clusters are at the forefront of new technological developments and constantly generate new knowledge. Other clusters mainly absorb knowledge from outside the cluster (Roelandt *et al.*, 1999). An example is the Dutch construction cluster: although there are some innovative segments such as bridge and road construction, in general this cluster does not have an impressive track record in terms of innovation (den Hertog and Brouwer, Chapter 10 of this volume). The ability to generate new knowledge from within is only just developing. Therefore, new knowledge tends to come from outside the cluster. This new knowledge is often embodied in new materials, thus making the suppliers of these materials very important to the construction cluster. The location of these suppliers is less relevant and many are established outside the Netherlands (Belgium, Germany, the United Kingdom).[5] From a spatial point of view, clusters can be concentrated in a region or even a city but can also cross national borders (Enright, 1999).

Thus, the context in which a cluster operates may vary substantially. Following the innovation systems approach, various types of systemic imperfections may occur in these learning and innovation processes. Systemic imperfections refer to hindrances in the interactions both among firms and among firms and non-firm actors such as knowledge institutions. Such obstacles may constrain the sharing of knowledge and its further development, and hence the innovation performance of a cluster. The systemic imperfections concept is broader than the "old" market imperfection argument since its focus is on the functioning and performance of an innovation system as a whole (OECD, 1999). The goal of cluster policy is to address these systemic imperfections and in particular:

- *Limited interaction.* "Firms do not innovate in isolation": interaction among them or with non-firm actors is important. When firms do not interact, potentially fruitful opportunities for learning and innovation may be missed. There may be various reasons why firms do not interact. First, firms may not be aware of the presence of other firms relevant to their business lines. In this case, an appropriate policy answer is to improve the cohesiveness of the innovation system. Relevant instruments in this respect are, for example, a brokerage event or a Web site allowing firms to learn about one another's existence and capabilities. Another reason may be that firms are cognitively too distant, *i.e.* their knowledge bases do not overlap. In this case, an appropriate policy response may be to facilitate the development of a common infrastructure to enable firms to communicate. Elements of such an infrastructure include, for example, technical standards or norms, joint experimentation platforms or joint programming of scientific research.[6]

- *Informational imperfections.* These imperfections refer to a situation in which lack of insight into future technological and/or business developments constrains the innovative potential of a cluster. An appropriate policy response is to develop, together with industry and knowledge institutes, a common understanding of these key trends,. This can be done by means of foresight studies, technology radars, market studies or technology roadmaps.

  Technology roadmaps are a relatively recent tool which allow future technological developments to be planned. By portraying how a specific technology is likely to develop over the next five to ten years, the cluster actors can develop the "path" along which future R&D should be aligned and can identify the (private and public) partners with which they could usefully co-operate. To be effective, a roadmap must be developed in close interaction with all relevant actors. This implies that the above-mentioned "process role" for government is critical and must address the issue of trust among cluster participants. However, trust cannot be built from scratch but needs to exist already. This makes technology roadmaps more suitable for clusters where a certain level of trust already exists than for clusters which consist of arms' length relations between wary parties. In addition, technology roadmaps appear to be more suited to relatively established technological fields which can, to a certain extent, be projected into the future based on technical standards and/or rules of thumb. We view roadmaps as being especially appropriate for clusters such as energy,[7] semiconductors or catalysis.[8]

- *Mismatch between knowledge infrastructure and business needs.* This type of systemic imperfection refers to a situation in which (public) knowledge institutes develop knowledge which is too far removed from business needs. This imperfection may manifest itself in various ways. Knowledge may be too "scientific", making it very difficult to judge whether there is a potential market for it. Knowledge may also be sufficiently application-oriented but not commercially promising. A more complicated case relates to the institutional set-up of the knowledge infrastructure, especially when knowledge is systemic as is the case for ICT. The systemic ICT knowledge base means that the development of a new ICT application requires the integration of various key technologies, such as information technology, database technology, (mobile) telecommunication technology, language technology, intelligent agent technology, content management technology, etc. The systemic character of ICT means that changes in one technology often require adaptations in others. The incentive structure of the public knowledge infrastructure rewards academics who achieve excellence within a specific functional area rather than those who build multidisciplinary bridges. Firms often do not have the capabilities (*i.e.* expertise, time) to integrate these technologies in order to be able to develop new products and services. Therefore, even in the case of excellent public knowledge on a particular technology, it may be economically unsound to invest time, money and expertise in adopting the new technology due to the required integration with adjacent technologies. In order address this situation, a number of "Technological Top Institutes" were founded in 1997. These are bridging institutions which aim to connect the worlds of academia and business in order to secure knowledge transfer. They cover technology areas offering potential growth opportunities for Dutch firms. The bridging institutions allow firms, often by means of a consortium or as a cluster, to ensure that research is aligned with business needs.

- *Lack of demanding customers.* Demanding customers are customers who are ahead of the mass market for a particular technology or novel application and in this respect may provide powerful incentives for firms to innovate. Such requirements may speed up internal learning processes and enable firms to build up a strong competitive position as soon as a mass market develops. Especially in public areas such as construction, environmental issues, education, safety and energy, there may be substantial potential for such procurement policy. The usefulness of such policy in boosting the innovation performance of other clusters is certainly worth exploring.

In conclusion

What becomes clear is that differences among clusters can be substantial. The existence of cluster specificities suggests that neutral policies are likely to be so abstract and general that they will have little effect within the distinct structures of a specific cluster (Edquist *et al.*, 1998). We argue that cluster policy should be designed with such specificities in mind, thus calling for substantial analytical capabilities. A framework is therefore needed to allow us to better understand these system specificities and to identify possible "systemic imperfections" in order to design policy. In the following section, we will elaborate on the insights gained and lessons learned, placing them into a policy perspective which will result in such a strategic framework.

From 2000 and onwards: a strategic framework

A strategic framework has been constructed to act as a support tool for policy makers in developing policy for a particular cluster. This framework serves three specific goals. First, it serves a managerial goal, allowing policy makers to decide on the resources to be allocated to a particular cluster. Second, it serves as a support tool to policy makers who are operationally involved in cluster policy by enabling them to systematically develop policy tailored to the cluster in question. It is important in this respect to ensure that the strategic framework does not become a straight-jacket but rather functions in a flexible way. Third, the goal of the strategic framework is to make the cluster policy process transparent to the outside world, not only to firms but also to Parliament. In this respect, it responds to the growing demand for public sector accountability. The model comprises three sequential phases (shown in Figure 2).

The information phase implies identifying those clusters that are of importance to the Dutch economy. In the initiation phase, involvement grows as interaction with business is needed to help solve the question of how systemic imperfection(s) may be addressed. If there is a case for governmental involvement, the third phase represents the implementation of that role. Following the lessons learned earlier, the appropriate policy tools will differ depending on the phase.

Phase 1: Information

In this phase, the policy question is how to identify which proposals or initiatives for clustering are of potentially strategic importance to the Dutch economy. Most topics will have a certain bias towards new technologies, although newly emerging market trends will also be covered. Due to the focus on new-to-the-world technologies or business trends, limited information will be available and a value chain is often not yet clearly delineated. Networks of firms will barely exist, if they exist at all.

Figure 2. **Strategic framework**

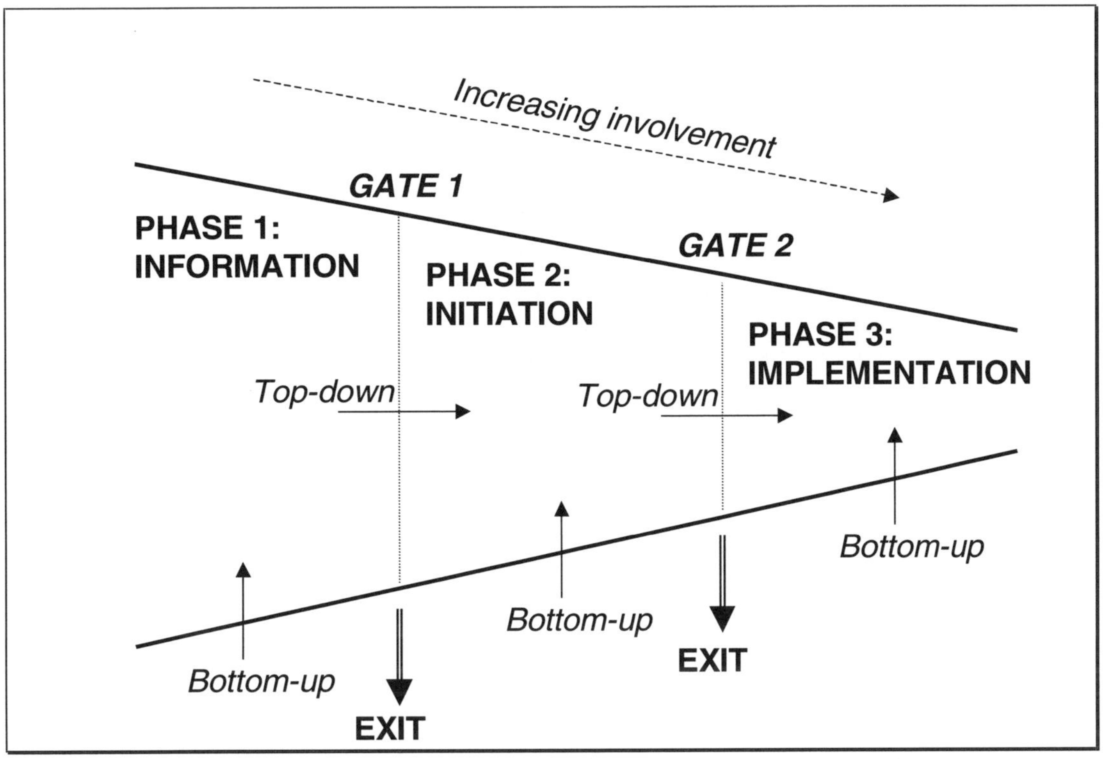

The main responsibility from a policy point of view is to systematically collect strategic information from a variety of information sources. The use of different information sources is very important in order to ensure that the new topics are covered as broadly as possible as well as to undertake the necessary cross-validations. The various information sources are the following:

- Scientific, technical and economic literature.
- Internet.
- Strategic market studies (*e.g.* benchmarks).
- Technology studies (*e.g.* technology radar, technology foresight exercises).
- Knowledge position studies.
- Cluster studies.
- External contacts with industry associations, platforms, consortia, etc.

The input for this phase can be generated either by policy makers (top-down) or by industry. An example of a top-down approach is the technology-radar study undertaken by Rand Technology in 1997 (Rand Corporation, 1997). This study reviewed 15 key technology areas of relevance to the Dutch economy. A bottom-up example is the foresight study conducted by STT on EMVT in 1998. This study extensively examined the emerging field of a new energy technology and highlighted its economic potential. As a result, EMVT was added to the policy agenda.

Although the exploration of these new trends may be a rewarding experience, it is not a goal in itself. In the end, it should give policy makers a better understanding of the key elements and provide a sufficient basis for moving on to the first "gate".

Gate 1: Towards the initiation phase

The objective of this gate is to assess the ideas and trends generated by the initiation phase for which the following criteria are used:

Cluster assessment. A first set of criteria relates to the potential importance of a cluster. This importance is determined by its potential size and its ability to generate or enable innovations in other clusters. More specifically, we use the following criteria:

- Potential:
  - Size;
  - Value added;
  - Growth rate.
- Innovative contribution to the overall Dutch economy:
  - Strengthening of Dutch competitiveness;
  - Innovative impact on other clusters.
- Interaction with other clusters:
  - Spin-offs to other clusters;
  - Enabling innovation in other clusters.

Assessment of the potential role of government. This second set relates to the question of whether government should legitimately become involved in a cluster. More specifically:

- A cluster is not able to capitalise on its innovative potential, *i.e.* a systemic imperfection occurs:
  - Informational failures;
  - Limited interaction;
  - Mismatch between knowledge infrastructure and business needs;
  - Lack of demanding customers.
- Potential ability of government to address the identified systemic imperfection through the use of one or more policy tools.

The two sets of criteria are cumulative; therefore, a cluster proposal should meet both sets. If a (potential) cluster is assessed positively on both these criteria, it will move on to the initiation phase. However, where a cluster initiative is positively assessed as a cluster but does not show systemic imperfections, it will *not* be selected to move on to the next phase. In other words, where a cluster has economic potential but no imperfections exist, there is no added value to be had from governmental involvement. Such a cluster does not merit further policy attention at this point in time and is thus taken off the policy agenda.

Phase 2: Initiation

This phase aims to assess how the identified systemic imperfection(s) can best be tackled by the relevant stakeholders. There is an emphasis on the collection of strategic information. Although this information still covers some explorative search, there is an increasing focus on whether the participating firms and institutions are ready and willing to invest. This requires close interaction between all the relevant actors; involvement by policy makers as well as by firms is growing. This means that the appropriate tools are different from those used in the information phase:

- Knowledge and technology instruments:
  - Studies for innovative research programmes;
  - Technology workshops;
  - Technology roadmaps;
  - Next-generation analysis and scenarios.[9]
- Business tools:
  - Business roadmaps;
  - Benchmarks;
  - Focused market studies (especially for bottom-up studies);
  - Benchmarking studies.
- ClusterMonitor[10]

An important element of these tools is their interactive nature. They form a combination of (quantitative) desk research and direct interaction with business.

As has been said, designing policy for a cluster requires substantial analytical capabilities. In this respect, the tools aim to bring an in-depth understanding of the key issues for policy implementation regarding a specific cluster and the potential role for government. Based on this, the step towards gate 2 follows logically.

Gate 2: Towards the phase of implementation

This gate aims to assess the initiatives, projects, etc., from the previous phase. The goal is to assess whether the removal of systemic imperfections requires government involvement. The following criteria are used:

- Level of urgency:
  - Declining cluster performance.
- Associated risks:
  - Sufficient commitment by all involved;
  - Willingness to invest (funding, time, resources);
  - "Damage" in the event of failure.
- Required input:
  - Level of financial resources;
  - Level of human resources.
- Sufficient return on public investment:
  - Collective interest should clearly outweigh the interest of (groups of) individual parties.
- Professional, well-thought out approach:
  - Appropriate business plan.

In this case, too, the sets of criteria are cumulative: a cluster that does not meet all the criteria will not be selected and will disappear from the policy agenda.

Phase 3: Implementation

This phase aims to remove the systemic imperfections identified earlier. These include: limited communication, lack of information, restrictive regulations, etc. Removing such imperfections may require substantial government involvement, with various roles for government:

- Brokerage function.
- Foundation of platform organisations:
  - Twinning as an incubator for new ICT start-ups;
  - Bio-platform Life Sciences;

  - Steering Group for Catalysis.

- Providing strategic information:

  - Through the Web site: www.clustermonitor.nl;

  - Through conferences, such as those on "Competing with ICT Capabilities" or on "Life Sciences".

- Removal of constraining regulatory conditions.

Here, cluster policy is directly connected with policies from other ministries. Examples include heavy regulation in the construction field and the lack of a regulative, normative framework in the field of biotechnology.

Present status of the framework

In the above, we have outlined the objectives and elements of our strategic framework. At the moment of writing, the framework has been officially adopted as the future direction for cluster policy. As a next step, it needs to be implemented in order to prove its value. Obviously, it needs further fine-tuning which can only be attained by using it in the daily practice of cluster policy. In this process of fine-tuning, some elements of "first-generation" policy will remain, others may be phased out and new elements added. An in-depth evaluation of the effectiveness of cluster policy over the past five or six years is scheduled to start soon and will be finalised by the end of 2001.

Conclusions

In this chapter, we have described how the cluster concept of the early 1990s has grown to incorporate a substantial range of policy tools. The development process has co-evolved with experimentation and policy learning, resulting in a number of lessons from which we have been able to learn. Putting these lessons, and our insights on how clusters can differ, into a policy perspective, we have been able to construct a strategic framework.

This newly developed framework marks the transition between a period of variety and experimentation towards a more structured and integrated approach, *i.e.* a second-generation cluster policy framework. The framework starts from the notion that clusters are different and that a "one-size-fits-all" approach is not effective. A solid, in-depth understanding of a cluster is required to enable a sound assessment of the specificities of the cluster and its possible systemic imperfections to be made. Based on this, policy can be developed and specific tools selected.

As we have shown in this chapter, the strategic framework offers an integrated and structured approach to cluster policy. It functions as a support tool in "managing expectations", enabling policy makers to communicate clearly with industry through explicit assessments resulting in a "go or no-go" decision.

The policy framework itself is an important outcome of policy learning. At the same time, it offers a challenging opportunity to deal with future policy learning in a more explicit way. The three policy phases provide a sound basis for systemically comparing various policy approaches, identifying key success factors for effective policy making and highlighting opportunities for continuous upgrading. Based on the lessons learned, we think that some of the lines along which this upgrading should take place could include the following:

- *Non-technological innovations.* The growing importance of non-technological innovations calls for serious consideration of how to incorporate this "soft side of innovation".

- *National vs. international.* How to assess cross-border clusters and how to develop policy for these clusters?

- *Interministerial relationships.* Do other ministries need to become involved in clusters in their respective policy areas? If so, how should this be co-ordinated?

- *The learning policy maker.* How to switch "intelligently" between the various policy roles and how to combine the required analytical capabilities with a "bias for action"?

To conclude, the strategic framework needs to prove its potential in practice and can certainly be further improved upon. In this respect, policy learning will remain an important ingredient of cluster policy in the years to come.

NOTES

1. For further information, see www.ecp.nl.

2. For further information, see www.twinning.nl.

3. For a more detailed analysis, refer to Chapter 10 of this volume which describes the Dutch construction cluster.

4. For a more detailed analysis of the Dutch multimedia cluster, see Chapter 7 of this volume.

5. Refer to Chapter 10 of this volume.

6. IOP = Innovatieve Onderzoeksprogramma's (Programming of Innovative Research).

7. For further information, see www.epri.com.

8. For further information, see www.technologyroadmapping.com.

9. Development of various possible application scenarios based on which a technology can be upgraded, often used in ICT. It differs from technology roadmaps in terms of time horizon (around 2-3 years) and the smaller number of different parties involved.

10. As an example, see the chapters by den Hertog *et al.* on the Dutch multimedia and construction clusters (Chapters 7 and 10 of this volume).

REFERENCES

den Hertog, P. and E. Brouwer (2001), "Innovation in the Dutch Construction Cluster", Chapter 10 of this volume.

den Hertog, P., S. Maltha and E. Brouwer (2001), "Innovation in an Adolescent Cluster: The Dutch Multimedia Cluster", Chapter 7 of this volume.

Edquist, C. (ed.) (1998), "The ISE Final Report: Scientific Findings and Policy Implications", http://www.tema.liu.se/tema-t/sirp/publi.htm.

Malerba, F. and S. Breschi (1997), "Sectoral Systems of Innovation", in C. Edquist (ed.), *Systems of Innovation*, Pinter, London.

Ministry of Economic Affairs (1993), *Competing with Knowledge*, Ministry of Economic Affairs, The Hague.

Ministry of Economic Affairs (1993), *Industrial Policy in the 1990s*, Ministry of Economic Affairs, The Hague.

Ministry of Economic Affairs (1995), *Knowledge in Action*, Ministry of Economic Affairs, The Hague.

OECD (1999), *Boosting Innovation: The Cluster Approach*, OECD, Paris.

Porter, M.E. (1990), *The Competitive Advantage of Nations*, Macmillan Press, London.

Rand Corporation (1997), *Technology Radar*, The Hague.

Roelandt, T.J.A., P. den Hertog, P.J. van Sinderen and N. van den Hove (1999), "Cluster Analysis and Cluster Policy in the Netherlands", in OECD (1999), *Boosting Innovation: The Cluster Approach,* OECD, Paris.

Willems and van den Wildenberg (2000), *Evaluatie technologie-radar*, The Hague.

Chapter 20

THE CLUSTER APPROACH IN FINNISH TECHNOLOGY POLICY

by

Jari Romanainen
Tekes, Finland

Introduction

This chapter discusses the role of cluster approach in Finnish technology policy.[1] It first provides a brief overview of recent history to show how mainstream policy concepts such as innovation system and clusters have affected science and technology policy thinking and how policies have evolved over the last few decades. This is followed by a description of some of the specific features of Finnish policy making to provide a better understanding of why and how approaches like national innovation systems (NIS) and clusters were integrated into policy making.

The chapter then discusses in more detail how the concept of clusters was introduced into industrial and technology policy during the 1990s. This is followed by an example of a recent policy approach which specifically targeted the interface between industrial clusters and public sector activities and actors.

This chapter seeks to explain how the cluster approach has become an integral component of technology policy rather than remaining a set of well-defined cluster-specific policies. Thus, it does not provide a detailed discussion of specific industrial clusters, although some references will be made to the forest, ICT and well-being clusters. In Chapter 2 of this volume, Laura Paija discusses in greater detail one of the main Finnish clusters, namely ICT, and provides a complementary view of the cluster approach in the Finnish context.

Finally, the chapter draws some conclusions from the experience gained in Finland and sets out future directions for an integrated cluster approach in technology policy.

A brief history of science and technology policy

The birth of modern science policy in Finland dates from the 1960s, at which time the role of R&D and education in economic growth was first recognised more widely.[2]

The 1970s were dominated by a strong emphasis on regional policy. During this period, the roles of industry and academia were quite distinct: the universities were not permitted to collaborate with industry; a number of regional universities were established; industrial policy relied on direct subsidies, government ownership and selected technologies; and SME policy was predominantly regional.

At this time, policy makers gradually became aware of the fact that education, science, industrial and economic policies could not be viewed in isolation and that many policy measures were linked. There was a general understanding that policies needed to be balanced and that science, industrial and social needs had to be considered simultaneously.

In the early 1980s, the policy environment began to promote R&D. Industrial R&D activities started to grow faster and, as a result, the role of technological and industrial R&D became more visible. Universities were now encouraged to collaborate with industry and Tekes was established in 1983 to enhance collaboration among industry, universities and research institutes. The first national R&D programmes were launched. At this time, policy favoured industrial restructuring from low-technology industries to high-technology industries, and information technology was considered to be a key technology.

Integration of science and technology policy

The Science Council had been established in 1963 as an advisory body for government on science policy. However, by the 1980s, the role of technology had become so important that it was time to develop the role of the Science Council to include technology. In 1987, the name of the Science Council was changed to that of Science and Technology Policy Council (STPC).

Towards the end of the 1980s, the main issues in science and technology policy were internationalisation, technology diffusion, collaboration and a number of key technologies such as ICT, biotechnology and materials. Although private investments in R&D had doubled from the levels of the 1970s, the need to increase national spending on R&D was highlighted. Consequently, in 1984, the Science Council set the target of increasing the share of R&D to 2% of GDP by 1990. This was confirmed by the STPC in 1987, with an additional target to increase the R&D share in GDP to 2.45% by 1995 and to 2.7% by 2000.

In the 1987 outline, the STPC argued strongly that all increases in resources would have to support the development of the system as a whole. The target at that time was "*improving integration, collaboration and division of labour within the whole research system*". Although the concept of the national innovation system (NIS) had yet to be introduced, NIS policy had begun to emerge.

Renewal of industrial policy: from intervention to framework conditions

The late 1980s saw a major economic boom resulting in the wake of the liberalisation of financial markets. This was followed in the early 1990s by a recession – one of the most severe in Europe.[3]

It was realised that more industrial activity and enterprises were needed. In addition, there was a recognised need for diversifying the economy away from the forestry and metals industries towards new high-technology industries such as IT. Industrial policy was revised and a number of policy actions were taken, the most important of which included a new competition policy, privatisation of government-owned companies, liberalisation of markets and increased emphasis on technology policy. In addition, during the 1990s, efforts were made to improve the firm- and business-related regulatory environment and institutional structures.

Venture capital markets were virtually non-existent until the early 1990s, when they were launched with public money. Since then, and especially during the second half of the 1990s, there has been a major shift towards private venture capital. In recent years, foreign direct investment has also increased, especially in the new high-technology sectors.

Higher private investment and a strengthened focus on commercialisation was reflected in an increasing emphasis on technology policy. The implementation of technology policy shifted from a "science-push" to an "industry-pull" focus and an increasing number of national technology programmes had a stronger industry and problem orientation than in the past. Technology programmes emphasised vertical collaboration.

The first half of the 1990s was characterised by Finland's upcoming EU membership and by globalisation, which had started to become an issue in the late 1970s. Most of the current international collaboration platforms and networks, such as EUREKA, COST, OECD/IEA and EU Framework Programmes, were established or included the participation of the first Finnish companies during the late 1980s and 1990s.

Introduction of NIS and clusters

The concept of national innovation systems was introduced in policy discussions at the beginning of the 1990s although, as stated above, such policies had already started to emerge in the late 1980s. At the same time, the aim of science and technology policy shifted from intervention to facilitation. Industrial policy was drawn from market-failure thinking and proned the benefits to be gained from externalities (R&D), investments in new factor conditions (knowledge and skills) and support of restructuring processes. There was a call for national strategies and competition, and technology policies became the most important components of industrial policy. The main focus of industrial policy was the facilitation of economic growth and specialisation through innovation and competitive markets.

Clusters were first introduced on the Finnish policy scene in 1992 as a result of an ETLA study which identified the major industrial clusters in Finland. There were no big surprises as to the clusters identified, but the approach itself was novel.

At the same time, competitiveness and networking were increasingly emphasised in policy discussions. Towards the mid-1990s, long-term national and global challenges, such as environmental issues, ageing population and energy, took on greater visibility in industrial and economic policy. In addition, the social dimension of science and technology policy received increasing attention during the 1990s.

An era of fast growth

Since the mid-1990s, the Finnish economy has been growing at an exceptionally fast pace, mainly due to the growth of export-based high-technology industries, especially ICT.[4] The performance of the high-technology industries has clearly shown that future economic growth will be based on innovation. Today, business models and value chains are changing rapidly in all sectors of industry and the role of services is increasing. Only companies that embrace change and innovate will be successful.

The private sector is increasing its investments in R&D at an unprecedented rate. The public sector faces a major challenge in developing its role in this new situation. Science and technology policy is struggling with issues like the supply of skilled workers, getting innovations rapidly to market and how best to support the restructuring of existing industries to enable them to cope with new value chains and global markets.

These changes in industrial and economic structures inevitably result in both winners and losers. In addition to major global challenges, identifying future growth areas and ensuring their continued growth while at the same time supporting the reform of existing and traditional sectors through innovation are the main challenges faced by policy makers today. This means that the social dimension of science and technology policy will demand more attention in the coming years.

The current policy challenges are listed in the recent STPC Policy Review (STPC, 2000),[5] which strongly emphasised both NIS and clusters.

Specific features of policy making in Finland

The main actor in designing science and technology policy is the Science and Technology Policy Council. Its role is to act as an advisory body to the government, but in fact it is more than that. A strong representation[6] from all key stakeholders makes the Review of science and technology policy, prepared once every three years, a very strong basis for policy.

Another characteristic of the policy process in Finland is the level of decision making. In the Review, the Council discusses the main policy challenges facing the country and makes general suggestions concerning all actors. The Review makes recommendations as to how resources, *i.e.* public funding of R&D, should be allocated. The actual implementation of these suggestions is left to the ministries and agencies. Individual research or technology programmes are not decided by the Council or even by the ministries, but by the implementing agencies. This makes it possible for the system to react relatively quickly to new industrial and societal challenges as they are identified.

An additional feature of the system is related to its size. Since the key actors are few and easily contacted, a great deal of informal interaction takes place between different actors at all times. Important issues are continuously discussed in an informal way, and major documents such as the Review originate from these discussions. Thus, the main purpose of the Review is not so much to identify new issues, but rather to discuss and set priorities and communicate these to a wider audience and to decision makers.

The cluster approach

The Finnish cluster approach can be argued to have originated from two main sources. First, the development of industrial structures and the subsequent renewal of industrial and economic policies in the late 1980s and 1990s geared the policy environment for the cluster approach. Second, the early adaptation of "NIS thinking" in science and technology policy made it possible to adopt the cluster approach very quickly in an integrated way.[7]

The cluster approach in industrial and economic policy

The principal industries in Finland, the forestry and metals industries, developed during the industrialisation era as the two mainstays of the economy. These industries were based on natural resources, wood and mineral resources. The metals industry, in particular, received an additional boost, when according to the peace treaty signed with the former Soviet Union after World War II, Finland agreed to provide various types of machinery to its eastern neighbour.

Industrial and economic policy followed the interests of these export-oriented industries, which attracted a great deal of associated industrial activity to interact with these industries. The resulting structure consisted of large forestry and metals companies linked with numerous smaller companies, typically subcontractors – the core and essence of a cluster was born.

However, policy did not recognise these emerging clusters. Instead, industrial and economic policy treated large companies separately from small and medium-sized enterprises (SMEs). Large companies' interests were a national issue while SMEs were seen as a predominantly regional issue.

During the second half of the 1980s, policy began to evolve, and linkages within and between industries were identified as important. Large companies actively built up subcontractor networks and began to collaborate more with other companies. Technology and competition policies began to replace traditional industrial policy and there was a need to find a new perspective. It was in this policy environment that the cluster approach was introduced at the beginning of the 1990s.

The *National Industrial Strategy* published in 1993 featured the first cluster analysis implemented by ETLA. The key message highlighted the realisation of clusters as the source of national competitiveness. This did not, however, result in strong cluster-specific policy. Instead, in accordance with the new role of government as a facilitator, it was decided that policy should not favour any particular cluster over others. Rather, policy should facilitate growth of all clusters, both now and in the future.

Forestry, metals, ICT – what next?

Clusters represent a natural way of understanding industrial structures in Finland. The major industrial strength after World War II was in the forestry-related industries. As the Finnish forest industry internationalised, an increasing number of manufacturing companies began to redirect their businesses towards products and services for the forestry industry. This ensured that the forestry industry benefited from a good supply of new ideas and innovations. When the cluster approach was introduced in Finland in the early 1990s, the forest-related industries represented an excellent example of an industrial cluster.

One of the original policy aims dating from the 1980s was to stabilise the economy and render it less vulnerable to economic fluctuations in the pulp and paper industry. The policy was to enable the development of new export-based industries to complement the two existing ones, forestry and metals. In the early 1990s, the telecommunications industry emerged as a potential third pillar of the Finnish economy.

Today, ICT has become a major export cluster. However, the cluster continues to rely heavily on the ICT companies themselves. The main challenge will be to strengthen the linkages between the ICT companies and the rest of the economy, thus resulting in an even stronger cluster. One answer to this challenge has been to complement the typical approach of adapting new information and communication technologies used in various other industries with programmes whereby these other industries provide new technologies for the ICT industry.[8] In addition, the ICT industry association, together with public sector actors, has prepared an outline of future needs for skilled resources in the ICT sector.

A major policy challenge was the identification of potential growth areas and how best to stimulate their performance. Some areas have already been identified through existing strategy processes and are listed in the STPC Review for 2000. At the more scientifically advanced end of the spectrum is the biomedical cluster that emerged towards the end of the 1990s. Energy is one of the more traditional clusters that has real growth potential in the coming years as energy markets in Europe are deregulated.

The cluster approach in science and technology policy

"NIS thinking" had been gradually entering into policy discussions over the course of the late 1980s. It was taken up in industrial and economic policy a few years before the introduction of the cluster approach.

Since both the NIS and cluster approaches are characteristically systemic, one is relatively easy to adopt once the other has been implemented.[9] Therefore, the early adaptation of NIS thinking supported the adaptation of a cluster approach. In fact, not only did it support the approach, it also strongly influenced the way in which the cluster approach was introduced into Finnish policy making.

The specific characteristics of policy making and the model of earlier NIS implementation resulted in a fairly pragmatic adaptation of the cluster approach at a very practical level of policy implementation. The STPC policy Review identified clusters at a general level, leaving a great deal of leeway in the design of the implementation, with the result that the cluster approach is integrated close to implementation of science and technology policy rather than being a dominant policy dimension.

The only visible exception to the integrated approach is the inter-ministerial cluster-based programmes that were introduced in the 1996 STPC policy review. These were part of the additional appropriations programme implemented during 1997-99 (STPC, 1996). However, the motivation underlying the design and implementation of these programmes was not simply to introduce the cluster approach into science and technology policy. Instead, one of the main objectives of these programmes was to improve collaboration between the various ministries and sectors of government.

Although a number of ministries are represented in the Science and Technology Policy Council, mainstream policy implementation included companies and universities, but only the most technologically oriented government research institutes. The remainder of the government research institutes receiving most or all of their funds directly from ministries were not properly linked to the mainstream of research and development.

The following section describes how the inter-ministerial cluster-based programmes were designed and launched.

The inter-ministerial cluster-based programmes

The inter-ministerial cluster-based programmes are a major new concept of Finnish technology policy of the late 1990s. The industrial cluster-based approach has existed for some time, but these programmes represented something new. The novelty was to gather all the stakeholders – not only universities, research institutes and companies, but also sectoral government research laboratories and the most relevant users – together to plan and execute joint projects aimed at increasing the competitiveness of the whole cluster.

The 1996 science and technology policy review identified the need for increased collaboration among the various policy sectors. It identified a number of promising sectors for collaboration: economic, financial, industrial, education and science, employment, regional, social and health, environmental and cultural policies and the work that had been carried out on developing an innovation society. It noted that the work that had already been initiated within industrial clusters, particularly in the sectors of well-being and telecommunications, should be extended to other sectors.

To implement this collaboration, the Council suggested additional funding for the Ministries of Trade and Industry, of Education and Science, of Agriculture, of Transport and Communications, of Social and Health, of Labour and of Environment. The additional funds were to be used for research and development, in collaboration with Tekes, the Academy of Finland, universities, research institutes and industry.

Once the government had agreed to the proposals recommended in the policy review, there was a planning period during which decisions were taken as to how the funds were to be used. A number of planning groups were established, the composition of which varied, but which were all led by the ministries that had received the additional funding.

As a result, seven inter-ministerial cluster-based programmes were identified. One already existed, others were based on or combined earlier activities, the remainder were new. The main feature of these programmes was that they did not try to duplicate work that had already been done. They added to or complemented the existing set of policy actions that were being implemented at that time.

The management of these programmes was carried out by programme steering committees, comprised of the funding organisations and the major stakeholders. Although funding decisions were made independently by the funding organisations on a project-by-project basis, calls for proposals, preliminary evaluation and programme monitoring were the responsibility of the programme steering committee.

An informal body chaired by an STPC chief planning officer was set up, consisting of the programme managers of the inter-ministerial cluster-based programmes. Tekes and the Academy of Finland were invited to join this informal group, together with a representative of the additional appropriations evaluation committee.

The additionality principle that had been identified during the planning phase ensured that the new programmes targeted areas where no similar policy action existed at that time. For example, the Ministry of Transport and Communications did not select the ICT cluster, since several large technology and research programmes had already been implemented by Tekes and the Academy of Finland in that sector. Instead, the new programme targeted transport, where the development of new logistics systems required extensive collaboration among private and public actors. The ICT sector was targeted with a small, focused programme which aimed to facilitate access of small firms to the Internet and e-commerce.

The forest cluster had benefited from a series of technology programmes during the late 1980s and 1990s. These had served to develop a good working collaboration between industry, universities and research institutes. Earlier programmes, however, had not targeted the initial stages of the value chain, in which collaboration between forest research and industry was not sufficiently highly developed. Over the last three years, the inter-ministerial cluster-based programme has significantly improved collaboration between these parties.

In the well-being sector, the inter-ministerial cluster-based collaboration has enabled the implementation of an exceptionally large and comprehensive healthcare system pilot project, Macropilot, which combines the efforts of a number of actors.

One of the specific objectives of these programmes was to create and improve linkages between government (ministries), research (universities, research institutes and especially sector research institutes) and industry. These first-generation programmes were versatile in nature and were dictated by current activities and needs. The second generation of inter-ministerial cluster-based programmes is currently in the planning stage; new areas and new challenges in existing sectors are being identified. The second-generation programmes will draw on the experiences of the first-generation programmes, some of which are still ongoing.

Developing the cluster approach

The original cluster studies have been recently updated. The new studies have typically been closely linked with practical policy design activities, *e.g.* technology programme evaluation or design. In this way, cluster studies are more closely linked to and will feed back into policy design and implementation, and policy implementation will be better tailored to the needs of each cluster. Identification of the needs of individual clusters has resulted in some specific policy actions, *e.g.* special measures to enhance ICT education in universities.

The cluster approach is far from being utilised to its full extent. Recent policy discussion has introduced the cluster approach to other than traditional industrial contexts, *i.e.* there are now knowledge clusters and regional clusters. Whether these should be termed clusters or simply knowledge concentrations and regional innovation systems is not relevant, since the main issue is that a systemic approach is useful in analysis.

In the more traditional industrial clusters, one area especially requires attention. There is a need for proactive measures in cross-disciplinary networking. Products and services in today's markets combine many different technologies. One of the main factors in facilitating the birth and growth of clusters is the selection of efficient policy actions which can help people from different disciplines to interact in such a way that they result in innovations. It is easy to provide platforms where people who speak the same language can meet, but the real challenge is to create platforms where people from different disciplines, speaking more or less different languages, can interact efficiently.

Conclusions

The adoption of the cluster approach in Finnish technology policy was strongly affected by "innovation systems thinking" and the specific features of the Finnish policy context. The NIS concept had been adopted at an earlier date, and this facilitated the adoption of a similar approach. In the industrial policy context, clusters were introduced as a new basis for policy, while in the science and technology policy context, clusters provided a complementary view to the existing policy basis founded on NIS. This meant that in the technology policy context, the cluster approach was fully integrated into the NIS approach and taken up mostly at the level of policy implementation.

Technology policy implementation in Finland involves both proactive and reactive measures. Proactive measures – programmes and schemes – provide an interactive top-down (hands-on) element, which is balanced by a more reactive, bottom-up (hands-off) element. The cluster approach is integrated mainly in proactive measures which try to create platforms where different people with different ideas can interact and innovations can be initiated.[10] Both the innovation systems approach and the cluster approach provide tools whereby the various features of these platforms can be understood and analysed at several levels.[11] Both approaches emphasise the importance of understanding the system as a whole: actors, framework conditions and interactions.

So far, the experiences of the cluster approach have proved useful. Technology programmes targeting industrial clusters have been found to improve networking and bring together new actors. The inter-ministerial cluster-based programmes have proved to be an interesting and potentially promising way to target fields in which the innovation process needs to be extended into areas and actors which have not been traditionally involved in collaboration with applied research and industry. In the Finnish context, these are, in many cases, areas where innovation activities need to be introduced, strengthened and linked to public sector research institutes and other public sector organisations.

The adaptation of the cluster approach is relatively new and some activities have been more successful than others. There is always a risk that some adopt the name, but not the approach. While there are inevitably examples of this, there are more examples of how systemic thinking is being adopted in many contexts through the cluster approach. Still, it remains to be seen, for example, to what extent forthcoming foresight activities will feature clusters, or what new forms the second-generation inter-ministerial cluster-based programmes will take.

So far, the benefits of the cluster approach include the following:

- *It helps to identify networks and linkages among industries.* An understanding of the various networks helps, for example, in the design of technology transfer measures and measures enhancing diffusion.

- *It provides the possibility to identify emerging growth areas and the role that various organisations can play in enhancing performance.* Understanding new types of growth helps in the design of measures to support this growth.

- *It is useful for identifying technologies and problems common to several industries.* This helps create platforms (*e.g.* programmes) to enhance cross-disciplinary and/or cross-industry interaction.

While the cluster approach provides a useful tool for analysing the economic system, it must constantly evolve if it is to be able to grasp the changing dynamics of the economy. New ways of looking at networks and businesses are constantly called for. Strongly internationalised clusters and changing value chains are examples of the current challenges faced by policy makers.

Policy design and implementation must be innovative and able to experiment with different approaches and tools in order to meet the challenges of the changing innovation environment. This is possible only if the theoretical framework and methodologies continue to evolve and are able to provide a better understanding of the complex interactions and linkages within the innovation environment. Understanding how the system works is the key to successful policy design and implementation.

NOTES

1. This chapter refers to "technology policy", since that is the term currently used in Finland. However, the content of the current policy would warrant it being termed "innovation policy" rather than "technology policy".

2. Towards the end of the decade, industry underwent, and continues to undergo, major structural change, the need for which largely arose from the forces of globalisation and innovation. Large companies, in particular, began globalising through foreign direct investment. In addition, companies' investments in R&D and other intangibles began to rise.

3. The recession helped to speed up industrial restructuring and resulted in a stronger industrial base. Those industries that survived the recession now outperform most of their foreign competitors in terms of productivity growth. The recession also resulted in unemployment which, by the end of the 1990s, was dominated by structural problems.

4. The flagship of this growth performance has been Nokia, which has become one of the most valued companies in the world. However, growth has not been limited to Nokia or even to the network surrounding the firm; many other companies have also performed exceptionally well.

5. The recent science and technology policy outline from 2000 identified five major challenges for the coming years: *i)* How to enable growth of the ICT cluster in Finland? The major challenge is to enhance the supply of skilled workers; *ii)* How to ensure innovation and growth in social, cultural and other sectors of the economy? Society as a whole must be transformed into a knowledge economy; *iii)* How to identify potential clusters and how to enable their growth? The ICT cluster will continue to grow at a fast pace for a number of years. Unless other strong and fast-growing clusters are identified, the economy runs the risk of becoming too dependent on a single cluster; *iv)* How to enhance technology transfer and diffusion to enable widespread use and benefits of new technologies throughout society? The innovation system must provide high-quality innovation services to all organisations, regardless of location and capability; *v)* How to strengthen the science base? Basic funding for universities has not increased sufficiently in recent years to make up for the rise in the number of students, and the introduction of competitive funding has not sufficed to induce the necessary structural changes in the universities.

6. The Council is chaired by the Prime Minister and comprises ministers from the Ministries of Trade and Industry, of Education and Science, of Finance, of Transport and Communications, of Environment and of Defence. Other members of the Council come from industry, industry associations, labour unions, universities, Tekes, the Academy of Finland and the Finnish Environment Institute.

7. A more comprehensive description of the adaptation of the cluster approach and it's relation to policy making can be found in a recent PhD thesis by Jari Jääskeläinen, ETLA, 2001.

8. For example, the "Polymers are Building the Future" Technology Programme. Further information on Tekes programmes can be found at http://www.tekes.fi.

9. Both NIS and cluster approaches look at whole systems and try to identify the key actors, framework conditions and interactions that characterise the system. Clusters have, in fact, been referred to as

innovation systems – not at the national level, but on a smaller scale. However, this argument applies only when the cluster concept is extended to cover flows of knowledge and skills; the original concept of industrial clusters can hardly be argued to be an innovation system. There are a number of similarities between the two approaches and it is fair to predict that the development of these approaches will result in increasing degree of integration. Both approaches are moving towards the analysis of structures and interactions and their relation to outcomes, whether measured directly in economic terms (clusters) or indirectly as competitiveness (NIS). Both of these systemic approaches are facing a particular challenge in globalisation and restructuring of industries and value chains. It is becoming increasingly difficult to clearly define the borders of a given system. What is a national system? Is there a regional system? When should the system be international and how far should it reach? Companies are becoming global, value chains are becoming global – what about clusters? One solution to this is to move away from clear-cut systemic thinking towards analysing and thinking in terms of environments. Instead of trying to pre-define the borders of a system and analyse its contents, a attempt should be made to define the core and analyse the environment around it. This does not mean that one should abandon the systemic approach, rather it is an attempt to improve it using a complementary approach from another viewpoint. Implementation of innovation policy in Finland today has adopted the concept of innovation environment as a complementary concept to the innovation system. This is an attempt to bring a new perspective to the discussion, one that originates from innovation rather than from the system.

10. "Innovation systems" thinking has shifted the focus in science and technology policy towards innovation. Today, innovation combines different technologies and requires a broad science base. This means that issues like the knowledge and skills base, technology transfer and diffusion, commercial utilisation and the business environment are at the very crux of policy. The innovation-centred approach emphasises the innovation process and the environment in which innovations are initiated and brought to market. The focus of policy implementation is moving towards the building of different platforms which encourage and facilitate innovation.

11. Macro, meso, micro.

REFERENCES

Federation of Finnish Electrical and Electronic Industry (1999), "Need of Skilled Resources in the Electrical and Electronic Industry 1997-2002" (in Finnish).

Jääskeläinen, Jari (2001), *Clusters – Between Science and Policy. From Industrial Policy to Social Policy*, PhD thesis, ETLA (in Finnish).

Ministry of Trade and Industry (1990), *Industry Policy Outlook 1990.*

Ministry of Trade and Industry (1992), *Industry Policy Outlook 1992.*

Ministry of Trade and Industry (1993), *National Industrial Strategy for Finland.*

Ministry of Trade and Industry (1996), *Government Report on Economic Policy.*

Ministry of Trade and Industry (1996), *Industrial Policy Vision.*

Rouvinen, Petri (ed.) (1996), *Advantage Finland – The Future of Finnish Industries*, ETLA.

Science and Technology Policy Council (1987), *Science and Technology Policy Review 1987.*

Science and Technology Policy Council (1990), *Review 1990: Guidelines for Science and Technology Policy in the 1990s.*

Science and Technology Policy Council (1993), *Towards an Innovative Society: A Development Strategy for Finland.*

Science and Technology Policy Council (1996), *Finland: A Knowledge-based Society.*

Science and Technology Policy Council (2000), *Review 2000: The Challenge of Knowledge and Know-how.*

Chapter 21

BRIDGING CLUSTER THEORY AND PRACTICE: LEARNING FROM THE CLUSTER POLICY CYCLE

by

Paul Benneworth and David Charles
Centre for Urban and Regional Development Studies, University of Newcastle

Introduction

The cluster approach to industrial support has recently increased dramatically in popularity, offering market-led solutions to innovation policy makers (Roelandt and den Hertog, 1998). It is undeniable that the idea of a "cluster policy" has become increasingly attractive to a number of national and regional governments because of its low cost and high potential returns.

A distinction must be drawn between "clusters" and "clustering", although policies work with both of these concepts (Charles, 2000). "Clusters" are economic phenomena in the form of agglomerations of firms with patterns of interaction which boost their competitive advantage. "Clustering" is a process whereby inter-firm linkages and cluster externalities are built up, and so what were hitherto disparate firms gain competitive advantage from their interaction (Lagendijk and Cornford, 2000).

There are two sets of policies which correspond to each of these concepts – clusters are supported by a stable macroeconomic environment, appropriate infrastructure, etc., while clustering is supported by direct (low-cost) intervention to map and facilitate cluster formation and encourage contacts between firms. The majority of active policy measures aim to improve innovative performance by encouraging linkages between firms through support to latent and potential clusters. Thus, in this chapter, we are primarily concerned with those policies that promote "clustering".[1]

In Part IV of this volume, a number of specific cases of cluster policies in practice have been presented. Each of those chapters begins with cluster policies as a *fait accompli*. However, a closer reading of these accounts demonstrates that differences in the way the policies were chosen ultimately shaped their implementation. In a number of cases, the first approach taken to government support for clustering was in some way deficient, and so subsequent policy changes were made. Cluster policies are not divorced from the other interests of government, and so in a number of cases, while cluster policies were favoured by ministries of industry, the policies of other ministries were far more significant to the cluster than the marginal interventions made in the pursuit of "cluster policy".

This chapter analyses the questions raised when academic conceptions of clustering are operationalised into policies to boost economic performance by encouraging inter-firm networking.

Common to each of the national accounts was that cluster policy in effect became *an answer to a particular problem* faced by governments. In each case, the local (regional or national) political context affected the process of policy formation, and the decision to use a clustering approach. What differed was the way in which these ministries defined their clusters and how they promoted clustering.

This chapter examines how the context affects policies to analyse different types of clusters, to develop appropriate policy forms and the key administrative hurdles to their introduction. To circumvent this problem, this chapter uses the idea of a policy cycle to examine the way cluster policies have emerged in a number of OECD countries. The cluster policy cycle is then used to ask a number of specific questions about what can be learned from the experiences of particular cluster policies detailed in this volume. In the final section, a categorisation is offered of the scale and scope of clusters, and the appropriate scale of policy formulation and delivery; which leads into the synthesis of theory, policy and administration of cluster policies offered in the concluding chapter of the book.

The cluster policy cycle – why governments choose clusters

The cases presented in Part IV demonstrate the heterogeneity of cluster policies in practice, but also highlight the benefits which cluster policies have over more interventionist industrial and economic policy paradigms. However, clustering policies are inevitably framed by the circumstances of adoption in terms of the political intentions and chosen instruments. To capture these formative elements, it is necessary to consider the full cluster policy life span, and in this chapter, we elongate the period over which the idea of a cluster policy is considered. Hogwood (1987) uses the idea of a "policy cycle" to describe the life span through which an idea becomes implemented and evaluated, corresponding to the translation of an idea into actions.

The key feature of a policy cycle is that the degree of innovation in the policy tends to increase with the strength of the driver of change; the earlier in the policy cycle the idea of clustering is developed, the greater an influence the idea can exert on the policy framework. In Denmark (Holm Dalsgaard, Chapter 18 of this volume) and Flanders (Larosse *et al.*, Chapter 6 of this volume), clustering was adopted at the outset, giving a coherent policy framework. Conversely, in the United Kingdom, clustering was only adopted very recently by national government, and so it has hitherto been within the remit of the very limited economic development powers of local and regional authorities.

Cluster policies as a rational policy choice

Cluster policies have become an increasingly popular governmental tool because of their capacity to boost innovative performance; in the 1980s and 1990s, analyses of successful regions and nations drew attention to the fact that these success stories were home to networks of collaborative firms which created a competitive advantage through continual innovation and market leadership (Porter, 1990). One element of these networks was that they built connections between firms and technological suppliers, and integrated technology transfer into innovation (*cf.* Piore and Sabel, 1984; Saxenian, 1994). It was later observed that this idea could be more widely generalised to less successful regions; they could improve their economic performance with an institutional/policy "fix" borrowed from a more successful region (Morgan, 1992; Hassink, 1993; Lundvall and Johnson, 1994).

Cluster policy fits neatly with the idea of a learning economy, creating territorial institutions and/or mechanisms to facilitate business-led interactive learning (Florida, 1995). Porter (1998) explicates that clustering is the outcome of a set of strategic choices made *by businesses* to generate competitive advantage with respect to their competitors. Cluster policy therefore seeks only to augment businesses' own strategies by encouraging them to collectively solve problems without individual solution. Under this interpretative framework, the process of clustering is driven by businesses and the marketplace, leaving only limited scope for government involvement (and importantly for expenditure).

There are a number of barriers to innovation which governments can encourage businesses to collectively overcome (Boekholt and Thuriaux, 1999). In some cases, governments themselves cause problems by implementing regulations that inhibit innovation or simply by making mistakes in the design of policy. More generally, the publicly owned knowledge infrastructure might lack incentives to engage collaboratively and strategically with clusters (Lawton Smith, 2000). Firms themselves might lack a common identity and awareness of similar firms, or particular elements of a coherent cluster might be missing. Small firms, in particular, face a number of specific problems; short-term financial pressures can obscure the potential of longer-term collaborative activity, and might also inhibit the adoption of new technologies which need to be adapted to their specific needs.

The limits to rationality and the policy process

The rationale for the adoption of a cluster policy marks the beginning of a lengthy policy process in which a hypothetical solution to a perceived problem is realised and evaluated. However, cluster policies are by no means the only solutions to these problems and the reasons why governments choose a cluster approach can have profound implications for the evolution of the policy. Regardless of the identification of the problems, the *first stage of a cluster policy process is the decision to use a cluster approach.*

Once that decision has been taken, the particular national meaning and state role in the approach has to be debated (*e.g.* direct intervention, research funding, collaboration facilitating). This *determination of the state role* further affects the policy, as it directs the set of tools which can be used. If clustering is an administrative approach, relatively removed from the centres of power, then many of these definitions may remain undetermined until late on in the policy process (*e.g.* the United Kingdom). If clustering is introduced as a high-profile policy, then there is much greater latitude to join up policy areas and produce a common government approach (as in Finland).

The next stage is the *selection and designation of clusters*. In some cases, this might be aspirational, which may be based on input/output criteria or sectors offering potential strategic national industrial advantage. In other cases, the designation of clusters has transcended industrial statistics to other criteria, such as platform technologies or even self-designation.

Once the decision has been taken to support a particular cluster, then the cluster policy enters a much more technocratic phase *from strategy formulation to programme delivery*. Willing participants in the cluster are identified, aims and targets for the cluster determined, and then actions are planned and delivered. Finally, because of its political nature, there is an *evaluation and reporting-back stage*, where lessons are learned, and the possibilities of subsequent policy phases evaluated. The policy finally re-emerges into the political sphere, where its appropriateness and efficiency *as a policy measure* can be democratically debated, and decisions taken over the future of cluster policies. This is represented in Figure 1 below.

Figure 1. **A stylised representation of the cluster policy process cycle**

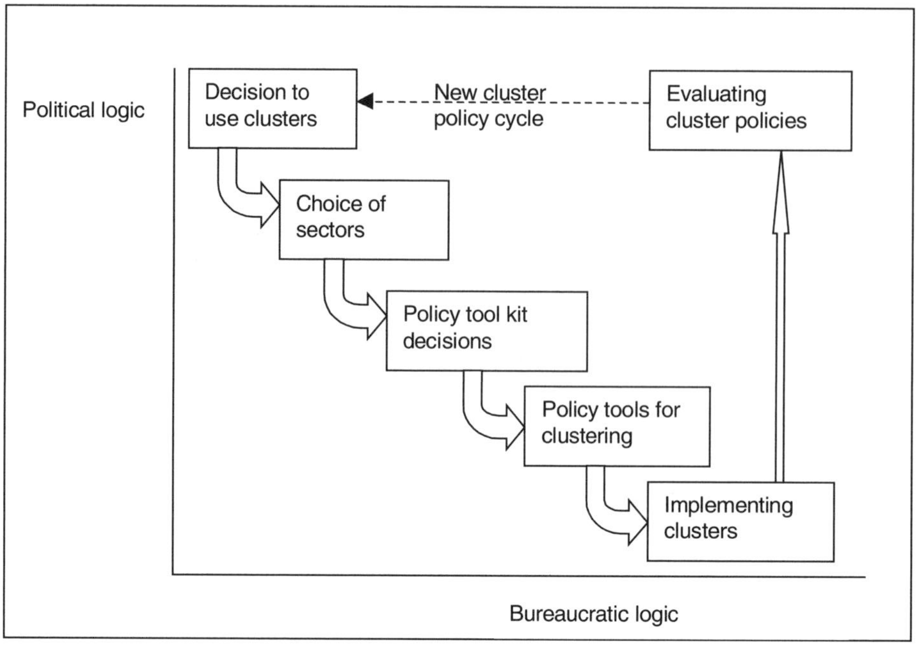

Source: After Hogwood (1987).

Cluster policies evolve over time, and lessons learned are carried forward into successive innovation policy generations. Regarding cluster policy as a cycle, it is first necessary for governments to have embraced the idea of a non-interventionist market economy; otherwise, clustering lacks appeal as a policy tool. *The cluster policy cycle has a number of stages, and in each of the stages, a set of decisions are taken which affect the outcome and ultimately the success of those cluster policies.* Although with the benefit of hindsight, the adoption of cluster policies in some country may appear smooth, at each stage there is a degree of uncertainty with respect to the development of the policy. It is clear that government decisions are important in shaping events, and this raises questions about the issues faced in the development of the policy between each of the stages. These issues differ at each stage of the process; at the start and end of the cycle, political interests are paramount, while at other times, bureaucratic interests are more important.

Cluster policies in practice: the OECD experience

We now look at six questions faced by governments at each stage of the process, the way choices are affected by different pressures, and ultimately the influences those choices have on the success of whatever becomes implemented in the name of "cluster policy". This forms the basis for analysing the different policy and spatial scales at which cluster policies operate, and of which policy makers need to be mindful. The six questions are as follows:

- Why do governments choose cluster policy?
- How do governments set the criteria for targeting a cluster?

- How are the cluster policy instruments chosen?
- How are general framework policies customised to specific clusters?
- How does government shape the development of clusters?
- What role do governments play in supporting innovation in clusters?

In this book, a wide range of clusters have been analysed across the gamut of the OECD's membership, and this provides the basis for answering those questions. Reference will be made in particular to the Finnish, Danish and Dutch policy chapters, as well as to particular clusters in Flanders, Scotland and England.

Why do governments choose cluster policies?

The choice of clusters as a framework for innovation and industrial policy is affected by political culture. Although past debates about industrial policy have addressed questions of state intervention, it is generally accepted that industrial policy is no longer concerned with *dirigiste* and planned approaches to economic development. The introduction of cluster policies can be relatively uncontroversial – in England, *de facto* cluster policies have been introduced by administrative *fiat* in the form of directions to the eight regional development agencies to include clustering measures in their regional economic strategies.

However, in other cases, cluster policy has been intimately associated with other (politically more sensitive) issues, and as the political popularity of these other issues has fluctuated, so the much less contentious cluster policies and those charged with their administration have also experienced shifting importance (Larosse *et al.*, Chapter 6 of this volume). This can have positive or negative implications – while association with a temporarily popular political programme might facilitate the introduction of new policy tools and methodologies, that same association can hinder the survival of cluster policies if their umbrella movement falters. Equally, in some cases, clustering began as a specific policy item and was later adopted into the mainstream.

How do policy makers identify clusters?

Methodologies for identifying clusters have been vital in establishing the existence of the phenomenon, but it is vital to distinguish between methodologies for proof of concept and tools for identifying clusters in practice.

What the policy chapters made clear was that the designation of clusters is a highly politicised process, often heavily dependent on the prior existence of sectoral groups to lobby and press government to support particular sectors. In England, for example, the introduction of the cluster policy for the regions was directly associated with support for particular industrial sectors that have experienced dramatic decline and whose survival is under threat (*e.g.* the automotive industry in the West Midlands). However, in a number of the cases, attempts were made, at least after the event, to try to rationally justify the choices made.[2]

It is in the interests of effective public administration that the selection procedure is transparent. However, because the benefits of clusters are not wholly measurable, they are not always amenable to quantitative analyses. This is illustrated rather neatly with the case of Denmark, which initiated its cluster policies with a rational selection of its mega-clusters, and later complemented this with a second layer of more strategically and intuitively chosen and self-defining cluster areas (Holm Dalsgaard, Chapter 18 of this volume).

How are cluster policy instruments chosen?

In tandem with the identification of clusters, it is necessary for governments to take a number of political decisions about what support will be given to those clusters, which can then be applied, either generally to all applicable clusters, or more specifically, offering a cluster tool-kit from which targeted cluster policies can be chosen. Obviously, the greatest influence on cluster policies is the economic policy framework of the particular country. For all OECD countries, there is a presumption of the operation of free markets and consequently competitive pressure, with the role of all government intervention being restricted to pre-competitive support and overcoming market failure rather than the establishment of new state-owned firms with monopoly positions.

However, on a more micro level, cluster policy is often implemented with the intention of joining up existing policies. In these cases, the choice of instruments is limited by established governmental practice. This means that what actually takes place under the rubric of "cluster policy" is tightly conditioned by a wide range of government systems from the technology, economic, utilities and scientific policy areas.

In Denmark, to avoid the lead role of the Ministry of Trade and Industry limiting the applicable policy areas, small-scale clusters and close inter-departmental relations have characterised the use of cluster policies to "combine competitiveness of Danish trade and industry with the general welfare of society". The general government agreement that clusters were to be the focus of an industrial modernisation programme meant that there was a great deal of flexibility to include a range of activities relevant to each of the cluster groups (Holm Dalsgaard, Chapter 18 of this volume).

England demonstrates that the choice of instruments within "cluster policy" is strongly dependent on a set of other political choices. Although the Treasury's own policies are critical to the success of clusters in terms of providing macroeconomic stability, in the English case a bureaucratic rule change rather than a development in the real economy has been one of the key drivers in permitting the development of regional cluster policies which are vital to a mature economy as regionally differentiated as England. Since then, clusters have increased their importance, with DTI launching a national mapping and regional strategy exercise with EUR 80 million of dedicated funding.

Policy innovation is a prerequisite for the introduction of successful cluster policies. In a sense, this reduces the attractiveness of the clustering rationale which is intended to be a low-cost light-touch alternative to more *dirigiste* forms of economic policy. These decisions determine the generic tool-kit available to support clusters; this raises the question of how to support particular clusters, a question to which this chapter now turns.

How are framework policies tailored to specific clusters?

Once a cluster has been identified, and the limits to the extent of the policy fixed, it is then necessary to apply those tools to particular clusters. This is strongly constrained by the prior political choices through which cluster policies were selected as a generic solution. If the government has taken a "one-size-fits-all" mechanistic approach to clusters, then once econometric analysis has determined the existence of a cluster, the generic set of institutions, determined above, can be applied.

However, mechanistic selection processes are rare, and it is much more likely that there will be a mix of minimum qualifying standards (such as size and strategic importance) alongside more opportunistic and political factors. This means that what constitutes clusters within one policy system can be very varied; in the case of Denmark, there is the parallel existence of the mega and small-scale clusters, but even within the small-scale clusters there is a mix of territorial agglomerations and national aspirational/strategic sectors, such as wind technology and children's play and learning. In such cases, cluster policy sets out a framework for the form of government intervention, without prejudice to the eventual determination of the particular instruments to be used.

The lesson in the selection of policy instruments is that different contexts require different tools, particularly if cluster policy is to reflect the diverse rationality of cluster selection. If there is already co-operation between firms, then it is necessary to bring the cluster into the policy cycle at a half-way point. The main problem is in mixing administrators' expertise in process management and firms' needs for specific outcomes. Well-minuted participatory meetings are vital in the orientation phase to determine a common direction, while in the intensification or delivery phase, they can be definitely harmful by focusing activity on continually debating and revising the purpose of the group rather than developing that purpose into a set of meaningful targets (Benneworth, 2001).

Although national context affects the set of policies available, this suggests that different instruments may be appropriate at different stages of the clustering process, as emphasis shifts from self-identification to collective problem solving. This problem is exacerbated by the fact that "clustering" (the bottom-up process), is neither entirely rational nor mechanistic, which means it is difficult to produce a generic policy framework which may have to deal with a diverse array of development trajectories. The case of the Netherlands (Gilsing, Chapter 19 of this volume) thus demonstrates the need for sensitivity to the needs of the cluster; not all clusters will have the same needs in each of the phases, but there are likely to be more commonalties between clusters in similar phases. Similarly, the relationships between the firms in a cluster will not develop in line with a prior timetable. The challenge for government is to accurately identify how the parallel processes of cluster and policy development correspond.

How does government shape the development of clusters?

The role of government can often be critical in boosting the activity of clusters. Although "cluster policy" is one area in which governments can support clusters, encouraging firms to work closer together, there are a range of other policies which governments pursue that are essential to the operation of clusters. The Dutch and Finnish Governments, for example, both place a premium on the smooth operation of the market mechanism to increase the competitiveness of their businesses and by implication their clusters; in the 1990s, deregulation and privatisation rather than direct intervention were a critical means of boosting innovation within clusters (Gilsing, Romanainen, both this volume). Indeed, the Dutch classify government's roles in policy intervention into framework conditions, brokerage/*animateur* activity, and as a sophisticated customer.

However, this framework is deliberately focused upon clustering policies, and overlooks the range of policies which affect industrial sectors, from whence clusters ultimately originate. As clusters mature and latent clusters are activated, so the range of policy tools that can affect the clusters grows. However, governmental policy can still, in this phase, affect "clustering" of the common activities as well as providing a suitable macroeconomic environment for those clusters.

Although clustering is a relatively recent innovation in England, it has been used more extensively in Scotland. In Scotland, a specific clustering approach was chosen for electronics (Charles and Benneworth, Chapter 17 of this volume), principally because despite the presence of many electronics firms, there were believed to be comparatively low levels of innovation. However, there is a limit to the benefits that improved co-operation can bring, and so more traditional industrial policy tools have played a supporting role for the cluster. In particular, in response to global restructuring of the semiconductor industry, large amounts of investment support were made available to particular flagship investors who, it was believed, would lay the basis for the next iteration of the cluster.

What role do governments play in supporting innovation in clusters?

Cluster policies are popular because clusters boost innovation, a fact observed in a number of free-standing clusters not driven by direct government intervention. Ultimately, the key to the success of a cluster policy is the degree to which innovation is stimulated, and the way this provides territorially-based firm agglomerations with the capacity to compete more successfully. From a policy perspective, innovation is certainly a desirable outcome, but it is necessary to continually ensure that the cluster support process is actually encouraging useful innovation by member firms.

Unfortunately, the importance of intuition and opportunity to innovation mean that a deterministic policy framework has difficulty in ensuring that activities actually are boosting innovation. In the case of the Netherlands, this is overcome by ensuring that there are two points at which intermediate outcomes are assessed to make sure that there is some correspondence between activity and innovation (Gilsing, Chapter 19 of this volume). In Denmark, the small scale of the cluster activity allows a more intuitive appreciation of the progress in developing an innovation-boosting cluster (Holm Dalsgaard, Chapter 18 of this volume).

A recurrent theme of the national experiences has been that clustering *is* a useful policy approach, which raises a question as to the degree to which policies are encouraging groups of similar and complementary companies to co-operate to improve their innovative performance. The Finnish case study highlights a key feature of the interrelation between government activity and the boosting of innovation (Romanainen, Chapter 20 of this volume). Policy can readily be influenced by successful examples of practice, and the way success is defined can have significant consequences for subsequent policy developments. In Finland, cluster policies became associated with technological dynamism and industrial diversification, factors of absolute importance for as open and internationalised an economy as Finland. Similarly, in the United Kingdom, the success of the Cambridge bioscience industry has led to an association between cluster policy and the impacts of university spin-outs and graduate labour markets.

In these cases, clustering was a partial explanation for the success of existing economic activities where strong leaders have driven a process of change outside the government mechanisms. When they have chosen to interact with government, it has been under much more equal terms that the generally less successful top-down attempts to encourage non-associating firms to engage in clustering activity. Such leaders are critical to the success of government attempts to encourage clusters; where they spontaneously emerge, they prevent policy focusing exclusively on process rather than outcome. If such leadership does not exist, then the first phase of the policy must concentrate on creating leadership capacity through strong individual firms and their networks.

Concluding discussion: cluster policies or policies for clustering?

There is an intimate interrelation between the policies used to initiate and support clustering and the (successful or otherwise) experience of governments with particular industrial sectors. At the heart of the problem experienced by governments is the tension between the need for government to represent all its constituency (which is easily done in welfare and education policy areas) while supporting excellence without favouritism. In each of the cases there has been a lengthy learning process as governments have sought to understand what can and what should be done, and to replicate, generalise and consolidate particular past successes.

In some cases, this has been explicit. Denmark deliberately reviewed and amended the mega-clusters policy to include small-scale clusters (Holm Dalsgaard, Chapter 18 of this volume). The Netherlands recently reviewed their policies and moved from a generic to a phased (orienting, intensifying, targeting) approach (Gilsing, Chapter 19 of this volume). Scotland phased the introduction of cluster policies, with a first phase of very general and broad clusters and a much more specific second phase. In other cases, the development has been less formal; the Finnish Science and Technology Policy Council adopted the language of clusters in 1999 to give coherence to previous activities (Romanainen, Chapter 20 of this volume), while what has really galvanised the potential for regional cluster policy in the United Kingdom (vital because of very divergent regional innovation systems) has been the creation of regional development agencies which have asked for the remit to draw up regional cluster strategies.

Clustering and the cluster policy cycle

Cluster policies may be chosen for a range of reasons, to fulfil a political need, to integrate existing programmes, or to channel government support to technologically strategic industries.[3] The actual policies adopted may be constrained by existing programmes and instruments, while clever and sophisticated users may manipulate the policy environment to the exclusion of less sophisticated but potentially more dynamic businesses. These factors all limit the effectiveness of the implementation of the cluster policies. In each of the cases presented, however, governments have worked to circumvent the problems and to ensure that the policies encourage co-operation, innovation and competitiveness from the participating firms.

Each of the policy chapters demonstrated how, in particular cases, policies have been used to stimulate businesses to behave more like the cluster archetypes analysed and advocated by Porter. Porter's emphasis is firmly upon the clustering process, centred on the firms, while we have demonstrated that it is insufficient to regard "government" merely as a conditioning factor, particularly given the rise of purposive cluster policies. However, it is also insufficient to regard cluster policies instrumentally, identifying what firms need to achieve and then taking a series of steps to achieve that end. There is a clear difference between the progress of a cluster and a cluster programme. It is clear that when cluster programmes fail to continually focus on their impact on firm behaviour, the evolution of the cluster and the policy may become asynchronous and the value-adding potential of the policy is diffused.

Successful clusters are not necessarily underpinned by good cluster policy, but the cases suggest that the "best-value" cluster policies are those which are most highly targeted to the needs of the particular groups comprising the cluster. The Netherlands emphasises this through the use of three phases, through which it is anticipated that not all of the clusters will pass. The paradox is that success requires quite senior individuals with a policy-making as well as an implementation brief to work continually with small groupings of companies to ensure that the activities funded and encouraged are continually useful. As important as the policies are the refinements between policy cycles, through which the framework for innovative activity is created.

Linking the cycles — sequences of policies for clustering

One fact demonstrated by the case studies is the influence of successful policies on subsequent development. In a number of the cases, cluster policy has been shaped by past successes; in Finland, the past success of clustering by the forestry sector had helped to shape a vulnerable economic sector under the pressures of global competition. In England, the success of the Cambridge city-region complex supported government policies encouraging technology transfer and commercialisation by universities as the basis for high-technology growth. Much of the policy literature (*e.g.* Lagendijk, 1999) concentrates on the evaluation aspect of the final learning phase, sensibly attempting to determine whether particular cluster policies have added value to the industrial development process. However, the ramifications of policy success are wider, and relate to the influencing of subsequent developments.

A key problem is that evaluations are context-dependent and often lack a long-term perspective. The exemplar in this regard is the Silicon Valley cluster in the United States. Silicon Valley has gone through a number of phases, and in each phase, commentators failed (for essentially short-term reasons) to notice the features which would lead to the end of that phase and a radical change of direction.[4]

For political reasons, the majority of policy learning takes place within one policy cycle – the Netherlands, Denmark and Scotland all evaluated and reviewed their cluster policies and amended their operation in the light of those findings, while crises which entail seismic political shifts often produce wholesale institutional amnesia. Concern must be raised so that if (or when) the notion of cluster loses its kudos, the important innovation policy lessons learned through the cluster policy cycles are not themselves forgotten.

Targeting policies for clustering: the importance of scope and scale

Early analyses of clusters as a policy tool (*cf.* OECD, 1999) established the multiple ways in which governments can influence cluster behaviour (Boekholt and Thuriaux, 1999). This is further complicated by the multiplicity of entities which can be described as clusters, from highly inter-trading industrial sectors through self-sustaining corporate research networks to industry associations facilitating and encouraging co-operation. Given such a wide range of objects for policy intervention, and at times the inappropriate use of the term, it is perhaps unsurprising that policy has seemed unfocused and uncoordinated. An alternative way of coming to terms with this is to begin to categorise the dimensions on which things called "clusters" can be differentiated.

Charles (2000) characterised clusters along two dimensions, the geographical scale[5] and the density of the bonding between the elements in the cluster. This is presented below in Figure 1 (which deals only with the industrial groupings, not the features of the policies by which those clusters have evolved).[6] It is possible to characterise the policy style along similar axes, to consider how policies are directed at different types of cluster (Figure 2).

Figure 2. **Geographical scale and intensity of interaction**

National	Danish small-scale clusters	Finnish telecommunications cluster	Danish mega-clusters
	Japanese–style supply chains	Scottish Enterprise clusters	Motor Sport Valley (UK)
		Northern Offshore Federation (NE England)	Silicon Valley, California
Local	Micro-clusters or company networks	Flanders Language Valley	Cambridge biotech cluster
	Tight interaction		Loose agglomeration

Figure 3. **Variations in policy style with the geographical scope and intensity of interaction of the "cluster"**

	Tight interaction	Loose interaction
National	Strategic Interventionary	Enabling, facilitating Foresighting/programming
Local	Daily work and group involvement	Dealing with specific issues (*e.g.* planning regulations)

When the above table is taken in conjunction with the policy-cycle model, it becomes possible to see how the eventual style of policy making is affected by political factors influencing the adoption of "cluster policies". National policies which begin from input-output or other aggregate analyses are best equipped to deal with clusters as a means of futurological control, to provide strategic information for firms who are not necessarily aware of their parallel existence. A range of cases demonstrated that sensitivity was required to ground-reference the quantitative data against the real economy, and that decisions made about access to subsequent panel meetings could profoundly influence the allocation of benefits at a national scale.

Conversely, when national policies are targeted at self-aware and self-defining groupings, policy tends to centre around meeting the needs of those firms, and civil servants tend to build much closer and personal linkages with firm members than in enabling policy regimes where a few roundtables, rather than repeated industrial intervention, is the goal. The case of Denmark's small-scale clusters illustrated this very well (Holm Dalsgaard, Chapter 18 of this volume), although there are problems to this – over-identification by civil servants with the "cluster" can blight the evaluation and reduce its overall added value.

Local tight clusters often lack direct contact with policy makers and are constrained by existing instruments, meaning that the officials become guides to ensure that national policy instruments are relevant to the needs of local clusters (which does not always proceed smoothly). This was not the case in Flanders, where knowledge of local needs was congruent with the relevant political powers, thus greatly improving the added value of the policy measures (Larosse *et al.*, Chapter 6 of this volume).

Finally, in local loose agglomerations, which lack policy measures and co-ordinated industrial groupings, the role for the public sector is to establish the local needs and to ensure that the general conditions are met for the success of the cluster. In Cambridge, England, for example, the greatest drivers for the success of the biotechnology industry are national and international R&D investment flows, so local (and to a degree national) cluster policy has defaulted to sustaining the prestige of the location through restrictive spatial planning policies.

Boekholt and Thuriaux finished their analysis with the observation that there was a great deal of diversity among clusters and the appropriate policies. In this chapter, we have argued that not only is there diversity between clusters, but also between the policies used to support them, and that much value can be added to cluster policy by tailoring the entire *cluster policy style* to the characteristics of the cluster, which in turn depends on the way "clusters" have been defined. This does not diffuse the overall applicability of the cluster notion because in each case the competitiveness of the firms is being enhanced by encouraging networking to improve co-operation.

Greatest policy value can be added if the cluster policy style is congruent with the cluster policy cycle; national and loose clusters fit best with lengthy/long-term policy cycles which in turn need a bi-partisan commitment to the approach in order to ensure its long-term adoption. Local/tight clusters can be promoted locally, and indeed there can be problems with governments promoting particular local clusters (even if of national competitiveness) because of the requirements of impartiality. Thus, the best starting point for cluster policy making lies in identifying from the outset the style of the clusters which agencies are interested in promoting, and tailoring the length of the policy cycle appropriately.

NOTES

1. In this chapter, the term "cluster policy" is generally used, although the emphasis is primarily on those policies that actively promote clustering.

2. The choice of experts has not been without its own political significance; the case of Flanders above, as well as the Catalonian, Scottish and Quebecois cases, show that each chose Porter's Monitor consultancy, not just because of expertise in cluster policies, but in order to make a statement that regional industrial policy was autonomous from the respective federal state.

3. The idea that the policy-making process has its own logic is not a new idea; Heclo and Wildavsky's (1974) famous study observed the behavioural patterns and cultures which built up in the United Kingdom around the annual Civil Service budgetary negotiations with the Treasury. Although recent trends in public administration have stressed the new managerialism/privatism of civil services and governments (providing focused services for customers rather than environments for citizens, *cf.* Williams, 1999), each of the cases illustrates the importance of contextual determinants on the evolution, development and ultimately the implementation of policy.

4. The first analysis, in the 1970s and 1980s, regarded Silicon Valley as an exemplar, for its commercialisation of research activities but overlooked the dependence of the economy on a few large defence firms. Consequently, in the late 1980s, during lengthy federal defence restructuring, Silicon Valley was regarded almost as a senile cluster which had failed to adapt to changing circumstances. These analyses themselves overlooked the intrinsic adaptability of the individuals and firms of Silicon Valley, besides the explosion of the Internet which prefigured the rebirth of Silicon Valley and its re-admittance into the select grouping of industrial districts. It is not inconceivable that some external stock market or oil shock could once more invert the position of current success.

5. Geographical scale here does not preclude the possibility of strong international links, but focuses on the primary geographical unit identified by the cluster members themselves or by the level at which policy is applied.

6. The Northern Offshore Federation (NOF) is the central institution in the North East offshore cluster with a membership of several hundred firms in offshore construction, supply and related services. It plays a role in promoting skill development, innovation and other competitiveness-enhancing developments.

BIBLIOGRAPHY AND FURTHER READING

Benneworth, P.S. (2001), "Innovation and Modernisation of a Peripheral Industrial Region: The Case of the North East of England", unpublished PhD Thesis, University of Newcastle upon Tyne, Newcastle.

Boekholt, P. and B. Thuriaux (1999), "Public Policies to Facilitate Clusters: Background, Rationale and Policy Practices in International Perspective", in OECD, *Boosting Innovation: The Cluster Approach*, OECD, Paris.

Charles, D. (2000), "Clustering and its Contribution to Economic Development: Implications for Tyne and Wear", paper presented to Tyne and Wear Partnership Cluster Workshop, North Shields, England, 19 September.

Charles, D. and Benneworth, P. (2001), "Clustering and Economic Complexity: Regional ICT Clusters in the United Kingdom", Chapter 17 of this volume.

Dankbaar, B. and J. Cobbenhagen (2000), "In Search of a Regional Innovation Strategy for Flanders", in F. Boekema, K. Morgan, S. Bakkers and R. Rutten (eds.), *Knowledge, Innovation and Economic Growth: The Theory and Practice of Learning Regions*, Edward Elgar, Cheltenham.

Danson, M. and G. Whittam (1998), "Networks, Innovation and Industrial Districts: The Case of Scotland", in M. Steiner (ed.), *Clusters and Regional Specialisation*, Pion, London.

Florida, R. (1995), "Towards the Learning Region", *Futures,* Vol. 27, No. 5, pp. 527-536.

Gilsing, V.A. (2001), "Towards Second-generation Cluster Policy: The Case of the Netherlands", Chapter 19 of this volume.

Hassink, R. (1993), "Regional Innovation Policies Compared", *Urban Studies*, Vol. 30, No. 6, pp. 1009-1024.

Heclo, H. and A.B. Wildavsky (1974), *The Private Government of Public Money: Community and Policy inside British Politics*, Macmillan, London.

Hogwood, B. (1987), *From Crisis to Complacency: Shaping Public Policy in Britain*, OUP, Oxford.

Holm Dalsgaard, M. (2001), "Danish Cluster Policy: Improving Specific Framework Conditions", Chapter 18 of this volume.

Keating, M. (1998), *The New Regionalism in Western Europe*, Edward Elgar, Cheltenham.

Lagendijk, A. (1999), "Good Practices in SME Cluster Initiatives: Lessons from the Core Regions and Beyond", *ADAPT Project Report*, CURDS, Newcastle.

Lagendijk, A. and J. Cornford (2000), "Regional Institutions and Knowledge: Tracking New Forms of Regional Development Policy", *Geoforum*, Vol. 31, No. 2, pp. 209-218.

Larosse, J., P. Slaets, J. Wauters, S. Bruninx, P. Simkens and R. Wintjes (2001), "ICT Cluster Organisations in Flanders: Co-operation in Innovation in the New Networked Economy", Chapter 6 of this volume.

Lawton Smith, H. (2000), Technology Transfer and Industrial Change in Europe, Routledge, London.

Lundvall, B.-Å. and B. Johnson (1994), "The Learning Economy", *Journal of Industrial Studies*, Vol. 1, No. 2, pp. 23-42.

Morgan, K. (1992), "Innovating by Networking: New Models of Corporate and Regional Development", in M. Dunford and G. Kaflakas (eds.), *Cities and Regions in the New Europe*, Belhaven, London.

Munro, A. (2000), "Experiences of the Food and Drink Cluster", paper presented to the OECD Cluster Group Workshop on "Do Clusters Matter in Innovation Policy?, Utrecht, the Netherlands, 8-9 May.

Paija, L. (2000), "Industrial Network Relationships in the Finnish ICT Cluster", paper presented at the OECD Cluster Group Workshop on "Do Clusters Matter in Innovation Policy?", Utrecht, the Netherlands, 8-9 May.

Piore, M.J. and C.F. Sabel (1984), *The Second Industrial Divide*, Basic Books, New York.

Porter, M.E. (1990), *The Competitive Advantage of Nations*, MacMillan, Basingstoke.

Porter, M.E. (1998), *On Competition*, Harvard Business School, Boston.

Roelandt, T.J.A and P. den Hertog (1999), "Cluster Analysis and Cluster-based Policy Making in OECD Countries: An Introduction to the Theme", in OECD, *Boosting Innovation: The Cluster Approach*, OECD, Paris.

Romanainen, J. (2001), "The Cluster Approach in Finnish Technology Policy", Chapter 20 of this volume.

Rosenfeld, S.A. (1997), "Bringing Business Clusters into the Mainstream of Economic Development", *European Planning Studies*, Vol. 5, No. , pp. 3-23.

Rouvinen, P. and P. Ylä-Anttila (1999), "Finnish Cluster Studies and New Industrial Policy making", OECD, *Boosting Innovation: The Cluster Approach*, OECD, Paris.

Saxenian, A.-L. (1994), *Regional Advantage: Culture and Competition in Silicon Valley and Route 128*, Harvard, Cambridge, Mass.

Scottish Enterprise (1993), *Improving the Business Birth Rate: A Strategy for Scotland*, Scottish Enterprise, Glasgow.

Williams, N. (1999), "Modernising Government: Policy-making within Whitehall", *Political Quarterly*, Vol. 70, No. 4, pp. 452-459.

Wintjes, R. (1999), "Flanders Language Valley", *MERIT Working Paper*, MERIT, University of Maastricht.

Chapter 22

CREATING AND SUSTAINING INNOVATIVE CLUSTERS: TOWARDS A SYNTHESIS

by

Pim den Hertog, Dialogic Innovation & Interaction and Department of Innovation Studies, Utrecht University,
Edward M. Bergman, Vienna University of Economics and Business Administration, and
David Charles, Centre for Urban and Regional Development Studies, University of Newcastle*

Introduction

Clusters are about interdependencies and innovation

The cluster approach, as presented in all its variety in the preceding chapters, is part of the growing family of innovation systems approaches (Edquist, 1997; Malerba, 2000[1]). Clusters reflect the systemic character of modern innovation and interactive innovation processes; innovation increasingly depends on market- and non-market-induced interactions among interdependent actors (firms as well as other types of organisations). Interactions between actors in clusters are based on trade linkages, innovation linkages, knowledge flows in various forms or the sharing of a common knowledge base or factor conditions. Although clusters are in principle market-based, non-market-based relationships do play a role. These interactions and interdependencies, by definition, transcend the borders of individual sectors and industries. The cluster perspective offers useful insights into how these linkages and interdependencies are shaped, how they evolve over time and how they affect innovation, and defines the scope that exists for policy action.

Clusters as frameworks for new forms of governance

The cluster approach is more than just an analytical instrument for passive use, clusters also offer a robust organising framework for addressing or removing systemic imperfections in the functioning of these reduced-form national innovation systems (NIS). In practice, the cluster approach has proven to be a useful framework for developing and applying new forms of governance, moving away from direct intervention towards forms of indirect inducement. This approach focuses upon facilitating networks and creating the institutional setting that provides incentives for market-induced cluster formation and for the revitalisation of existing clusters. For policy makers, it can be seen as a tool for

* The authors wish to thank Svend Remoe, co-ordinator of the overall OECD NIS project, for generously providing feedback on drafts and for taking part in a substantial part of the editorial deliberations.

knowledge and innovation management that can pinpoint those actions that are most needed to overcome barriers to innovation and to customise these to the specific cluster.

We cannot emphasise enough that the contributions included in this volume convince us that there is neither a standard cluster approach, nor a fixed policy recipe for implementing the cluster approach in practice. Below we will, without doing full justice to the richness of the individual contributions, briefly flag what we see as the main overall lessons to be drawn from an analysis of innovation clusters and cluster innovation policies.

Overall lessons from analysing innovation clusters

Cluster analyses reveal that every country/region has a unique cluster blend

Every country or region has its own selection of clusters and specialisations with different characteristics and role in the economy. Cluster studies are useful analytical instruments for understanding better how individual economies are structured, their specialisations and the role that various clusters play in the wider economy. The input-output analysis as performed by Peeters *et al.* (Chapter 12 of this volume) for Flanders and Switzerland illustrates – in addition to offering a good starting point for cluster analysis in an international, comparative, context – that every country has its own collection of clusters and specialisations. Even at the level of mega-clusters, individual clusters have different characteristics and play a different role in the economy. In addition, an examination of the profiles of the various clusters reveals the heterogeneity of the economic activities, in terms of size, connectedness, R&D intensity, share of innovative products, etc. Luukkainen (Chapter 13 of this volume) sees the action of identifying mega-clusters and their characteristics basically as a technology policy tool that can be useful in analysing and developing industrial structures and for assessing the economic impacts of R&D investments in advance. He also notes that knowledge-intensive business services (KIBS) have an important role as an external knowledge transfer mechanism in the clusters.[2]

Clusters are thought to represent a useful level of analysis between the national and the sectoral levels, as they are a recognisable level of analysis for most firms and also for the relevant actors in the knowledge infrastructure formalising a representation of the "world in which we operate and function".

Clusters are variation and selection environments that are inherently different

Clusters, perceived as reduced-NIS, are the relevant selection and variation environments in which firms and other type of organisations such as intermediaries or knowledge institutes operate and innovate. National, and for that matter regional, economies can be thought of as consisting of various mixes or blends of these innovation clusters. Selection and variation processes in clusters proved to be inherently different between countries (or regions), between technologies, between policy systems and ultimately between individual clusters. Not only do innovation and innovation processes differ between construction, agro-food and ICT clusters, the way innovation is taking shape in, for example, the Finnish, the British or the Flemish ICT clusters – or between the various ICT clusters within a single country – are inherently different.

Therefore, the notion of "ideal" innovation clusters is a fallacy

History, the types of knowledge, the particular stage in the cluster's life cycle and networking practices are among the most important factors responsible for cluster specificity and for differences between clusters in what has been labelled "innovation style". These are substantiated below.

Historic background and country specificities. Current innovation patterns in, for example, the Finnish ICT cluster cannot be appreciated fully without understanding the long history of liberalisation of telecommunication markets and the rigour with which Nokia reinvented itself (Chapter 2 of this volume). Similarly, the examination of three UK ICT clusters reveals them to have been differentially shaped by varying combinations of inward investment, local entrepreneurship and defence, regional planning and science policies over a number of decades (Chapter 17 of this volume). The innovation dynamics in the Norwegian agro-food cluster calls for an understanding of the historic role that co-operatives have played (Chapter 8 of this volume). Building an Austrian new media cluster requires an understanding of the rich content base that Austria has developed over centuries and which can be redeployed using new media (Chapter 15 of this volume). In construction, typical localised strengths such as those of the Swiss in tunnelling technology (Chapter 11 of this volume), the Dutch in civil engineering in soft soils and water (Chapter 10 of this volume), or the way the Danish construction industry is affected by the Scandinavian or Danish design tradition (Chapter 9 of this volume), all point to the need to appreciate historical and country specificities before discussing innovation performance and possible policy interventions. History matters, also when assessing innovation in clusters.

Type of knowledge. Clusters also differ according to the characteristics of the knowledge base and the way in which knowledge is diffused. The type of knowledge may differ between clusters in the sense of the relative importance of tacit and codified forms of knowledge or the degree to which knowledge is science-based or more application-oriented. ICT knowledge is generally thought of as being fairly well codified, which may be true for the underlying fundamental knowledge base. Yet, the various case studies included in this volume show that application knowledge is key and leads to numerous specialisations in ICT.[3] ICT clusters in different regions were found to be highly dissimilar! Quite often, ICT clusters emerged in those regions in which the exchange of tacit and implicit forms of knowledge was abundant and where specific knowledge bases were in place. In the Flemish case, for example, this resulted in various specialised ICT valleys (Chapter 6 of this volume). Similarly, in the Dutch multimedia cluster, clear regional specialisation could be found (Chapter 7 of this volume). In the Spanish case, two quite different ICT-related clusters could be identified, a telecommunication-based cluster in the area around Madrid and a consumer electronics cluster in the Catalan region (Chapter 5 of this volume). The Japanese contribution showed that industries that spatially cluster might be categorised into industries whose key technologies depend on implicit or tacit knowledge such as design or know-how, industries with a fairly substantial service component (Chapter 14 of this volume).

Stage of cluster development. An important variable that should not be overlooked is the stage of cluster development. The life-cycle approach suggests evolutionary patterns that contain adaptation and variation. As Peneder (Chapter 15 of this volume) rightfully indicates, a cluster life cycle can be said to exist. Cluster development progresses over various stages and the constellations of actors, as well as the particular barriers with which a cluster is faced over time, might vary. Getting a cluster off the ground might imply investing in new knowledge, cross-disciplinary platforms, new networks and alliances. In more mature clusters, for example, codification of knowledge is more important, non-technological knowledge gains in importance and in most cases the number of links with end users increases. The whole idea of a cluster life cycle points to the fact that the form of facilitation by policy makers will vary over the lifetime of a cluster. Facilitating an emerging cluster requires different actions from policy makers than revitalising an existing mature cluster.

Networking practices. Finally, an important variable that often makes clusters so different from one another is the considerable variation in networking practices. Networking not only requires a certain level of trust and preparedness but also the ability to define common goals, to engage in partnerships across industrial sectors, to set up relationships with research institutions as needed, as well as establishing links with end users or their representatives. The study on industrial clustering in Japan found that firms had frequent and high-intensity external links with business partners across industrial groups. However, significant differences were found in the linkages with business infrastructure and research infrastructure (Gonda and Kakizaki, Chapter 14 of this volume). The need for networking, the ability to engage in networking and the actual shape of networking in practice are unevenly spread over clusters, and over countries and regions since culture plays a role in this as well. As pointed out by Pentikaïnen (Chapter 16 of this volume), these differences need to be understood before specific cluster initiatives are implemented; the more so as networking should not restrict market competition.

In conclusion, thinking of the management of innovation at the cluster level in terms of *ideal types* and *best practices* is destined to fail, as individual clusters are likely to differ on many aspects. Cluster specificity implies the need for customised sets of policies to promote innovation in clusters (see below).

Industrial clusters may transcend geographical levels

Clusters are economic phenomena that can be perceived as selection and variation environments that operate at various geographical levels.[4] Value-chain-based clusters represent the relevant set of actors and inter-linkages that firms and other cluster actors recognise as "the world in which we operate and function".[5] Sometimes very localised (industrial district) clusters operate on world markets. Sometimes, localised markets are served by clusters that are intrinsically international in terms of production networks or value-chain-based clusters. Even within a single cluster, clear international as well as regional or even localised elements can be identified, as was shown in many of the case studies on ICT clusters. In that respect, the notion of clusters may transcend geographical levels. Increasingly, national and regional cluster performances are linked to and form part of international value chains. This implies that clusters of which all individual elements are to be found in a confined area are the exception rather than the rule. In some trades of business, it might even be counterintuitive to expect "complete" clusters at the regional or national level as the relevant knowledge base is strongly internationalised. Linking into the international knowledge base and tapping into an international reservoir of knowledge may be essential in these cases.

The scale of individual countries and regions is an important factor here as well. Charles and Benneworth (Chapter 17 of this volume) noted that in relatively large economies like the United Kingdom, cluster-based innovation systems are likely to emerge at a regional scale, and indeed the concentration of particular clusters in particular regions may be an unforeseen consequence of public policy. These regional innovation systems have a particular set of characteristics that differentiate them from scaled-down national innovation systems. They are specialised around focused clusters, they draw heavily on national inputs that may lie outside the region, and their boundaries may be less determined by administrative territories and more by interactions and flows between firms. Thus, the mapping of such clusters cannot be restricted to a specific region, but is often trans-regional.

How then do clusters relate to concepts like RIS and NIS?

Regional or national economies cannot be automatically defined as regional innovation systems or national innovation systems if the only qualifying criteria is a geographical boundary. Some regions are recognised as having attributes which foster the development of clusters but not every region functions automatically as an innovation system. The biggest risk is that every administrative region is called an innovation system or every collection of firms and industries linked in a value chain an industrial cluster. This brings us to the question of what are the systemic properties of these selection and variation environments? How "complete" do sets of interdependent (and networked) firms, linked in value chains, have to be before one can say that they belong to a cluster? How close need user-producer links be? To what degree are specialist suppliers a prerequisite? How developed do strategic links with universities, research institutes, knowledge intensive business services and bridging institutions need to be? To what degree have these actors to share resources such as available technological competence, specialised labour, a client base, specialised knowledge institutes, reputation and a specific innovation culture before we can speak of a cluster? This is, of course, a continuum. The danger of working with an analytical concept such as clusters – as with any concept – is that its flexibility is put to the test and wishful thinking comes into play. Sometimes the concept is abused such that random selections of firms are labelled clusters just for the sake of it. If a hammer is the only tool available, then everything looks like a nail!

Emerging, technology-based clusters are more likely to be part of wider international clusters

Technology-based clusters, such as the ICT clusters described in this volume, seem to be international in character from the outset. Green *et al.* (Chapter 3 of this volume) discuss the ICT cluster as a boundaryless cluster and rightfully pose the question "can there be 'national/regional' clusters for a technology that is developing by definition at a global scale?" In a similar fashion, Larosse *et al.* (Chapter 6 of this volume) present some localised ICT clusters and describe these as mixed local and global phenomena, introducing what they label the "global-local paradox". California's Silicon Valley or, if focusing more on ICT-based services for the New Economy, New York's Silicon Alley, are usually seen as the ultimate examples of rather complete ICT clusters. However, although often politically motivated to build similar powerful ICT clusters, most countries are not capable of replicating these kinds of value chains within their national borders. This is most obvious for the smaller, open economies. This implies that most ICT clusters are more or less incomplete or, to put it more positively, specialise in certain segments. The Finnish telecommunication cluster is a relatively complete cluster although it relies on foreign sources for certain knowledge and technology inputs. Its key player, Nokia, excels not only in using the considerable Finnish knowledge base, but also taps into foreign sources for its R&D and non-technical knowledge (*e.g.* on product branding). As Paija (Chapter 2 of this volume) indicates, international ICT firms are increasingly using Finland as a test-bed for new technologies and services. Green *et al.* (Chapter 3 of this volume) show how Ireland, lacking a typical national ICT champion, has moved away from an exogenous development strategy towards a strategy of embedding these firms into a local context and developing more indigenous firms, to eventually create an indigenous growth cluster. Denmark (see Dahl and Dalum, Chapter 4 of this volume) cannot be said to possess a complete ICT cluster, but rather seems to be part of a wider Nordic ICT cluster in which it has traditionally played a role as advanced user. The Dutch study revealed that, notwithstanding the presence of a (regionally distributed) multimedia cluster, a player like Philips sees itself more as part of a wider ICT cluster (den Hertog *et al.*, Chapter 7 of this volume).

More mature clusters typically function at a national or regional scale

In general, more mature clusters, such as agro-food and construction, are likely to function primarily at a national or regional level. However, even within these mature clusters, there are parts of the value chain that are not present at the local or regional level. So, while there certainly is a Norwegian agro-food cluster, some of the necessary machinery is imported. Most of the construction clusters, especially the residential housing parts, are typically national or regional clusters, yet some specialised supplies (electro-mechanical equipment or ceramics) might be supplied from abroad.

Innovation in so-called low-tech clusters is more advanced and complex than is often depicted

The cluster studies undertaken by members of the cluster focus group indicate that innovation in so-called low-tech/mature clusters is more advanced and complex than often depicted. Innovation in these clusters is knowledge-based as well. They often have highly specific innovation styles which may be, but need not be, less advanced compared to the so-called high-tech clusters. In clusters such as agro-food and construction, in which sources of knowledge might prove to be exogenous and not as clearly dominated by one (fancy) technology such as in biotech or ICT, quite knowledge-intensive production systems were seen to be present. The knowledge base of these clusters was observed to be relatively more application-oriented and more process-oriented. Notwithstanding the fact that technological knowledge was seen to derive mainly from supplying manufacturing industries, these clusters have remarkably complex sets of institutions (in the widest sense) and historically developed ways of innovating, influencing the way in which knowledge is created, diffused and applied.

However, this requires an open mind towards the role that non-technological knowledge plays in innovation

The inter-linkages present in clusters are not necessarily focused on advanced technologies. Of course, applying or diffusing technological knowledge along the value chain requires a minimal knowledge base at the receiving end. However, these inter-linkages equally extend to other types of knowledge needed for innovation such as organisational and marketing knowledge. The construction cluster cases show the considerable organisational and market knowledge required to bring about innovation there. Much of the "innovation problem" in this cluster can be said to relate to the non-technological aspects of innovation. Major barriers which hinder innovation in the construction cluster include co-ordination across disciplines and between phases in the building process, a limited sensitivity to client-led or market-pull innovations, detailed procurement, and a current weakness in some market segments concerns market pressure for innovative products. This is a further indication that looking at R&D intensities does not suffice when facilitating innovation.[6]

Main challenges for cluster analysts

Apart from further closing the gap between analytical cluster studies and the day-to-day reality of policy makers using the cluster approach in policy processes, further efforts are needed on at least four issues. In the first place, micro data and especially regular innovation and production data need to be available at the level of clusters. This requires some persuasion, especially of statisticians who conventionally organise their data at sectoral level. Second, new analytical work is needed in order to better identify clusters in their infancy, *i.e.* clusters that have not developed fully, but could mature into important clusters. Possibly, the life-cycle concept can be used to identify new constellations of economic activity and identify "what will be the next pillar of the economy". This also requires a greater sensitivity to clusters that are not purely built on (a combination of) technologies as there seems to be tendency to define clusters around key technologies, leading to a bias towards technological knowledge and clusters dominated by manufacturing firms. Third, a major challenge for cluster analysts will be to measure international clusters or the ways in which national clusters are linked to parts of the value chain that go beyond national borders. Finally, there is a further need for a closer link between cluster analysis and cluster policy practice in order to make the whole process of cluster identification, cluster selection, cluster actions (and knowing when to stop) more transparent and verifiable. This is clearly reflected in the strategic framework introduced recently in the Netherlands (see Gilsing, Chapter 19 of this volume), which aims to make the trade-offs in policies towards clusters more objective and allow, in principle, for a better assessment of cluster policy making.

Overall lessons from cluster innovation policies[7]

Megacluster analyses need to be combined with action-oriented small-scale cluster approaches

If the cluster approach is to succeed, it is important to achieve the right mix between analysis and action. Although cluster analysis aimed at identifying mega-clusters has also been used to identify barriers that hinder the more general development of these clusters, the general belief is that this type of analysis is a fairly crude mapping device. Much of the time, the relevant cluster context for policy is confined to small-scale clusters. Mega-clusters are useful for general background information but not for formulating policies and not for immediately developing a cluster strategy for a particular cluster. Identifying mega-clusters as a step in the policy formulation process will therefore have to be supported by other, less formal and more qualitative methods at lower levels of aggregation. This is what has happened in most countries and regions that have implemented policies aimed at increasing the innovativeness of clusters. They have developed and used more detailed, and practical, action-oriented tools that are applied at lower levels of aggregation. Up until now, these two trajectories (mega-cluster and small-scale clusters) have developed too much in isolation. There is a need for bridging the gap between extensive cluster identification and mapping exercises (approaches put forward by researchers and policy analysts), and the practical cluster actions taken at lower levels of aggregation (put forward by policy implementation bodies, cluster organisations and consultants). A fine attempt to do so are the recent "clusters of competence" recently introduced in Denmark (see Holm Dalsgaard, Chapter 18 of this volume). The introduction of "ClusterMonitor" studies in the Netherlands is aimed at identifying barriers to innovation and proposing practical measures to overcome them (see den Hertog *et al.*, Chapters 7 and 10 of this volume). This type of analysis, more so than the mega-cluster exercises, shows that innovation is as much an organisational and even cultural phenomenon and cannot be treated as a technological phenomenon only.

Cluster policies can benefit from cluster analysis in several ways

Cluster policy is about creating the right framework conditions for innovation, it is about identifying barriers to innovation, and it is about building relationships and networks. As such, it requires an appropriate mix of analysis and action. Adopting the cluster approach implies that governments need to be able to engage in pragmatic, problem-solving processes, including a willingness to learn through experimentation and institutional adaptation. Cluster studies can in practice be used as a working method for systemic innovation policy making.[8] They can be a way of opening up the dialogue on how innovation takes place in a particular cluster and of learning how policy making can contribute to this process. How this is realised differs across clusters and across countries. Important variables here are the way in which dialogue among industry, research and governments is institutionalised, the size of the economy, the policy culture, the level of government intervention and the degree of industrial and technological specialisation. Romanainen (Chapter 20 of this volume), discussing the Finnish approach towards cluster policy, lists as the major benefits of the cluster approach, the assistance it provides in identifying networks and linkages between industries; the possibility to identify emerging growth areas and the roles that various kinds of organisations play in it (which helps to design measures to support this growth); and the practical use it offers for identifying technologies and problems common to several industries. The latter mainly refers to the fact that, as much as an analytical tool, the cluster approach provides a working tool for policy makers to (proactively) create platforms and programmes for cross-disciplinary and/or cross-industry interaction.

Innovation clusters are shaped by all kinds of policies (including non-innovation policies)

The case studies included in this volume have proven without exception that policy matters in the way clusters emerge and in the way their innovative capacities develop over time. This is especially apparent when the history of certain clusters is analysed in more detailed and shown to be affected by deliberate and less deliberate policy initiatives. What is important, though, is that many of the policies that shape clusters lie outside formal "cluster policy" or even industrial and technology policies. Each cluster is affected by a complex interplay of policies influencing the trading environment, sources of innovation, the nature of places where cluster resources come together and the regulation of the cluster. A focus purely on innovation or industrial policy, without such a broad and historically contextualised perspective, will yield a narrow and myopic view. The development of the UK ICT clusters described by Charles and Benneworth (Chapter 17 of this volume) was noted to be heavily affected by various types of policy as diverse as defence procurement policies, science and regional development policies (among others). In a similar vein, the history of the Danish construction cluster showed that policy intervention played a substantial role at all levels in shaping the cluster and the scope for innovation (see Dahl and Dalum, Chapter 9 of this volume). The direction of innovation in the construction cluster was seriously affected by macroeconomic policies (the construction cluster was used to stabilise the overall economy), but also by detailed specifications and building standards in private homes, housing policies, energy-saving policies and fiscal policies. Some of the resulting innovative strongholds in the Danish construction cluster, such as energy-efficient construction, can be said to have been actively shaped by government. However, many policy initiatives and regulations that affected and direction of innovation in clusters were not originally initiated to support innovation. In conclusion, clusters and innovation in clusters are as much influenced by other types of policy making as by explicit cluster policies.

Some policies might have perverse effects on innovation in clusters

The foregoing also implies that some policies might damage the emergence or further development of clusters. This can be a matter of political choice or the unintended result of policies that generally have not been put in place to facilitate innovation in clusters.[9] Think, for instance, of how regulation of housing markets has affected the construction cluster, how trade policies and policies targeted at co-operatives have traditionally affected market dynamics and innovation in agro-food markets, and how education and science policy affect the development of ICT clusters. As policies aimed at stimulating innovation in clusters mostly start from analysing the systemic imperfections in clusters, it might well be that some forms of policy making are counterproductive from an innovation point of view. Governmental procurement policies in construction are sometimes not sufficiently used as a lever for facilitating innovation, *e.g.* by using detailed specifications instead of formulating functional requirements, often in combination with a clear cost focus (see, for example, Vock, Chapter 11 of this volume; den Hertog and Brouwer, Chapter 10 of this volume). Even specific cluster policy tools can hamper innovation if they are not sufficiently customised to the specific needs of individual clusters. Not all clusters might benefit from policies aimed at bringing knowledge institutes and industry closer together. Similarly, a government that is too active in facilitating strategic studies and platforms can in some clusters lead to cluster actors leaning back too far, waiting for government to make the next step.

Innovation clusters can be facilitated by varying customised mixes of innovation and non-innovation policies

Looked upon differently, policies aimed at increasing innovation in clusters should not be perceived as a well-defined and specific set of policies (cluster policy in its narrow sense). In fact, policies aimed at supporting innovation in clusters should determine, on a case-by-case basis, the factors that influence (hamper, trigger) innovation in clusters (cluster policy in its widest sense). This implies that policies that are not specifically aimed at innovation can be used as levers to support innovation in clusters. Hence, there is a need to look at a wider array of policies and their interactions in policy systems. In practice, policies aimed at facilitating innovation in clusters require appropriate and customised doses of intervention by policy makers that more often than not consist of fairly limited actions. It is important to understand that bringing about changes in government regulations that act as a barrier to innovation in a specific cluster can quite often be as effective in spurring innovation as more specific innovation policy tools. This also implies that cluster policy makers might need to intervene in policy areas and policy domains that may not be immediately associated with innovation policy. As Holm Dalsgaard (Chapter 18 of this volume) notes, cluster policy is not a separate type of policy but rather represents an approach that impinges upon other policy fields as well and therefore requires interdepartmental co-ordination. Cluster policy then is a rather general approach to policy making and an action tool enabling framework conditions to be optimised for innovation in clusters. This requires an almost missionary orientation on the part of cluster policy makers to convince others who are not primarily interested in innovation to take innovation on board as one of the relevant steering criteria. This places high demands on the capabilities of the individual cluster policy maker.

Clusters are useful frameworks for co-ordinating policies and reducing complexity

Governments must assess how different policies have affected the cluster in the past and how they might affect its future trajectory. Viewed from this angle, the cluster provides an integrating perspective within national innovation systems to turn the theory of systemic innovation into a more pragmatic and practical application. The cluster approach provides an integrative knowledge and innovation management tool or framework for spurring innovation in clusters and customising all policies affecting innovation in clusters. Clusters provide policy makers with a way of dealing with increased complexities and better targeting policy by addressing particular systemic failures that hamper innovation. Thus, the role of governments may be seen as one of providing selective responses to the needs of innovative clusters. In other words, leveraging innovation in clusters is in itself a challenge, calling for appropriate policy mixes to be designed in pragmatic ways. With further convergence in macroeconomic policies, such policies are likely to increase in importance at the regional and national level as they allow some differentiation *vis à vis* other countries and regions.

The cluster approach does not offer standard policy recipes[10]

From the foregoing, it will come as no surprise that the cluster perspective does not prescribe one or two ideal tools to increase the innovativeness and adaptation capabilities of a particular cluster. What might benefit one cluster in one country might be counterproductive elsewhere. How the cluster approach should be translated into practical policy tools is highly cluster-specific as different innovation dynamics call for different actions.[11] In one cluster, policy actions can be limited to making sure that competition is sound,[12] in other clusters, it might consist of various roles (demanding customer, technology foresight, creating the appropriate knowledge infrastructure, looking after competition practices, IPR problems, etc.). The main task of policy makers promoting innovation in clusters is to facilitate the networking process and create an institutional setting which provides incentives for market-induced cluster formation and forms of co-operation in both emerging and mature clusters. They have at their disposal a set of possible roles, instruments and analytical tools from which they can choose. Examples of innovation policy measures include: raising awareness of the benefits of knowledge exchange and networking; providing support and appropriate incentive schemes for collaboration; initiating network brokers and intermediaries to bring actors together; facilitating the informal and formal exchange of knowledge; setting up competitive programmes and projects for collaborative R&D; and ensuring that (public) institutions (especially schools, universities, research institutes) cultivate ties with industry. However, non-innovation policies might also be used to spur innovation, such as competition policies; education policies; cluster-specific deregulation; fiscal policies; environmental policies; regional policies, etc. This menu-approach is present in the new strategic framework for so-called second-generation cluster policies in the Netherlands presented by Gilsing (Chapter 19 of this volume).

Cluster policies are dependent on the stage in the cluster life cycle

The choice of what role, instrument and analytical tool to use, and when, depends not only on the particular needs of the actors in a particular cluster, but also on the stage in the cluster's life cycle. Over the life cycle of a cluster, various constellations of actors ask for various sorts of facilitation by policy makers. Furthermore, it depends on the degree to which policy makers are prepared to customise policies to particular clusters. The Finns, for example, have not chosen strong cluster-specific policy; instead, in accordance with a perceived government role primarily as a facilitator, it was decided that policy should not favour any specific cluster over others (see Romanainen, Chapter 20 of this volume). The danger is that by favouring existing clusters, less attention may be devoted to identifying and facilitating emerging innovative clusters. In many respects, the cluster approach requires a balancing act by policy makers (between new and established clusters, between technology- and non-technology-based clusters, between supporting co-operation while not destroying competition, etc.).

A major risk for further development of the cluster approach is "high-tech myopia"

Possibly the biggest risk in cluster analyses and cluster policies is that policy makers and researchers tend to focus on "high-tech" clusters and the obvious cluster success stories that abound. This is a major risk as it is usually forgotten that the rise of such clusters in the first place is the result of a combination of an often unique mix of mostly strongly localised factor conditions and development trajectories built up over decades that cannot be replicated overnight. The mechanisms and experience built up in clusters – no matter whether these are labelled as high-, medium- or low-tech – are valuable capacities. As long as clusters have built-in mechanisms to renew and re-invent themselves over time, this is a very precious asset. Therefore, characterising clusters as low- or medium-tech might be misleading. Hauknes (Chapter 8 of this volume) shows how knowledge-intensive a cluster like agro-food has become; the more so if one does not overlook the non-technological knowledge involved in innovation.

A second risk is adopting (counterproductive) standard cluster policy models and standard tools

One of the major pitfalls that follows from the above argument is that working with standard policy models and using a tool-push approach can be dangerous. Quite often, policy instruments and working tools are developed for particular vanguard clusters (for example, ICT or biotechnology) and subsequently applied to other innovation clusters without too much consideration for their specificities. This is a particularly dangerous development as cluster policies are precisely about customising sets of policy tools to the needs of a particular cluster and not about applying a standardised cluster approach. Similarly, at the level of individual cluster tools, there is the threat of a "tool-push" approach, which may lead to failures. Gilsing (Chapter 19 of this volume), for example, suggests that the application of particular analytical tools (*e.g.* technology radar) was used in too standardised a way in different clusters, leading to some disappointments. Both policy models and cluster policy tools need to be applied with great care and a "one-size-suits-all" approach is counterproductive. There is no magic recipe for the success of future cluster policies; it is a constant process of learning and improving. Most important for the cluster approach is to create awareness and a constructive debate about the situation in each cluster (Holm Dalsgaard, Chapter 18 of this volume). The ideal outcomes are a common agenda and initiatives and policies that strengthen innovation in clusters.

Policies aimed at clusters should strike a balance between creating and sustaining innovative clusters

The creation of a knowledge-based economy should not be equated with an exclusive focus on high technology. Switching at great cost to artificially created new (mostly high-tech) clusters that are *en vogue* could turn out to be ill-advised from both a policy and an industry point of view. It is recommended to build on existing strengths – even if these are in so-called low-tech clusters – and not to concentrate too much on creating new clusters from scratch. An additional argument for not neglecting opportunities to increase the innovativeness of low-tech clusters is that such neglect might prove costly as a consequence of their "weight" in the national or regional cluster blend. Even within individual clusters, a focus on high technology and R&D intensity is no guarantee that innovation will be supported most effectively.

Clusters as policy tools require experimentation and learning

For cluster policies to be successful, a "new breed" of policy makers is called for. They need to combine the analytical skills required to obtain an in-depth understanding of the innovation dynamics and innovation style of a particular cluster and the flexibility to decide on their most appropriate role (including the decision that there might not be a role to play!) to foster innovation. This mainly requires a trajectory of experimentation and constant policy learning as policy making related to clusters involves a great deal of "trial and error". Increasingly, this learning and experimentation will have to take place in an international context. Not only is it useful to reflect on how policies for facilitating innovation in clusters are dealt with in other countries, but strongly internationalised clusters require appropriate policy responses as well. As noted by Romanainen (Chapter 20 of this volume), one of the most pressing challenges for the cluster approach is how best to deal with strongly internationalised clusters. For policy makers dealing with clusters, it is advisable to look at what might be called supra-national innovation systems and consider cross-border policy co-operation for cross-border clusters (see also Holm Dalsgaard, Chapter 18 of this volume). Another way of working on the continuous improvement of cluster-based innovation policies is to evaluate experiences as systematically as possible, not only in order to increase the accountability of cluster-based innovation policies, but also as a way of codifying "good practices".

Cluster policy makers must be able to mix and switch between various roles …

Policy for innovation, as much as innovation itself, is a learning process. This implies that those responsible for executing cluster policies (or for co-ordinating in a cluster framework the complex set of policies that affect innovation in a cluster) need to possess certain capabilities. Adopting a cluster approach implies that the policy-making process – the day-to-day operations of innovation policy makers – is changing. Today, policy makers are less and less likely to implement policy measures and policy programmes using a top-down approach; they have become an integral part of the decision making which takes place at network level. Creating favourable conditions for straightforward interaction between the relevant actors (Paija, Chapter 2 of this volume) is one of the most logical outcomes of this. Cluster policies are demanding for policy practitioners in many ways as they have to be able to: *i)* close the gap between cluster analysis and daily policy practice; *ii)* understand the whole set of policy tools available; *iii)* understand which tools to use when; *iv)* handle complex cluster programmes and projects; *v)* withstand the pressure to use cluster policies for traditional industrial policy purposes; *vi)* act in various capacities, ranging from sparring partners, programme managers, stimulating dialogue, etc.; *vii)* combine a fairly pragmatic action-oriented approach, but at the same time be able to reflect on policy making as well; and *viii)* switch between various clusters that might have very different needs in terms of support and steering. The mix of roles needed to accomplish these tasks will vary, and governments must find their individual paths in line with national traditions and constraints. However, it is clear that cluster-based policies aimed at furthering innovation require intelligent, flexible and creative policy makers capable of engaging in a trajectory of policy learning and experimentation, constantly switching between analysis and pragmatic action.

… and need to be armed with persuasiveness and perseverance

The process of using and operationalising the cluster approach is a process of experimentation, variation, customisation and codification. The cluster approach cannot be applied overnight: for various reasons, such an approach requires considerable persuasive abilities and persistence to become established in policy circles. Not only is the cluster approach demanding in terms of the time and skills it requires of individual policy makers as indicated above, it is also demanding because it has to fit into the framework of existing policies and budgets. The cluster approach is an organising framework in which various types of policies that affect innovation in clusters come together. Automatically, a need for co-ordination between different political jurisdictions (various departments and levels of policy making) emerges, while not offering an immediate policy fix or immediate standardised policy programme. Perceived in this way, cluster-based policy making is a multidimensional balancing act between analysis and highly practical cluster policy actions; between allowing for bottom-up initiatives and top-down steering; between looking after the needs of established mature clusters and the needs of newly emerging clusters; between international, national and regional levels of policy making; between various policy roles or capacities; and, finally, between the needs for customisation of policy actions and a need to increase accountability of cluster-based policies. However, we never said that embarking on the cluster approach was going to be easy!

NOTES

1. Malerba's definition of what he terms "sectoral systems of innovation" resembles quite closely our own understanding of clusters.

2. This type of information can be useful in the hands of policy makers who wish to maximise the social rate of return on R&D subsidies However, to assess this at the micro level would require a fairly detailed level of information.

3. A comparison with biotechnology, for example, would reveal that in the biotech cluster the importance of fundamental research is relatively more pronounced and value chains relatively shorter.

4. Some clusters operate at a regional or even local scale. These are mostly the well-known examples from the industrial district literature in which clusters are depicted as strongly localised phenomena in which craft-like capabilities are relatively important as well as the presence of demanding customers, related sectors and other strongly localised specialised production factors. Examples used are mostly "complete" clusters operating in strongly regionalised or localised settings.

5. As Jacobs *et al.* observed earlier (1996, p. 427), a clear differentiation needs to be made here between the scale of trade and the scale of production clusters.

6. This argument also has consequences for the policies aimed at one individual (mega-)cluster. Peeters *et al.* (Chapter 12 of this volume) rightfully conclude that even though it is easy to accept the idea that the core sectors or clusters characterised by high levels of R&D expenditures should be favoured when designing industrial and technological policies, high levels of R&D expenditures are not necessarily a prerequisite for innovativeness, and vice versa. Interpreting innovation policies as mainly consisting of policies aimed at supporting and facilitating R&D would be a gross oversimplification!

7. For more general overviews of the implications of "innovation systems approaches" on innovation policy, see Edquist *et al.* (1998) and Cowan *et al.* (2000).

8. In fact, many of the contributions to this volume have been written by policy makers themselves or by analysts on behalf of policy makers.

9. A popular example outside the three clusters discussed in this volume is the biotechnology cluster. Here there is significant regulatory restriction on new applications of genetic technologies, often due to societal distaste and a fear of the unintended consequences of innovation.

10. See also Chapters 12-16 in OECD (1999) and Jacobs *et al.* (1996).

11. See, for example, Chaminade (Chapter 5 of this volume) for a discussion of appropriate policies aimed at furthering innovation in the Spanish ICT cluster.

12. In fact, balancing competition and supporting co-operation is one of the major balancing acts in facilitating innovation in clusters. Pentikaïnen (Chapter 16 of this volume) suggests that one of the major risks of cluster policy, especially if the initiatives are highly focused, is that they may stimulate collaboration with only minor additionality, if any, or that they may create market distortions.

REFERENCES

Chaminade, C. (2001), "Innovation Dynamics in the Spanish Telecommunication Cluster: Policy Implications", Chapter 5 of this volume.

Charles, D. and P. Benneworth (2001), "Clustering and Economic Complexity: Regional ICT Clusters in the United Kingdom", Chapter 17 of this volume.

Cowan, R. and G. van de Paal (2000), *Innovation in a Knowledge-based Economy. A Study Commissioned by the European Commission*, DG Enterprise, Publication No. EUR 17023, Luxembourg.

Dahl, M.S. and B. Dalum (2001), "The Construction Cluster in Denmark", Chapter 9 of this volume.

den Hertog, P., S. Maltha and E. Brouwer (2001), "Innovation in an Adolescent Cluster: The Dutch Multimedia Cluster", Chapter 7 of this volume.

den Hertog, P. and E. Brouwer (2001), "Innovation in the Dutch Construction Cluster", Chapter 10 of this volume.

Edquist, C., Hommen, L. Johnson, B., Lemola, T., Malerba, F., Reiss, Th. En Smith, K. (1998), *The ISE Policy Statement – The Innovation Policy Implications of the 'Innovations Systems and European Integration' (ISE) Research Project*, Linköping, Sweden.

Edquist, Ch. (Eds, 1997), *Systems of Innovation. Technologies, Institutions and Organization*, Pinter, London and Washington.

Gilsing, V.A. (2001), "Towards Second-generation Cluster Policy: The Case of the Netherlands", Chapter 19 of this volume.

Gonda, K. and F. Kakizaki (2001), "Knowledge Transfer in Agglomerations: A Regional Approach to Japanese Manufacturing Clusters", Chapter 14 of this volume.

Green, R., J. Cunningham, I. Duggan, M. Giblin, M. Moroney and L. Smyth (2001), "The Boundaryless Cluster: Information and Communications Technology in Ireland", Chapter 3 of this volume.

Hauknes, J. (2001), "Innovation Styles in Agro-food Production in Norway", Chapter 8 of this volume.

Holm Dalsgaard, M. (2001), "Danish Cluster Policy: Improving Specific Framework Conditions", Chapter 18 of this volume.

Jacobs (1996), "Clusters, Industrial Policy and Firm Strategy", *Technology Analysis & Strategic Management*, Vol. 8, No. 4, pp. 425-437.

Larosse, J., P. Slaets, J. Wauters, S. Bruninx, P. Simkens and R. Wintjes (2001), "ICT Cluster Organisations in Flanders: Co-operation in Innovation in the New Networked Economy", Chapter 6 of this volume.

Luukkainen, S. (2001), "Industrial Clusters in the Finnish Economy", Chapter 13 of this volume.

Malerba, F. (2000) "Sectoral Systems of Innovation and Production", Working paper EU TSER Programme ESSY Project, Milan.

OECD (1999), *Boosting Innovation: The Cluster Approach*, OECD, Paris.

Paija, L. (2001), "The ICT Cluster: The Engine of Knowledge-driven Growth in Finland", Chapter 2 of this volume.

Peeters, L., M. Tiri and A. Berwert (2001), "Identification of Techno-economic Clusters using Input-Output Data: Application to Flanders and Switzerland", Chapter 12 of this volume.

Peneder, M. (2001), "Dynamics of Initial Cluster Formation: The Case of Multimedia and Cultural Content", Chapter 15 of this volume.

Pentikainen, T. (2001), "Trade-flow-based Networks in the Finnish Economy: Heterogeneous Practices", Chapter 16 of this volume.

Romanainen, J. (2001), "The Cluster Approach in Finnish Technology Policy", Chapter 20 of this volume.

Vock, P. (2001), "An Anatomy of the Swiss Construction Cluster", Chapter 11 of this volume.

OECD PUBLICATIONS, 2, rue André-Pascal, 75775 PARIS CEDEX 16
PRINTED IN FRANCE
(92 2001 08 1 P) ISBN 92-64-18706-5 – No. 51909 2001